Key Technologies for Tunnel Construction under Complex Geological and Environmental Conditions

Series Editor

Qihu Qian, Nanjing, China

This series is written on the basis of several difficult projects, aiming at complex geological conditions such as complex topography and landform types, complex geological structure and poor geotechnical engineering geology, systematically describes the Key Technologies for Tunnel Construction under Complex Geological and Environmental Conditions like mountains, underwater, soft and uneven strata, from the aspects of geological exploration, early warning and risk control, design, construction and digital application. It's a monograph that introduces innovative achievements and core technologies formed in the construction of complex geological tunnels in recent years. The series contains both texts and illustrations. It can be used as a reference for engineers, technicians, scientific researchers, teachers and students of related specialties in tunnels and underground engineering.

This series systematically describes the Key Technologies for Tunnel Construction under Complex Geological and Environmental Conditions like mountains, underwater, soft and uneven strata, from the aspects of geological exploration, early warning and risk control, design, construction and digital application.

Contents of the series are mainly from a number of key projects, such as Yuelongmen Tunnel of Chenglan Railway in Sichuan Province, Jinan Yellow River Tunnel Project, Qiyueshan Tunnel of West Hubei Expressway, Jiaozhou Bay Tunnel of Qingdao, Nanjing Yangtze River Tunnel and Wuhan River Crossing Tunnel, as well as the 973 Projects of the State-Mechanism, Prediction, Early Warning and Control of Major Water Inrush Disasters in Deep and Long Tunnels, Science and Technology Project by the Ministry of Transport "Study on Key Technologies of Underwater Large Section Shield Tunnel Construction under Complex Geological Conditions", National Defense Research Project "Three-Dimensional Advance Prediction Technology for National Defense Underground Engineering".

More information about this series at http://www.springer.com/series/16378

Jian Chen · Fanlu Min

Shield Tunnel Cutter Replacement Technology

Jian Chen
China Railway 14th Bureau Group Co., Ltd.
Jinan, Shandong, China

Fanlu Min
College of Civil and Transportation Engineering
Hohai University
Nanjing, Jiangsu, China

ISSN 2662-2904 ISSN 2662-2912 (electronic)
Key Technologies for Tunnel Construction under Complex Geological and Environmental Conditions
ISBN 978-981-16-4109-1 ISBN 978-981-16-4107-7 (eBook)
https://doi.org/10.1007/978-981-16-4107-7

Jointly published with Shanghai Scientific and Technical Publishers
The print edition is not for sale in China (Mainland). Customers from China (Mainland) please order the print book from: Shanghai Scientific and Technical Publishers.

Translation from the Chinese Simplified language edition: 盾构隧道刀具更换技术/*Dun gou sui dao dao ju geng huan ji shu* by Jian Chen, and Fanlu Min, © Shanghai Scientific & Technical Publishers 2019. Published by Shanghai Scientific & Technical Publishers. All Rights Reserved.
© Shanghai Scientific and Technical Publishers 2022
This work is subject to copyright. All rights are reserved by the Publishers, whether the whole or part of the material is concerned, specifically the rights of reprinting, reuse of illustrations, recitation, broadcasting, reproduction on microfilms or in any other physical way, and transmission or information storage and retrieval, electronic adaptation, computer software, or by similar or dissimilar methodology now known or hereafter developed.
The use of general descriptive names, registered names, trademarks, service marks, etc. in this publication does not imply, even in the absence of a specific statement, that such names are exempt from the relevant protective laws and regulations and therefore free for general use.
The publishers, the authors, and the editors are safe to assume that the advice and information in this book are believed to be true and accurate at the date of publication. Neither the publishers nor the authors or the editors give a warranty, express or implied, with respect to the material contained herein or for any errors or omissions that may have been made. The publishers remain neutral with regard to jurisdictional claims in published maps and institutional affiliations.

This Springer imprint is published by the registered company Springer Nature Singapore Pte Ltd.
The registered company address is: 152 Beach Road, #21-01/04 Gateway East, Singapore 189721, Singapore

Foreword by Jun Sun

In recent years, the development of infrastructure construction in China has been developing vigorously. The construction of "The Belt and Road" has been continuously promoted. Many major projects have mushroomed vigorously, such as the Three Gorges Project, the Qinghai-Tibet Railway, the South-to-North Water Diversion, the three vertical and four horizontal high-speed rail network, the Hong Kong-Zhuhai-Macao Bridge, Shanghai Center Tower, and the construction with China's aid of the Jakarta-Bandung high-speed railway, the China-Laos railway, the China-Thailand railway, the Gwadar Port, the Piraeus port and so on. There is no doubt that China has become one of the countries with the largest number of major projects in the world. In terms of construction scale, technical difficulty and capital investment, these major projects are among the top not only in China, but also in the world, and even in the first place. A series of major key technical problems emerged in the construction process of these projects, many of which have been well optimized and solved through analysis, exploration and innovation. Some even created new technical means and methods on the basis of the original theory and technology, and applied for a large number of technical patents. For example, Shanghai Center Tower, 632m, as the world's tallest green building, has achieved many scientific achievements in many aspects such as super high-rise design, green construction, construction supervision and building information modeling (BIM) technology. It has applied for 8 invention patents and authorized 12 practical new technologies. Only in the aspect of structural engineering, it has been applied to many innovative technological innovations, such as the support technology of super deep foundation pit, the technology of super high pumping concrete, the installation technology of complex steel structure and the technology of structural crack control, some of which have reached the world level. These optimizations, breakthroughs and innovations will be very valuable references for Chinese engineers and technicians.

During the plenary session of the National People's Congress held in early March 2016, many representatives said that a great deal of technological innovation and development is a key driving factor for realizing China's macroeconomic strategic adjustment during the 13th Five Year Plan period, and a fundamental support and key driving force for realizing the overall development of the country under the overall layout.

At the same time, in the face of the opportunity of a new round of scientific and technological revolution, only by making innovations and breakthroughs in key core technologies one by one can we realize the overall leap of social productivity, make the level and ability of our scientific research achievements and engineering technology control enter the ranks of developed countries as early as possible and as quickly as possible, so as to continuously improve the technological competitiveness in the world, and the national strength will be stronger! At present, many engineering technology innovations have been widely recognized, but there are still many problems in the promotion and application of innovation results. In the field of major engineering construction, after the key engineering technical problems have been broken through and solved in practice, it is necessary to further sort out and summarize the new theories or methods, and then widely apply them to production practice again, which in turn will promote further innovation and development of technology, which is a huge driving force for the sustainable development of technology. One of the most effective aspects is to systematically summarize the innovative achievements and publish a set of technical monographs. This is also the significance of publishing the series of "Research on Key Technologies of Major Engineering Construction". With the promotion of academic innovation as the main goal, the series of "Research on Key Technologies of Major Engineering Construction" has the following characteristics:

1. Focus on major projects and key projects. At present, China's infrastructure construction is booming in various fields, and various kinds of engineering projects continue to mount. From the perspective of project volume and technical difficulty, we have selected a number of major projects and key projects, based on which, we have summarized the professional theory and technology to make them into books. Due to the large number of fields and disciplines involved in various projects, and the intersection and integration of disciplines, it is difficult to set a series of books with a single discipline. Therefore, based on the main line of Engineering categories, a series of books have been written in eight fields, i.e., tunnel and underground engineering, bridge engineering, railway engineering, highway engineering, super high-rise and large-scale public buildings, water conservancy engineering, port engineering, urban planning and architecture, basically covering the main fields of engineering construction in China, with a view to providing professional and technical reference guidance for major engineering construction in the future. As there are many fields and specialties involved, there are similarities and differences between technologies. In the field of cross technology, they are dealt with according to the specific situation to avoid content duplication and disjunction.
2. Highlight common technology and innovation achievements, and focus on application technology theorization. Focusing on a series of key technical problems in major projects in recent years, and based on the innovation achievements

and technological breakthroughs made by the project, the series of books sort out pertinently the technical experience and scientific research achievements of common, key or significant promotion value in each series, and makes in-depth, systematic and detailed analysis and elaboration from the perspective of technical methods and engineering practice experience, and provides practical theoretical basis and application reference for the solution of similar problems and the improvement of technology. In the "Series of Key Technologies for Tunnel Construction under Complex Geological and Environmental Conditions" (Academician Qihu Qian is the director of the editorial board), the key problems in the current tunnel and underground engineering construction are systematically described and the corresponding professional technical theory system is formed, including the prediction, warning and risk control of major water inrush disaster in deep and long tunnels, the treatment of soft and hard uneven stratum and extremely soft stratum in shield engineering, a sort of rectangular shield method, underwater shield tunnel, ground access shield method tunnel, extra-long highway tunnel, three-dimensional geological detection of tunnel, rapid detection of shield tunnel disease, digitalization of tunnel and underground engineering, new anchorage material for large deformation tunnel in soft rock and so on, and the key problems are solved in different degrees in the research and effectively implemented in the follow-up project.

3. Pay attention to the practical value of the project. The technical achievements involved in the series have been adopted many times in China, which have been proved to be reliable and effective in practice, and some of them have obtained technical patents. The series emphasizes the theory as the guide, the application as the key, and the case as the illustration. All the technical achievements are required to take the engineering project as the background and the production practice as the support, so that the series not only has the academic connotation, but also has the important engineering application value. For example, the "Series of Key Technologies for Long-span Bridge Construction and Maintenance" (academician Jielian Zheng is the director of the editorial board, academician Zhengqing Chen is the deputy director), focusing on major bridge projects such as long-span suspension bridges, long-span bridges across the sea, multi tower cable-stayed bridges, long-span concrete-filled steel tube arch bridges, long-span pedestrian bridges, large-scale spatial suspension bridges with variable width, etc., focuses on the design innovation theory, construction innovation technology, technical breakthrough of construction difficulties, bridge structure health monitoring and state assessment, maintenance and maintenance during operation period, etc. The main contents include the vacuum assisted pouring technology of large-scale concrete-filled steel tube structure, the spatial cable plane suspension bridge system with large proportion and variable width, the new eddy current damping vibration reduction technology, the cable hoisting and cable-stayed buckle construction of long and large bridges, the super large-scale deep-water foundation super high

composite bridge tower, intelligent deformation monitoring, and the integration of construction and maintenance based on BIM, etc. These technologies are based on major construction projects, including the first bridge of Hejiang Yangtze River, the second bridge of Hejiang Yangtze River, Wushan Yangtze River Bridge, Nanpanjiang bridge of Guangxi Guangzhou railway, Zhangjiajie Grand Canyon bridge, Xihoumen Bridge, Jiashao bridge, Hong Kong-Zhuhai-Macao Bridge, Humen second bridge, etc. in the book, the relevant contents of specific project cases are analyzed in detail, with good application reference Test the value.

4. Focus on the hot spot, focus on risk analysis, disaster prevention and mitigation, health detection, engineering digitization and other emerging branches in recent years. Under the guidance of the principles of green and sustainable development, in recent years, technological innovation in the field of infrastructure construction has many achievements and examples in energy conservation and emission reduction, low-carbon environmental protection, green civil engineering, risk analysis, disaster prevention and mitigation, health detection (remote wireless video monitoring), safety and economy in the life cycle of the project, reliability and durability, construction technology organization and management, digitization, etc. Series of books are also reflected in these aspects, in order to give full play to the role of series books in promoting the long-term, green and sustainable development of major project construction.

5. Establish an open framework. As a result of the above characteristics, the progress of each volume of the series is different, so the open framework is adopted, and the flexible way of publishing in stages is adopted in the setting of each volume of the subsequent series.

6. The editor in chief has first-class academic level, which lays a solid foundation for the academic quality of the series. The chief editor of each series is the academic authority in this field and has an important academic status and influence in this field. For example, Prof. Zhengqing Chen, academician of Chinese Academy of engineering, chief scientist of Project 985, bridge structure and wind engineering expert, Prof. Jielian Zheng, academician of Chinese Academy of engineering, bridge design and construction expert, Prof. Qihu Qian, academician of Chinese Academy of engineering, protection and underground engineering expert, Prof. Zhiqiang Wu, academician of Chinese Academy of engineering, urban planning and construction expert, etc. The main writers who participated in the writing were all the front-line personnel active in the research, education and engineering of infrastructure construction in China, who had undertaken major engineering construction projects or national major scientific research projects, mainly from China Railway Tunnel Bureau Group Co., Ltd., CCCC Tunnel Engineering Bureau Co., Ltd., China Railway Construction Co., Ltd., CCCC highway planning and Design Institute Co., Ltd., and the army Institute of engineering design, Qingdao Metro Group Co., Ltd., Shanghai Urban Construction Group, China Merchants Chongqing Communications Research and Design Institute Co., Ltd, Tianjin Urban Construction

Group Co., Ltd., Zhejiang Communications Planning and Design Institute, Tongji University, Hehai University, Southwest Jiaotong University, Hunan University, Shandong University, etc. In addition to undertaking the heavy tasks of engineering construction and scientific research and teaching, all experts have contributed their wisdom, knowledge and sweat to the progress of engineering technology in China. On behalf of the General Editorial Committee of the series, I would like to express my heartfelt thanks and respect for your hard work.

At present, not only the domestic infrastructure projects are in the ascendant, but also under the initiative of "The Belt and Road", China's major overseas projects are booming, and the demand for high level engineering technology is becoming increasingly urgent. It is believed that the publication of the series of books can provide certain assistance for the development of major engineering construction and the progress of innovative technology in China.

Shanghai, China
December 2017

Jun Sun

Mr. Jun Sun first-class Honorary Professor of Tongji University, senior academician of the Chinese Academy of Sciences, well-known expert in geotechnical mechanics and engineering at home and abroad. Chief editor of the series *Research on Key Technologies of Major Engineering Construction*.

Foreword by Qihu Qian

Since entering the 21st century, with the continuous development of economy, the continuous improvement of comprehensive national strength and the continuous application of high and new technology, China's tunnel and underground engineering have achieved unprecedented rapid development. China has become the country with the largest scale, the largest number, the most complex geological conditions and structural forms, and the fastest development of construction technology in the world. At the same time, with the continuous increase of urban subway construction, the number of cross-river and cross-sea tunnel projects continues to increase. National key construction projects such as long-distance water supply, underwater transportation, West to East Gas Transmission and other projects will involve the problem of crossing rivers. The construction of railway, highway, municipal administration, water supply, gas supply, flood control, hydropower and other tunnel projects will greatly increase the number of tunnels.

In the aspect of tunnel construction technology, the technical system of high-speed railway tunnel has been basically formed; the construction technology of long large tunnels with complex geological conditions in difficult and dangerous mountainous areas has made continuous progress; the construction technology of large section soft surrounding rock tunnels has made great progress; the construction technology of urban large-span shallow buried tunnels and river crossing underwater tunnels has made breakthroughs; the research and development and manufacture of tunnel boring machines have made great progress, these marks the tunnel construction technology in China has reached a new level of development. In particular, China has a vast territory, complex geological conditions, and extremely complex geological conditions are the main factors restricting the safe and efficient construction of tunnels, which are recognized as the difficulties of tunnel construction. Relying on a large number of key and difficult projects, such as Guanjiao tunnel of Qinghai-Tibet railway, West Qinling Tunnel of Lanzhou-Chongqing railway, immersed tunnel of Hong Kong-Zhuhai-Macao Bridge, Gaoligongshan tunnel of Dali-Ruili railway, Yangtze River Tunnel of Sanyang road in Wuhan, etc., China has made great achievements and more innovative achievements in tunnel and underground engineering, especially in tunnel construction under complex geological and environmental conditions. Therefore, in view of the complex geological conditions such as complex terrain and landform type, complex geological structure and poor engineering geology of rock and soil mass, based on the breakthrough and innovation achievements of key and difficult projects, the editorial board of the series and Shanghai Science and Technology Press jointly planned the series.

From the perspective of geological exploration, early warning and risk control, design, construction and digital application, the series systematically sorts out the key technologies of tunnel construction under complex geological and environmental conditions, such as mountains, underwater, soft and hard uneven strata. In the form of academic monographs, it introduces the innovative achievements and core technologies formed in the process of complex geological tunnel construction in recent years. The innovative achievements and technologies involved and introduced in the series are at the leading level in China, some of which have formed core technologies with independent intellectual property rights, and have been applied in major projects, making the series have the characteristics of cutting-edge, original, innovative and leading. For example, Kairong Hong's *Shield Tunneling Technology in Hard-Soft Uneven Strata and Extremely-Soft Strata* expounds shield treatment technology of soft and hard uneven stratum and extremely soft stratum of underwater tunnel from theory, technology and engineering cases, which represents the latest theory and practice of soft and hard uneven stratum tunnel and underground engineering; based on the scientific research project "National Defense Engineering Geological Prediction and Quality Nondestructive Testing System" presided over by the author, and combined with the research and application results of the team for many years, Guohou Cao and Hao Liu's *Three-Dimensional Exploration Technology of Tunnel Geology* comprehensively introduced the new theory, new method and new technology of tunnel complex geological geophysical exploration; Shucai Li's *Monitoring Method and Early Warning Technology of Tunnel Water Inrush* relying on the national major scientific research instrument and equipment development project "Comprehensive Geophysical Exploration Instrument for Quantitative Advance Prediction of Tunnel Unfavorable Geology Used in the Roadheader Construction", the National Natural Science Foundation for outstanding young scientists project "Mechanism and Disaster Control of Tunnel Water Inrush and Mud Inrush" and other major scientific research projects, introduces the advanced prediction technology of water inrush and mud inrush disaster source in complex geological tunnels and its application in large tunnel engineering, etc. The application cases in the series also include yuelongmen tunnel of Chengdu-Lanzhou Railway in Sichuan, Yellow River Tunnel Project in Jinan, Qiyueshan Tunnel of West Hubei expressway, Jiaozhou Bay Tunnel in Qingdao, Yangtze River Tunnel in Nanjing, and river crossing tunnel in Wuhan.

Under the strategic background of vigorously promoting the "One Belt and One Road" construction, implementing the innovation-driven development strategy and building a powerful transportation nation, we hope that the publication of the series will not only better summarize the above technological achievements and promote the popularization and application of innovative technologies, but also play a catalytic role in the breakthrough of basic theoretical research and common key technologies, and can play a positive role in the cultivation of technological innovation mode and professional talents. In the process of discussion, planning, organization, compilation and review, the series has been greatly supported by relevant large enterprises, colleges and universities, research institutions, societies and associations. Many experts have put forward many very good suggestions and ideas for the series in their busy schedule. Thank you all.

Guangzhou City, P. R. China Qihu Qian
August 2018

Mr. Qihu Qian Professor of the Army Engineering University of the Chinese people's Liberation Army and academician of the Chinese Academy of Engineering. Director of the editorial board of "Series of Key Technologies for Tunnel Construction under Complex Geological and Environmental Conditions".

Preface

Since the 1990s, the shield construction method has gradually become the preferred method for excavating tunnels in soft soil areas such as urban subways, cross-river tunnels and municipal tunnels due to its construction safety, excavating efficiency and small impact on the environment. At present, China has become the country with the largest number of shield projects and shield machines in the world. In the large number of shield tunnel construction projects, Chinese engineering technicians and academic experts have made lots of precious and original achievements in the process of conquering world-class problems such as highly water pressure, strong water permeability, composite stratum and long distance. These technological achievements have greatly promoted the development of Chinese shield technology. However, due to Chinese vast territory, large differences in geological conditions and technical level of construction enterprises, as well as the technical closure or confidentiality among different enterprises, accidents might happen during the construction process of shield tunnels. Hazards including excavation surface instability, excessive wear of cutting tools, breakdown of shield tail, and excessive ground settlement still occur. Among them, the failure of excavation caused by the excessive wear of cutter tool in the sand pebble and soil-rock stratum is one of the most important problems that plague the tunneling of China's shield.

The cutting tool is often compared to "teeth" of shield machine. There are also many different classifications depending on its function, such as scrapers that cut soft soil layers, cutters and hobs that crush rock. Generally, each shield cutting tool will be equipped with cutters of different numbers, types and combinations according to the tunnel layers in order to achieve the purpose of smooth cutting. However, when drilling in sand pebble and soil-rock composite stratum with high quartz content, the tool often has excessive and eccentric wear and abnormal damage dueto the large differences in the properties of stratums, which lead to the decreasing tunneling efficiency or the failure of excavation. At present, solutions to this type of problems are mainly to optimize tool configurations, replace new tools or optimize the tool (change the tool size, welding process, etc.).

At present, most of the projects in China that have excessive tool wear or damage and need tool-changing operations are underwater shield tunnels that pass through sand pebble and soil-rock composites over long distances, such as Nanjing Yangtze River Tunnel and Nanjing Wei Three Road Crossing River Tunnel, Nanjing Metro Line 10 cross-river section, Wuhan Metro Line 8 cross-river section, Wuhan Sanyang Road cross-river tunnel. The risk of tool-changing operations in these projects is very high, and A little carelessness can lead to failures, casualties and collapse of tunnels. With the developing trend of Chinese underwater tunnels to complex conditions of larger diameter and higher water pressure, we can foresee shield tools will face test more severely in the future. Therefore, how to change tools safely and efficiently under complicated working conditions is an important problem to be solved urgently in the construction of shield tunnels.

Since the beginning of the construction of Nanjing Yangtze River Tunnel, the author has directed and completed lots of tool-changing techniques and research work, including tool-changing under pressure, tool-changing in the cutter arm under normal pressure and hob-tooth cutter interchange. In order to build a perfect shield tool-changing technology system, the author summarizes and compiles this book. This book is based on his own scientific research achievements and engineering experience accumulated over the past decade, referenced associated projects and technical achievements at home and abroad and combined with the relevant domestic typical engineering cases, and can be used as reference for engineering and technical personnel and researchers, in order to contribute to the development of Chinses shield industry.

All the contents of this book were fully discussed and negotiated by Prof. Jian Chen, the deputy chief engineer of China Railway 14th Bureau Group Co., Ltd., the Environmental Science and Engineering Doctoral Student of Ocean University of China, and Associate Professor Min Fanlu of the Institute of Tunnel and Underground Engineering of Hohai University. While writing this book, the authors often conduct several discussions, consult experts, and refer to dictionaries for the accuracy of a certain terminology. Especially in the shield machine about "pressure warehouse" and "pressure cabin", which one should be used as the most frequent discussions and most heated debates. Finally, the authors reached an agreement, thinking that because the "cabin" in the shield machine can be pressurized and withstand pressure, which is similar to the function of "cabin" in the aerospace and marine fields, but the meaning of "warehouse" is different. Therefore, one with the function of inflating and withstand pressure in this book is called "cabin", and here can be seen as a description and explanation. Jingbo Xu, Hangbiao Song, Yixin Bai, Jie Liu, Jiayu Du, Chaojie Yu, Hai Liu, Yucheng Wang, Huanjie Lu, Sheng Wang, Dengfeng Wang, Zhicheng Xu, etc., graduate students of Institute of Tunnel and Underground Engineering, Hohai University all offer great help in the process of writing this book. They participate in the preparation of relevant chapters, data compilation, charting and calibration, and I would like to express my heartfelt thanks. We cite some results of research cooperation units and refer to technical materials and literatures at home and abroad while writing, and I would like to express my gratitude along with all the others.

In the process of writing this book, some research results of many research cooperation units are also cited. Some technical materials and literatures at home and abroad are referred to, and I would like to express my gratitude.

This book is jointly funded by the China Railway 14th Bureau Group Enterprise Science and Technology Research and Development Project (2017–2006), the National Natural Science Foundation of China (51778213) and the Central University's basic research business fee (2016B01214).

Some parts of this book are the authors' personal understanding of the shield tool-changing technique. Due to the limited knowledge, there are inevitably shortcomings and errors in the book. Experts and readers are welcomed to criticize and correct them.

Jinan, China Jian Chen
Nanjing, China Fanlu Min

Contents

1 **Introduction** 1
   1.1 The Development and Application of Shield Tunneling
       Technology 1
   1.2 The Severity and Universality of Tool Wear in Shield
       Engineering in Complex Stratum 5
   1.3 The Key Technology of Restoring Shield Tunneling—Tool
       Replacement Technology 9

2 **Cutting, Wear and Replacement of Cutting Tools During
Shield Tunneling** 11
   2.1 Types and Arrangements of Shield Cutterhead and Tool 11
       2.1.1 Shield Cutterhead Type 11
       2.1.2 Type of Shield Cutter 19
       2.1.3 Arrangement of Shield Cutter Head 31
   2.2 Cutting and Wear of Tools 38
       2.2.1 Cutting Mechanism of Shield Tool 39
       2.2.2 Wear Types and Examples of Shield Cutting Tools 45
       2.2.3 Analysis on the Causes of Tool Wear of Shield Machine ... 54
       2.2.4 Tool Wear Monitoring Method for Shield Machine 58
   2.3 Shield Opening Technology 61
       2.3.1 Brief Introduction of Shield Tunnel Opening 61
       2.3.2 Common Opening Technology of Shield Tunnel 63
       2.3.3 Brief Introduction of Typical Shield Tunnel Opening
           Examples at Home and Abroad 68
       2.3.4 Summary and Prospect of Shield Opening Technology ... 73
       2.3.5 Tool Replacement Technology of Shield Machine 73

3 **Conventional Cutterhead and Tool-Changing Under Normal
Pressure Technology** 77
   3.1 Self-stabilizing Stratum Tool-Changing Technology 78
   3.2 Reinforcement Stratum Tool-Changing Technology 80
       3.2.1 Rotary Jet Reinforcement Technology 82
       3.2.2 Deep Mixing and Reinforcement Technology 93

		3.2.3	Freezing Method Construction Technology	102
		3.2.4	Shaft Reinforcement Technology	109
		3.2.5	Underground Continuous Wall Reinforcement Technology	118
		3.2.6	Comparison of Several Techniques	127
	3.3	Normal Pressure Tool Change Technology		128
		3.3.1	Tool Change Process	128
		3.3.2	Shield Cutter Tool Welding Process	129
		3.3.3	The Contents of the Cutter Repair	136
	3.4	Example of the Yellow River Tunnel in the Middle Route of the South-To-North Water Transfer Project		141
		3.4.1	Project Overview	141
		3.4.2	Introduction to Downtime	142
		3.4.3	Guarantee the Stability Measures of the Excavation Face	144
		3.4.4	Excavation Support	147
		3.4.5	Cutterhead and Tool Repair	150
4	**Normal Pressure Tool Change Technology Based on Basic Pressure and Changeable Knife Design**			157
	4.1	Introduction of Normal Pressure Changer Technology Based on Normal Pressure Replaceable Knife Design		157
		4.1.1	Principle of the Process of Changing the Knife Under Constant Pressure	157
		4.1.2	Features of the Normal Pressure-for-Knife Method	158
	4.2	Normal Pressure Replaceable Cutting Plate Structure Design and Layout		159
	4.3	Removal and Installation of Normal Pressure Replaceable Tools		163
		4.3.1	Safety Points for Tool Replacement	164
		4.3.2	Tool Inspection and Replacement Plan	164
		4.3.3	The Process of Changing the Knife Under Constant Pressure	165
	4.4	Example Swords-Off of Normal Pressure Opening in the Cross-river Section of Wuhan Metro Line 8		179
		4.4.1	Engineering Overview and Geological Conditions	179
		4.4.2	Shield Basic Parameters and Characteristics	184
		4.4.3	Shutdown Position and Knife Change Plan	186
		4.4.4	Atmospheric Pressure Changing Process and Program	189
5	**Stabilization Technology and Tool Change Technology for Open Face with Pressure Open**			197
	5.1	Pressure Opening and Opening Face Stability Technology		199
		5.1.1	Mud Preparation Technology	199
		5.1.2	Mud Film Forming Technology	208

	5.1.3	Stable Excavation Face Technology Based on Mud Membrane Closed Gas 212

- 5.2 Pressure-Opening Tool Changer 217
  - 5.2.1 Diving Operation 217
  - 5.2.2 With Pressure into the Cabin 220
  - 5.2.3 Pressure Change Tool Technical Produces 224
- 5.3 Nanjing Yangtze River Tunnel (Wei7Road) Example of Conventional Compressed Air Pressure Opening and Changing 225
  - 5.3.1 Overview of the Nanjing Yangtze River Tunnel Project 225
  - 5.3.2 Analysis Process of Shield Tunneling and Shutdown of Nanjing Yangtze Tunnel 236
  - 5.3.3 Control of Stability of Excavation Face when Opening 239
  - 5.3.4 With Pressure Open Cabin Changer Implementation 243
  - 5.3.5 Open Cabin Experience Summary and Discussion 244
- 5.4 Example of Saturating Method of Pressure-Based Open Caving in Nanjing Wei Three Road Crossing River Tunnel 246
  - 5.4.1 Nanjing Wei Three Road Crossing River Tunnel Project 246
  - 5.4.2 Application of Mud Film Closed Gas in Pressurized Open Compartment 252
  - 5.4.3 Saturation Method with Pressure into the Cabin Change Process 261
  - 5.4.4 Summary of Saturation Method with Pressure Change into the Cabin 264

6 Other Special Tool Changing Techniques 267
- 6.1 Tool Replacement Technology While Encountering Riprap and Bedrock Shield 267
  - 6.1.1 Project Summary 267
  - 6.1.2 Cutter Head Design 270
  - 6.1.3 The Influence of Riprap and Bedrock on Shield Tunneling and Its Location Survey 271
  - 6.1.4 The Construction Plan 275
- 6.2 Encounter Large Boulder Shield Cutter Protection and Replacement Technology 288
  - 6.2.1 Project Overview 288
  - 6.2.2 Analysis of Influence of Boulder and Bedrock Bulging on Shield Tunneling Construction and Blasting Construction Technology 289
  - 6.2.3 Construction Parameter Design 295
  - 6.2.4 Construction Guarantee Measures 300
- 6.3 Japan's Kumagai Group SunriseBite Method of Tool Change 302
- 6.4 Japan's Kashima Construction Company's Disc Hob Replacement Technology 304

6.5 Japan Dacheng Construction Company Central Hob
    Replacement Technology 305
6.6 Japan Feidao Construction Company Chameleon Tool
    Replacement Technology 306

About the Authors

Jian Chen Ph.D. Graduated from the Ocean University of China, Deputy Chief Engineer and Chief Expert of China Railway 14th Bureau Group, a well-known domestic large shield expert, mainly engaged in technical management and scientific research of the large shield cross-river tunnel project. He was awarded the honorary title of "China Civil Engineering Zhan Tianyou Innovation Collective", "Mao Yisheng Railway Engineer", "Shandong Province Excellent Inventor", "Jiangsu Province Top Ten Staff Operation Law" and "Shandong Excellent Young Intellectuals". Successfully completed the construction of large-scale underwater tunnels such as the Nanjing Yangtze River Tunnel, the Yangzhou Slender West Lake Tunnel, and the Wuhan Metro Line 8 Cross-river Tunnel. The scientific and technological achievements of the research and development have reached international advanced and domestic leading level. He received 6 provincial and ministerial-level scientific and technological progress first prizes, 2 second prizes; 1 national-level construction method, 5 provincial-level construction methods; 23 national patents authorized; 1 monograph published; published core and above scientific papers 14 Article.

Fanlu Min, Ph.D. associate professor of Hohai University, master tutor, visiting scholar of Hong Kong University of Science and Technology, selected as a "big scholar" program at Hohai University. He is mainly engaged in research and teaching of underwater shield tunnels, urban subways and other aspects. He has presided over 5 vertical projects such as the National Natural Science Foundation of China and the National 973 Program Sub-project, and presided over or participated in the Nanjing Yangtze River Tunnel, the Nanjing Weisan Road Crossing River Tunnel, the Yangzhou Slender West Lake Tunnel, the South-to-North Water Transfer Middle Line, the Yellow River Tunnel and the Wuhan River Crossing Subway. More than 10 large-diameter underwater shield tunnels and technological innovations in urban subways. He won the first prize of provincial and ministerial-level scientific and technological progress award; authorized 16 invention patents; published 1 monograph; published more than 60 papers, including 30 SCI/EI papers.

Chapter 1
Introduction

In recent years, with the improvement of urban density, the limited available ground space has gradually become an important reason restricting the development of urban construction. The infrastructure and underground pipelines in urban areas put forward higher requirements for creating underground space. In tunnel construction, the traditional open cut method because of its long time covering construction, cut off traffic, electric power, water and gas supply pipeline defects, no longer apply to the city within the tunnel construction, and has little impact on the environment, the construction of high efficiency, green environmental protection shield method gradually get the wide attention from various asp. The term "shield" in the shield tunneling method has two meanings: shield refers to the cutter head and pressure chamber that can maintain the stability of the excavation surface of the stratum, and the steel shell that supports the soil around the tunnel; the structure refers to the lining structure of the tunnel segments and the grouting body behind the segments. Shield tunneling is a set of tunnel construction methods with shield tunneling as the core.

1.1 The Development and Application of Shield Tunneling Technology

Shield method of birth can date back to 1818, when the French engineer (Mare Isambard Brunel) in London got the enlightenment from the behavior that moth in the planking wormhole paint around the hole of the discharge, found the driving principle of shield method, perfect the idea after the patent registration. Since then, the shield tunneling method has experienced open hand excavation, mechanical and closed pressurized construction. Since the end of the nineteenth century, shield tunneling has been introduced to the United States, France, Germany, Japan, the Soviet union and China, and achieved different degrees of development. Since the 1960s, shield method has made a number of technical progress, appeared a variety of modern shield method of closed thoracic, pneumatic type of shield method, grid extruding shield method, earth pressure balanced shield method, mud pressure type of shield method, etc., in which earth pressure balanced shield method and mud pressure type of shield technique technology invention, is the modern shield technology leap, soft soil has been widely used all over the world in the construction of the tunnel.

Since the 1950s, China began to try to use shield tunneling method in practical engineering. In the early 1950s, the shield tunneling method with small section was adopted in the drainage roadway project of Fuxin coal mine in northeast China and the sewer project of Beijing in 1957. After that, the mesh shield tunneling method was successfully used in Shanghai to construct the diameter of 10. 22 m Dapu road crossing, and the diameter reaches 11.3 m Yan'an east road north line across the river tunnel. On the basis of the successful application of shield tunneling method, the following projects of Shanghai metro line 1 and line 2 basically adopted the earth-pressure balanced shield tunneling method, and in 1996, the tunnel project of Yan'an east road south line was successfully adopted the mud and water pressurized shield technology for the first time. Meanwhile, shield engineering in Guangzhou, Nanjing and other areas started gradually, and shield technology in China entered a stage of rapid development. Figure 1.1 shows the development history of shield tunneling technology in China is presented.

After entering the twenty-first century, with the urban underground space development in China and a lot of big traffic tunnel construction, shield working method with the surrounding environment of construction disturbance is small, the construction layout is reasonable, more secure, short construction period and low cost a lot of

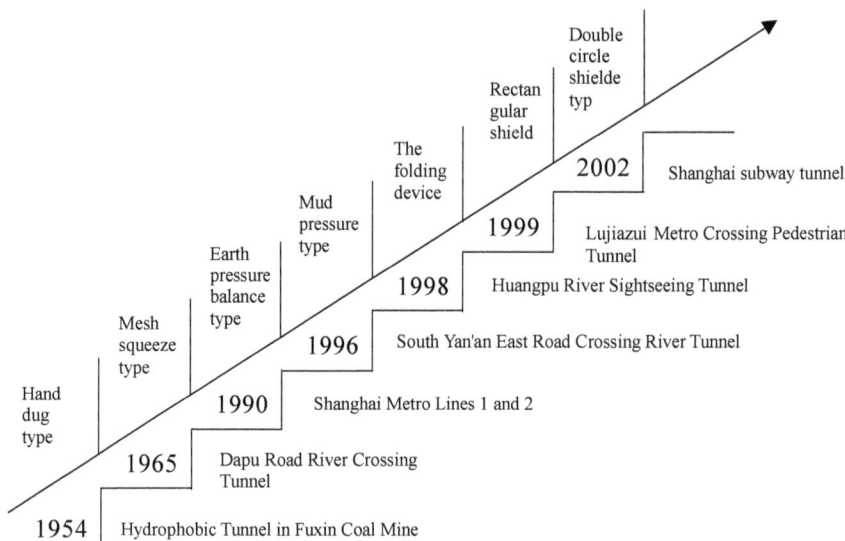

Fig. 1.1 The development history of shield tunneling technology in China

advantage, has become for metro in China, telecommunications, electricity, sewage and other urban tunnel construction, the main methods of shield tunnel technique into the leap in the development stage in our country. Under the requirement of the three-dimensional development of urban traffic, more and more cities begin to carry out subway construction. By the end of 2017, 33 cities had opened subways with a total mileage of more than 3,800 km, and 43 cities had been approved for subway construction. It is expected that by 2020, the number of cities with subway services will reach 40, with a total mileage 6000 km.

In the construction of urban traffic tunnel, the shield tunneling method adopted can be divided into soil pressure balance shield method and mud pressure balance shield method. The soil pressure balance shield tunneling method uses the cutting knife arranged on the cutter disc to cut the soil, and the soil under the cutting knife enters the soil chamber through the cutter disc. By controlling the rotational speed of the screw dump, the pressure in the chamber can be adjusted and controlled so as to keep the dynamic balance with the earth pressure on the excavation surface and maintain the stability of the excavation surface. The earth-pressure balanced shield tunneling method is suitable for almost all soft strata, and can effectively maintain the stability of excavation surface and reduce ground subsidence. Its overall performance has been proved by a lot of engineering practice in the subway construction of Shanghai, Guangzhou, Nanjing, Chengdu and Shenzhen, etc.

Mud film is formed on the excavation surface by injecting mud with appropriate pressure into the mud chamber to transfer mud pressure and maintain the stability of the excavation surface. Soil mass and mud under the cutting of shield cutter are fully mixed and then transported to the ground by mud discharge pipe for treatment. Mud-water pressurized shield tunneling method is considered to be applicable to almost all strata except hard rock, especially for the construction of river tunnel and submarine tunnel with large water content in the stratum and large water body above. In 1966 and 1984, Shanghai built the Dapu Road Crossing Tunnel with an outer diameter of 10.22 m and the Yanan Road North Line Crossing Tunnel with an outer diameter of 11.30 m, respectively, which opened the history of the construction of large underwater tunnels by shield method in China. Current Shanghai has built Dalian road tunnel, Fuxing east road tunnel, Xiangyin road tunnel, Shangzhong road tunnel and other large tunnels. Most of the strata through these tunnels are homogeneous silty clay or fine sand formation, and the construction control technology of mud-water pressure shield method is more mature in these strata.

Wuhan Yangtze River Tunnel, which was built and opened to traffic in 2008, is the first large highway tunnel built on the Yangtze River in China. Two mud-water pressurized shields with a diameter of 11.34 m are used to cross some fine sand, pebbles and other high permeability strata. The geological strip is complex, the strata are soft and hard, and it is the Jiangdi tunnel project with the highest technical content and the most difficult construction in China at that time. Subsequently, Shanghai Yangtze River Tunnel, Nanjing Yangtze River Tunnel (Wei Seven Road), Guangzhou Lionyang Tunnel, Hangzhou Qingchun Road Crossing River Tunnel (Qiantang River), South to North Water Transfer Center Line Crossing Yellow Tunnel, Hangzhou Qiantang River Tunnel, Yangzhou Lean Xihu Tunnel, Nanjing

Wei Three Road Crossing River Tunnel and other large diameter mud-pressure shield tunnels have been completed successively. And the construction of Wuhan Sanyang Road Tunnel, Nanjing and Yan passing through the Yangtze River Tunnel, Shantou Suai Bottom Tunnel, South Wuhu City Crossing River Tunnel, etc. marks the vigorous development of underwater large diameter mud-water pressurized shield tunnel in China.

Table 1.1 is a list of the main super-large diameter mud-water pressurized shield tunnels built in the world in recent years.

From Table 1.1, it can be seen that more than half of the super large diameter mud and water pressurized shield tunnels in the world are built in our country. In addition, the subway sections of major cities in China basically use mud and water pressure shield, such as Nanjing Metro Line 3 and Line 10 across the Yangtze River, Wuhan Metro Line 8 across the Yangtze River, Guangzhou Metro across the Pearl River section, Shanghai Metro across the Huangpu River section, Hangzhou Metro across the Qiantang River section, and so on. Obviously, mud-water pressure shield technology has become the key technology of underwater tunnel construction in our country. The Tuen Mun-Chik Wax Kok tunnel under construction in Hong Kong is constructed with a mud-water pressurized shield with a diameter of 17.6 m. It is the largest shield tunnel in the world at present. Both the Bohai Strait tunnel under planning and the Qiongzhou Strait tunnel will also be constructed by mud-water compression shield method. This also shows that shield tunneling technology in

Table 1.1 List of major large-diameter water-pressure type shield tunnels that have been built in the world in recent years

Name of tunnel project	Tunnel outer diameter	Shield diameter	Length of tunnel	Tunnel use	Year of completion	Country
Shanghai Chongsu Tunnel	15.00	15.43	8900	Highway	2008	China
Hangzhou Qiantang River Tunnel	15.00	15.43	4500	Highway	2012	China
Nanjing Yangtze River Tunnel	14.50	14.93	3020	Highway	2009	China
Nanjing Weisan Road Crossing Tunnel	14.50	14.93	4134	Highway	2015	China
Yangzhou thin West Lake Tunnel	14.50	14.93	1275	Highway	2014	China
Shanghai Yaohua Road Crossing Tunnel	14.50	14.88	2179	Maglev	2009	China

(continued)

1.1 The Development and Application of Shield Tunneling Technology

Table 1.1 (continued)

Name of tunnel project	Tunnel outer diameter	Shield diameter	Length of tunnel	Tunnel use	Year of completion	Country
Green heart tunnel	14.50	14.87	7155	High-speed railway	2003	Holland
Shanghai Shangzhong Road Tunnel	14.50	14.87	2802	Highway	2008	China
Yibeihe fourth Highway Tunnel	13.80	14.20	2561	Highway	2000	Germany
Moscow Ring Road Tunnel	13.80	14.20	3246	Highway	2003	Russia
Tokyo Bay Highway Tunnel	13.90	14.14	9100	Highway	1996	Japan
SMART traffic flood discharge tunnel	12.80	13.21	9400	Highway, flood discharge	2006	Malaysia
Hangzhou Qingchun Road Crossing River Tunnel	11.27	11.68	1766	Highway	2010	China
Wuhan Yangtze River Tunnel	11.00	11.34	3600	Highway	2008	China
Hill West Tunnel	11	11.34	6600	Highway	2003	Holland
Lion Ocean Tunnel	10.8	11.18	9340	Railway tunnel	2011	China

China is developing in the direction of high water pressure, long distance, complex strata and so on, which also indicates that shield tunnel technology will face greater challenges.

1.2 The Severity and Universality of Tool Wear in Shield Engineering in Complex Stratum

China has a vast territory and a wide area of shield tunnel construction, and there are great differences in geological and environmental conditions. With the exception

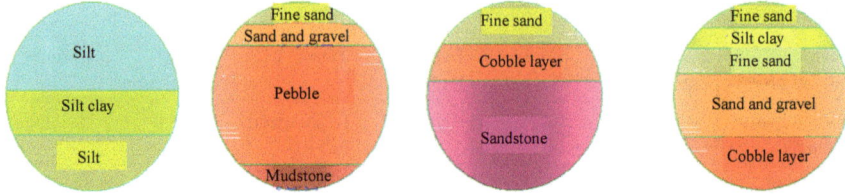

Fig. 1.2 Composite strata diagram of typical sections of Nanjing Weishan cross—river tunnel

of Shanghai and individual tunnels passing through muddy clay strata, the large diameter mud and water shield projects that have been built and are under construction in China have almost encountered composite strata of upper soft and lower hard, upper soil and lower rock due to the fact that most of the large diameter mud and water shield projects have crossed rivers and seafloor (Fig. 1.2), and Chengdu, Beijing, Shenyang, Lanzhou city subway all have a large size of pebble and gravel stratum. The durability of shield is very important for tunneling in complex stratum, and the durability of cutting tool is the most important factor affecting the durability of shield.

At present, the main problems about the durability of cutting tools are as follows: for the long distance tunneling of complex strata, there are many accidents such as abrasion, damage and shedding of cutting tools. In the process of tunneling, the wear and defect of the cutting head will reduce the tunneling speed and increase the rotating load of the shield cutter, which will stop the tunneling and even cause the excavation surface collapse and other accidents. Nanjing Yangtze (Wei Seven Road road) of shield tunnel under crossing the silty sand and sandy pebble on soft hard compound formation, due to the high sand strata quartz content was 70%, and the cutter disc and cutter excessive wear is shown in Fig. 1.3 lead to unsustainable development; a force for shield downtime with pressure tank to fill the knife dish and scraper surrounding the repair welding repair, lasted half a year, the losses were heavy.

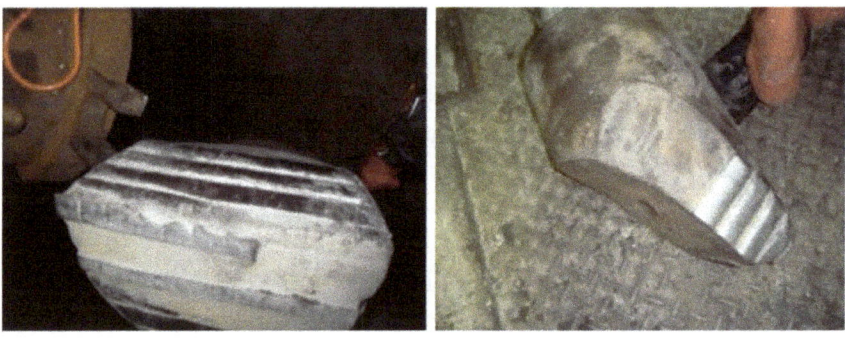

(a) Cutter tooth caving (b) Uneven wear of scraper

Fig. 1.3 Tool wear in Nanjing Yangtze river tunnel

1.2 The Severity and Universality of Tool Wear ...

During the shield construction of Nanjing Wei Three Road crossing river tunnel project, affected by the soil condition and the change of underground water pressure, with the continuous advance of the shield construction and the constant rotation of the cutter plate, the shield cutter is subjected to continuous and changing scraping, jacking, extrusion, wear, impact and other forces. The common forms of wear in the process of construction are: normal wear, eccentric wear, blade collapse and ring fracture and so on. Fig. 1.4 shows the typical tool wear during the construction of Nanjing Wei Three Road crossing tunnel. Among them, a total of 223 hobs,

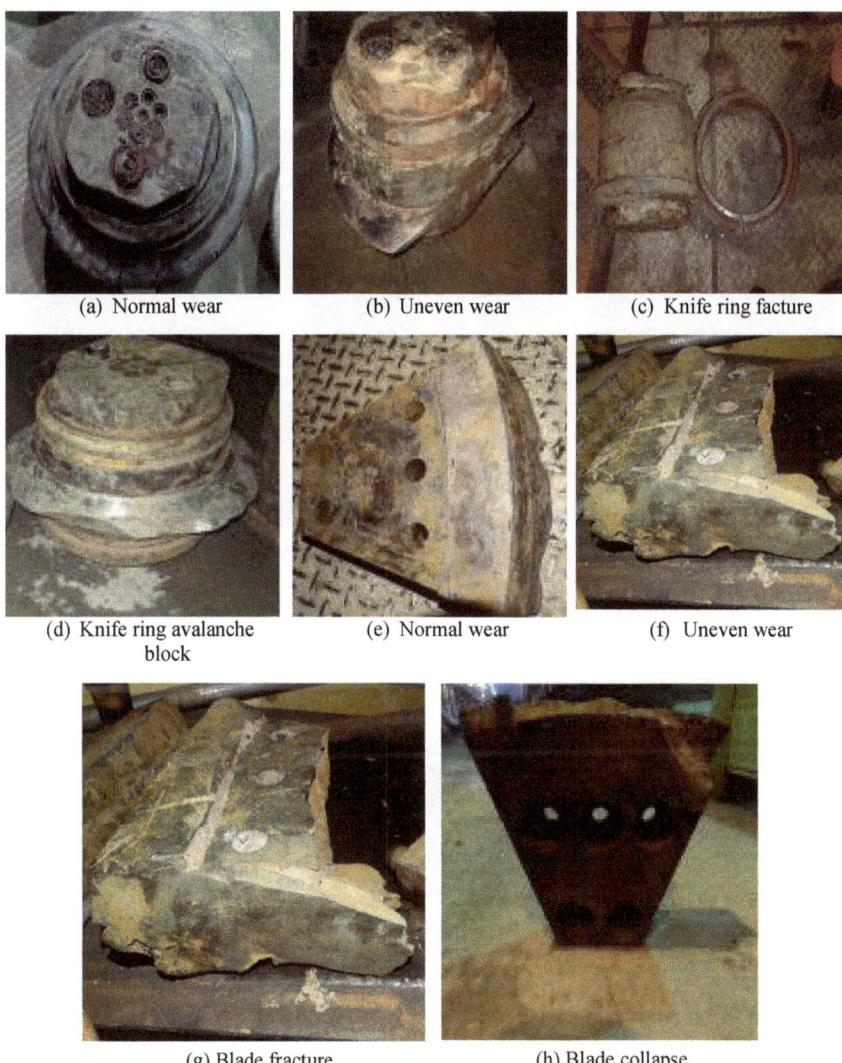

(a) Normal wear   (b) Uneven wear   (c) Knife ring facture

(d) Knife ring avalanche block   (e) Normal wear   (f) Uneven wear

(g) Blade fracture   (h) Blade collapse

Fig. 1.4 Wear condition of cutting tool in the Nanjing Wei Three Tunnel

accounting for 251% of the total, were replaced in the shield tunneling process. Replace 91 cutting tools, accounting for 15% of the total number of cutting tools. When the shield of the middle line of the south-to-north water diversion tunnel crosses the sand stratum with high quartz content and the hard calcareous concretion, all the 16 edge shovels are seriously damaged, accounting for 100% of the total shovels. 24 first knife alloy block fell, accounting for 100% of the total number of first knife; the outer edge of the cutter also suffered different degrees of wear (Fig. 1.5).

Wuhan metro line 8 a more Jiangduan tunnel excavating distance for more than 3000 m, in addition to strong weathering of the whole section conglomerate and cementation conglomerate formation, also on the cross section under the soft hard formation, the eccentric wear of shield cutting tools easily happened, up more than 1000 full replacement tool (Fig. 1.6), increased the difficulty of the shield tunneling. Shield tunnel technology has encountered serious tool wear problem in engineering construction, which has become one of the main factors disturbing shield tunnel

(a) 16 edge blades badly damaged   (b) The rotary shaft of hob ring is damaged

Fig. 1.5 Wear condition of cutting tool in the tunnel of south-to-north water diversion

Fig. 1.6 Wuhan metro line 8 crossing the river tunnel site six knife changes

construction, seriously restricting the development of shield tunnel construction technology.

Abroad, there is a lot due to excessive tool wear and affect the normal shield tunneling engineering cases, such as Singapore DTSS tunnel in hard rock with through the whole section, weathered rock, hard rock mixed with soft soil strata and so on three strata, due to the high corrosion resistance, hard rock cutter disc and cutter or the different degree of wear and damage. In fact, a considerable number of shield tunnel projects at home and abroad have experienced excessive wear of cutter heads when crossing complex strata. In serious cases, the project cannot be carried out and is forced to stop for a long time of replacement and maintenance, which seriously affects the smooth progress of the project. The severity and universality of the wear of shield tunneling cutters have become an important problem restricting the application and development of shield tunneling technology.

1.3 The Key Technology of Restoring Shield Tunneling—Tool Replacement Technology

In the process of shield tunneling, serious tool wear, abnormal wear and mechanical equipment damage generally occur, which leads to the forced shutdown of shield tunneling and the inability to continue normal tunneling. It is the main means to stop the engine to open the space, repair or replace the broken cutter, which is the operation that must be carried out to resume the shield tunneling under the condition of complicated stratum.

Up to now, many typical shield engineering projects at home and abroad have reported on the opening tool change or maintenance operations: the shield tunnel of the fourth tunnel of Elbe river in Germany stopped at the bottom of the river, and adopted two engineering methods of opening the cabin to change the tool; Weste schelde tunnel in the Netherlands adopts the method of ballast opening for tool maintenance and construction; The large-diameter underwater shield tunnel projects in China, such as Nanjing Yangtze river tunnel (Wei Seven Road), Nanjing Wei Road No. 3 crossing river tunnel, the middle route crossing yellow tunnel of south-to-north water diversion project, Wuhan Yangtze river tunnel, Beijing railway underground diameter line and Shizi Yang tunnel, have all carried out a variety of tool inspection, maintenance and replacement operations.

Similar to the development process of shield tunneling technology, in the early stage, Chinese engineering and technical personnel could only carry out relatively simple cabin opening and tool changing operations. Under high-pressure and underwater conditions, shield tunneling and tool changing technology was basically monopolized by large foreign companies. In China's engineering, foreign technical personnel should be hired with high salaries to carry out tool changing operations. Although at present our country has been independently completed a lot of underwater shield tunnel tube tunnels in the damaged tool replacement operation, make the project can proceed, also accumulated some experience, but most are engineering technicians should be organized according to the experience, combined with the engineering companies for technical secret, system of shield technology system has not been shaped into a tool changer. In the future, as the construction conditions of shield tunnel become more and more complex in China, the test of cutting tools in the process of shield tunneling becomes more and more severe, and the demand for innovative technology of shield tool replacement becomes more and more intense. Therefore, it is necessary to carry out the research and development of the innovative technology of tool replacement on the basis of the domestic and foreign examples of shield tunnel tool replacement operation, and form a new technology system of shield tool replacement, so as to meet the development requirements of shield tunnel technology in the new era.

Chapter 2
Cutting, Wear and Replacement of Cutting Tools During Shield Tunneling

In the process of excavation, the cutting of the front soil is carried out by the cutter head tool, and the adaptability between the tool and the soil is very important. If the cutter head tool does not adapt to the soil, it will make the shield cutting slow, accelerate the tool wear, and directly affect the working efficiency and the progress of the process. At the same time, the rationality of cutterhead structure design and tool placement is also one of the main influencing factors of shield service life. Therefore, it is necessary to clarify the relevant types, arrangement methods, cutting mechanisms, wear types and causes of shield cutters, and analyze and improve the tool change technology to improve the efficiency of shield tunneling and tool change, and reduce the safety risks during construction.

2.1 Types and Arrangements of Shield Cutterhead and Tool

2.1.1 Shield Cutterhead Type

The shield cutterhead is a cutting disc body with multiple feed slots, a panel for stabilizing the excavation face, a tool for excavating the formation, unearthed groove, the rotating drive mechanism and the bearing mechanism are partially composed. The cutterhead is located at the front of the shield and its main function is used as the installation of the tool for tunnel excavation, it also has the function of stabilizing the working surface, mixing the soil residue, blocking the extra large boulders. The working principle of the shield cutter can be simply compared to the razor cutter. In the process of excavation, the gravel soil is broken into small particles and placed in the cabin behind the cutter.

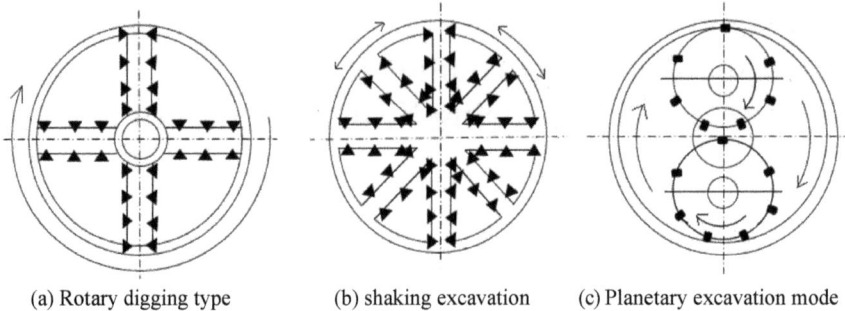

(a) Rotary digging type   (b) shaking excavation   (c) Planetary excavation mode

Fig. 2.1 Type of cutting mode of cutterhead

Shield cutters can be divided into various types according to the way of cutting, longitudinal shape and front shape.

2.1.1.1 Classification According to Excavation Method

It can be divided into rotary digging type, shaking excavation type and astigmatism type cutting type, as shown in Fig. 2.1.

2.1.1.2 Classification According to the Profile of Cutter Head

It can be divided into a vertical plane shape, a core shape, a dome shape, a slope shape, and a contraction shape, as shown in Fig. 2.2.

Figure 2.2a is a vertical planar cutterhead, which excavates and stabilizes the cutting surface in a plane form. Figure 2.2b is a protruding core cutter head. Compared with the vertical planar cutter head, the cutter head is equipped with a protruding tool in the center, which has good directivity and cutting property, and is beneficial to the mixing of additives and excavated soil. Figure 2.2c is a dome-shaped cutterhead. The cutterhead is designed with reference to the design principle of the rock roadheader. The cutterhead is mainly used in layers such as boulders and rock formations.

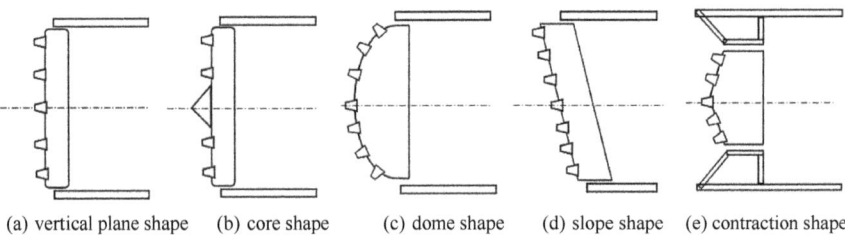

(a) vertical plane shape   (b) core shape   (c) dome shape   (d) slope shape   (e) contraction shape

Fig. 2.2 Shape type of longitudinal section of cutter head

2.1 Types and Arrangements of Shield Cutterhead and Tool

Figure 2.2d is a slanted cutter head, which is characterized by the inclination angle of the cutter head is roughly equal to the internal friction angle of the excavation soil layer, which is beneficial to maintain the stability of the cutting surface, and is mainly used for the excavation of the gravel layer.

Figure 2.2e is a shrink-type cutter head, mainly used for compressing shields.

2.1.1.3 Classification According to the Front Shape of the Cutterhead

It can be divided into two types: panel type (Fig. 2.3) and spoke type (Fig. 2.4). In addition, a web-type cutterhead between the two is now developed (Fig. 2.5).

1) Panel cutter head

The panel type cutter head is generally a welded box-shaped structure composed of spokes, a holder, a cutter, a notch and a panel. The cutter is arranged on both sides of the opening on the panel, and the hob is fixed to the panel by the holder. The panel type cutter head has a small opening ratio of about 30%. The construction of the above sea subway is the French FCB shield, The mud shields used in Shanghai Hu Chongsu Tunnel, Shanghai Shangzhong Road Tunnel, Wuhan Yangtze River Tunnel and the Qingchun Road Crossing Tunnel in Hangzhou are all panel-type cutterheads.

Cutter can be divided into slotted fixed panel cutters (Fig. 2.6) and slotted adjustable panel cutters (Fig. 2.7) according to whether the slot can be adjusted.

There are two types of notches on the panel: one is a strip-shaped notch that is always the same width from the center of the cutter head to the outer edge; the other is a fan-shaped notch whose width is gradually enlarged from the center to the outer edge. The width of the notch depends on the soil parameters of the excavation formation, such as the maximum particle size of rock and the soil particles, the

Fig. 2.3 Panel cutter

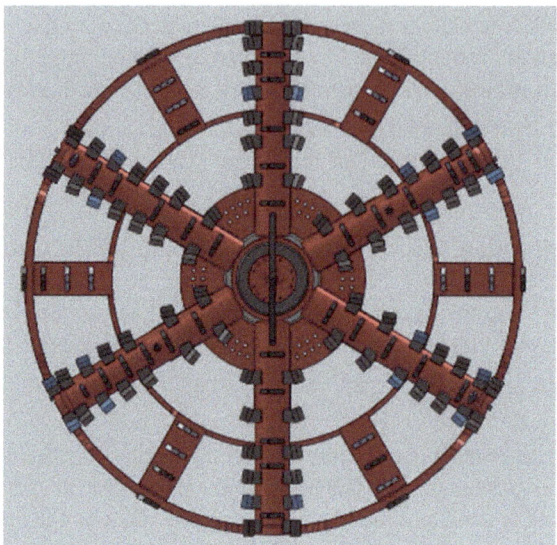

Fig. 2.4 Spoke cutter

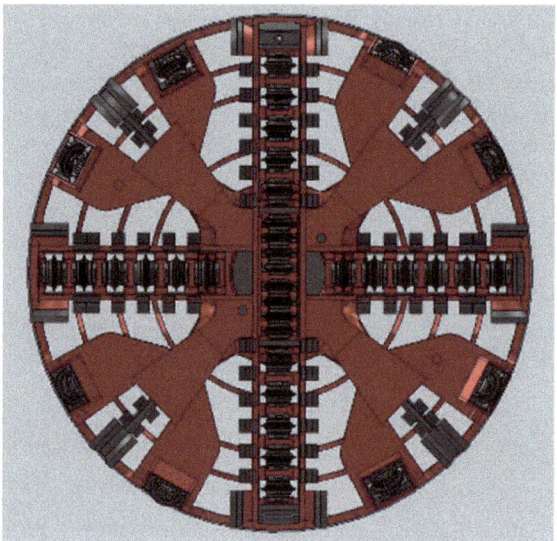

Fig. 2.5 Radial plate cutter

2.1 Types and Arrangements of Shield Cutterhead and Tool

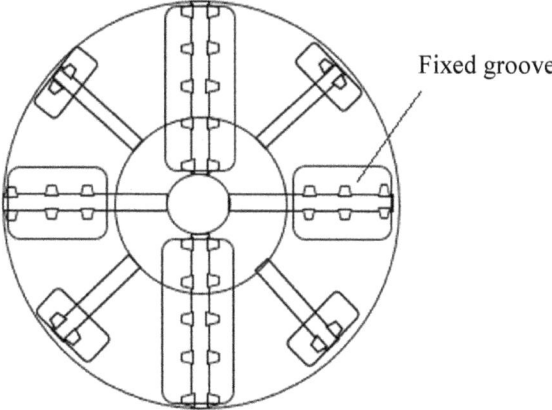

Fig. 2.6 Fixed panel cutter

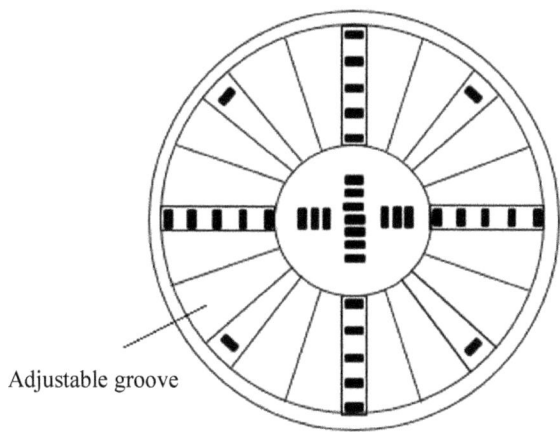

Fig. 2.7 Notch adjustable panel cutter

groove width is generally 20–30 mm, and the opening ratio 10–40%. The advantage of the panel cutter disc is that it directly supports the excavation surface through the cutter disc panel, which is beneficial to the stability of the excavation surface. The disadvantage is that due to the influence of the cutter disc panel, the earth pressure of the excavation surface is measured $\neq$ the earth pressure is measured, so the earth pressure management is difficult, and the cutter load is large and easy to wear.

2) Spoke cutter

The spoke cutter is mainly composed of a rim, a spoke and a cutter placed on the spoke. The cutters are placed on both sides of the spokes and it is generally difficult to arrange the hobs. The cutterhead opening rate is very large, ranging from 60 to 95%. The advantages of the spoke type cutter head are: low equipment cost; low torque resistance, good fluidity of the slag, weak secondary crushing, suitable for silt stratum and sand layer and sandy pebble stratum with small particle size; excavation Pressure = measuring earth pressure, thus enabling effective management of earth pressure. The disadvantages are: Can't limit large particle size stoneto enter the earthen cabin, the opening ratio is large, and the support effect on the stability of the excavation surface is poor. For example, Sichuan Island 6.14 m Shield (opening rate 95%), Komatsu 6.3 m Shield (opening rate 62%), Shanghai Metro Line 6, Sichuan Island 6.52 m double Circular Shield (opening rate 85%), are used in Beijing Metro Line.

3) Spoke plate cutter head

The spoke type cutter head has the characteristics of the panel type and the spoke type cutter head, and is composed of a wide spoke and a small piece of web. The cutter and the hob are respectively arranged on both sides and the inside of the wide spoke, and the opening ratio is 35–50%. For example, Mitsubishi heavy work 6.14 m shield used in Beijing Metro Line 4, Komatsu 6.32 m shield used in Tianjin Metro Line 1, Nanjing Yangtze River Tunnel, Nanjing Yangzijiang Tunnel, Yangzhou thin Xihu Tunnel, Guangzhou Shiyang Tunnel and other tunnel projects all use spoke plate cutter head.

2.1.1.4 Classification According to Whether There Are Directly Replaceable Cutters

It can be divided into conventional cutter head and atmospheric pressure change cutter head.

1) Conventional cutter head

Conventional cutter head means that there is no tool to be removed and replaced on the cutter head. If the cutter wear occurs during the tunneling process, in the bad phenomena, it is necessary to strengthen the soil in front of the excavation by techniques such as grouting and freezing, and the staff can open the cabin and enter the front of the cutterhead for tool replacement. This type of cutterhead has been used more in previous practical projects. In recent years, most of the shield machines used in China use atmospheric pressure tool changing tool head.

2) Atmospheric pressure change cutter head

Atmospheric pressure change cutter head means that some tools on the cutter head are equipped with tools that can be directly replaced or repaired under normal pressure, the tool is usually installed in the cutter gate. When the tool wears to a certain extent and needs to be replaced, the staff can enter the main arm of the cutter head through the center gate under the normal atmospheric pressure, and retract the tool through the cutter gate. After the cutter gate is closed, the worn tool can be removed and replaced with a new tool.

2.1 Types and Arrangements of Shield Cutterhead and Tool

The following is a brief description of the atmospheric pressure tool changer in several typical projects.

(1) Nanjing Yangtze River Tunnel Project. Nanjing Yangtze River Tunnel is one of the most direct fast passage connecting Nanjing urban area and Pukou District. It is located between Nanjing Yangtze River Bridge and third Bridge, with a total length of 5853 m, in which the shield tunnel is 3020 m long. The geological conditions of the Project are relatively complex. The composite stratum of silty fine sand, gravel sand and round gravel strata crossing the tunnel accounts for more than 43% of the whole tunnel length. The stratum has the characteristics of upper soft and lower hard, uneven soft and hard, strong water permeability and difficult construction. At the same time, the quartz content of sand and stone in the strata through which shield passes is generally high, which is a great test for the wear resistance of cutter head and tool.

The Yangtze River tunnel project of Nanjing is driven by the mud-water pressure shield produced by the German Heritage Co., and its cutter head is in the form of a center support and a spoke panel type. The shield cutter head is equipped with 6 main spoke arms and 6 auxiliary spoke arms, the cutting tools are composed of 189 scrapers (75 of which can be replaced at atmospheric pressure), 16 leading knives, 6 pairs of peripheral shovel knives, 6 pairs of small shovel knives and 2 copying knives. The front view of the cutter head is shown in Fig. 2.8.

The machine has the design of the original atmospheric pressure changing tool, which makes the shield cutter disc not seriously damaged during the complex formation tunneling process, and in order to completes the tunneling task more satisfactorily. The total number of atmospheric pressure changes in the construction of the Nanjing Yangtze River Tunnel exceeded 700, it can be said that the system contributed to the smooth progress of the Nanjing Yangtze River Tunnel. Figure 2.9 is the staff was replacing the tool under normal pressure.

(2) Wuhan Metro Line 8 Project. The project of Wuhan Metro Line 8 mainly includes one station, namely, Xujiahe Station and the Yujiang shield section of Xujiahe Station of the Huangpu Road Station. The project mainly passes through the soft-plastic silty clay layer, the fine sand layer and the medium coarse sand layer, and the 12.51 m composite mud-water balance shield is adopted, and the cutter head is in the form of a normal-pressure feed-in type. The cutter head is equipped with 61 normal pressure replaceable tools, of which 43 can be replaced by atmospheric pressure, 8 advance cutter can be replaced by atmospheric pressure, and 10 central cutter can be replaced by atmospheric pressure. In addition, there are 123 fixed scrapers, 20 fixed leading knives and 12 peripheral shovel knives. The front of the shield cutter head is shown in Fig. 2.10.

(3) Nanjing Wei Three Road Crossing Tunnel Project. Nanjing Wei Three Road Crossing Tunnel Project is located between the completed Nanjing Yangtze

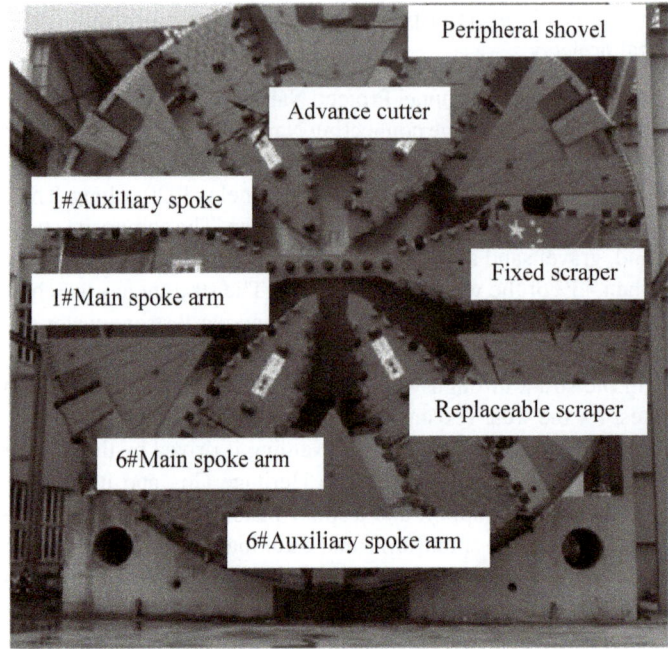

Fig. 2.8 Front view of atmospheric pressure change cutter head in Nanjing Yangtze River Tunnel

Fig. 2.9 Workers change tools under normal pressure

2.1 Types and Arrangements of Shield Cutterhead and Tool

Fig. 2.10 Front view of shield normal pressure change-tool plate for Wuhan Metro Line 8 Project

River Tunnel and Nanjing Yangtze River Bridge. It is a double-layer double-pipe tunnel and is constructed by mud-water pressurized shield with a diameter of 14.93 m. The shield cutter head of the project is also equipped with 83.19 in-double-edged hob and 6.17 in. single-edged hob. On the basis of the conventional tool configuration, part of the atmospheric pressure replaceable tool is arranged on the spoke arm, including 80 cutting tools at atmospheric pressure. In the process of crossing the gravel, pebbles and moderately weathered sandstone formation at the bottom of the river, there is tool wear and forced to stop and replace the tool. Figure 2.11 shows the front of the cutter head of the shield machine at atmospheric pressure.

Table 2.1 is for the adaptability of the formation of different cutter head structures.

2.1.2 Type of Shield Cutter

Shield excavation performance is mainly ensured by the selection and arrangement of the cutters. Different types of cutters are arranged on the shield cutterhead, so that the shied can adapt to the various layers of soft soil to hard rock. Mastering

Fig. 2.11 Front view of shield normal pressure tool changer in Weisan Road, Nanjing

the correct choices, how to use them, and making reasonable arrangements on the cutterhead is essential to improve the tunneling efficiency and create good economic returns.

In terms of the shape of the tool, Fig. 2.12 shows the face view and side view of four commonly used cutting tools (tooth-shaped cutter, roof-shaped cutter, insert-shaped cutter and disc-shaped hob). Among them, the tooth cutter and the roof cutter are mainly used for the excavation of soft formations such as sand, silt and clay; the insert-shaped cutter and the disc hob are mainly used for the excavation of hard layers such as gravel layers and rock formations.

The tool can also be classified according to the purpose of the cutting and the setting position, see Table 2.2.

2.1.2.1 Classification According to Whether It Can Be Disassembled and Replaced Under Normal Temperature Condition

Can be divided into atmospheric pressure replaceable tools and fixed tools. The fixed cutter head shall be replaced by opening the cabin under pressure. For mud-water pressurized shield, it is mainly through the formation of dense mud film on the excavation surface to realize the pressure opening, and the use of this method is not uncommon at home and abroad.

2.1 Types and Arrangements of Shield Cutterhead and Tool

Table 2.1 Selection of cutter structure types in different formation

Type	Aperture opening ratio	Suitable formation	Preparation note
Open type without cutter disc	–	Single sand pebble strata	The stability of excavation surface is not easy to control when the interval contains clay strata, and it is carefully used when the stability of excavation surface is not easy to control
Face plate cutter head	10–40%	Rich-water sludge formation and fine-fine sand formation	Usually only for mud-water balance shield
Spoke cutter	60–95%	sand-and-gravel formation	When the maximum particle size of pebble exceeds the slag discharge capacity of screw conveyor, it should be used cautiously
Spoke plate cutter	35–50%	Composite formation, common clay-gravel formation	The application scope is the most extensive, the earth pressure balance shield and mud balance shield are adopted

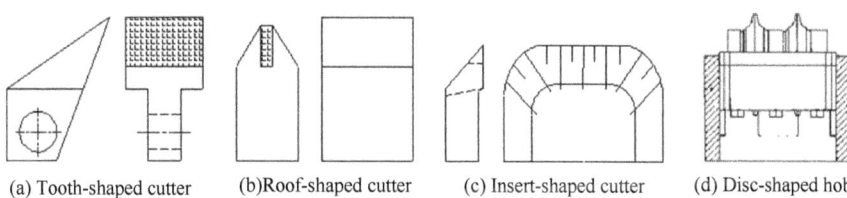

(a) Tooth-shaped cutter　　(b) Roof-shaped cutter　　(c) Insert-shaped cutter　　(d) Disc-shaped hob

Fig. 2.12 Cutting tool

The normal pressure replaceable tool does not need to be opened with pressure, and the tool change can be realized only by the switch gate. Replaceable under normal pressure tool is mainly composed of a cutter tooth, a cutter seat, a fixing bolt, a cutter cavity and a gate. The cutter teeth and the seat are integrally connected by fixing bolts, and are fixed in the cutter cavity, and the cutter cavity is welded on the cutter head. The special screw is used to move the cutter teeth and the tool holder back and forth in the tool cavity direction in the tool axis direction. During the tunneling, the cutter teeth and the seat are pushed to the front end and fixed. At this time, the cutter teeth extend out of the cutter to cut the face of the cutter. When checking and replacing, the cutter teeth and the cutter seat are fully retracted to the gate at the front

Table 2.2 Tool type, setting position, application

Category name	Use way	Set location
Fixed tool	Excavation	Front panel
Rotary cutter	Excavation	Front panel
Leading tool	Leading ahead	Front panel
Directional drill	Broken stratum	Front panel
Outer edge protection tool	Protect the outer edge of the cutter	Front panel extension
Trimming tool	Reduce propulsion resistance	Panel epitaxy
Back protection tool	Protective back panel	Back of panel
Plus mud nozzle protection tool	Protection plus mud mouth	Adding a mud mouth
Digging obstacle cutter of the front panel of the panel	Digging obstacles	Front panel center

end of the cutter rear. After closing the gate, the front part of the gate communicates with the excavation chamber, which is a high pressure chamber; the rear part of the gate communicates with the cutter head spoke, and is a normal pressure chamber. At the time, the inspection and replacement of the cutter teeth are performed under normal pressure. Figure 2.13 is the specific structure of the atmospheric pressure replaceable tool used in the Nanjing Yangtze River Tunnel Project.

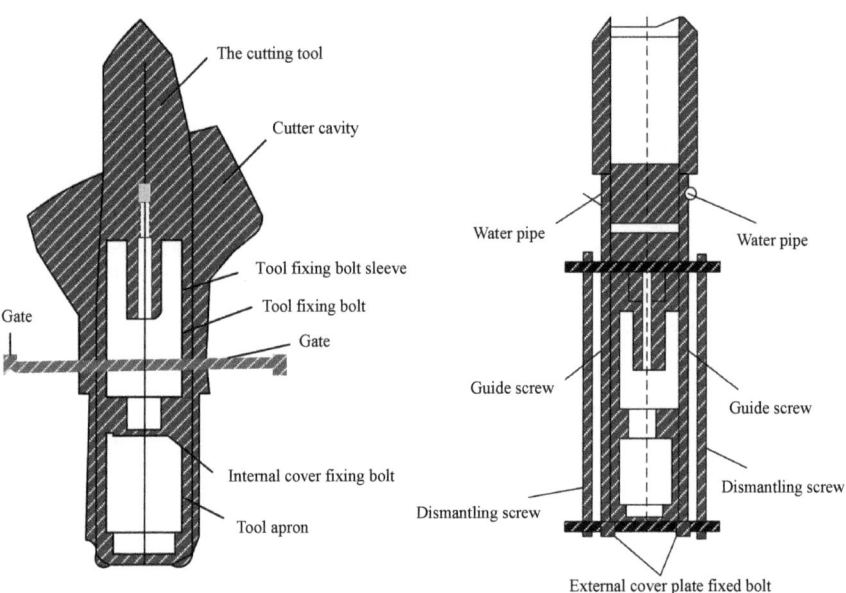

Fig. 2.13 Structure diagram of replaceable cutting tools at atmospheric pressure for Nanjing Yangtze River Tunnel Project

2.1 Types and Arrangements of Shield Cutterhead and Tool

Fig. 2.14 Removal of bolts at the installation position of the guide rod

The specific removal and replacement steps for the atmospheric pressure replaceable tool are as follows:

(1) the cutter head is rotated in place. Firstly, the cutter head spoke installed with the disassembled tool is rotated to a position of 6 points, so that the construction personnel can enter the cutter head spoke to work; then the hatch door on the central cone of the cutter head is opened for ventilation, the trachea, water pipe and lighting line from the shield to the spoke are installed, and the hoist of the hoisting tool is installed.

(2) Install guide rod and gate to open and close the cylinder.

① Remove the bolts at the end of the tool to install the guide screw, as shown in Fig. 2.14.
② Install the guide screw in the bolt hole of the bolt removed one step above, as shown in Fig. 2.15. Install the disassembly screw, as shown in Fig. 2.16.
③ Connect the special cylinder for opening and closing the tool gate to the two ends of the gate, as shown in Fig. 2.17.
④ Connect the manual hydraulic pump to the special cylinder for opening and closing the tool gate, and prepare to close the gate.

(3) Install the flushing water pipe. Connect the high-pressure water pipe on the shield to the tool flushing water pipe (the tool flushing water pipe has 2 pieces, One intake and one drainage, which are connected to the cutter body at the back of the tool gate and connected to the cutter cavity), as shown in Fig. 2.18.

(4) Remove the fixed bolt and withdraw the tool. Remove the remaining fixed bolts at the end of the tool. Use wrench to rotate two dismantling screw at the same time to make the cutter (cutter teeth and cutter base) exit along the guide screw evenly and slowly, as shown in Fig. 2.19.

(5) Flush the cutter cavity and close the gate. When the front end of the tool just retracts to the inside of the gate, open the water inlet valve of the flushing pipe

Fig. 2.15 Install guide rod

Fig. 2.16 Install the disassembly screw

Fig. 2.17 Install special oil cylinder for gate opening and closing

Fig. 2.18 Connect the flushing water pipe

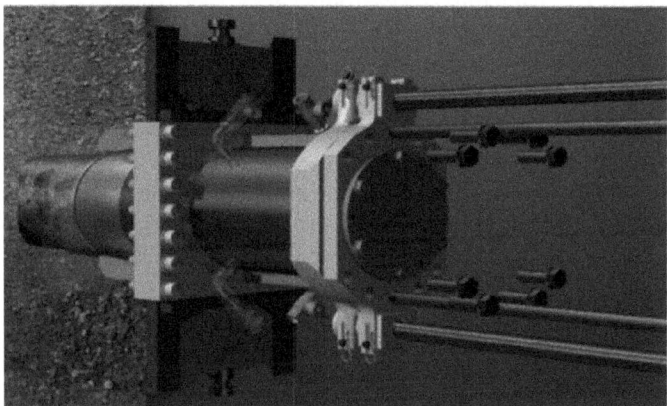

Fig. 2.19 Remove the remaining fixed bolts

       and rinse the cutter cavity with high pressure water. At the same time, use the manual hydraulic pump to supply oil to the special oil cylinder for opening and closing the gate, close the gate, and ensure that the gate is closed tightly. During the closing of the gate, the high-pressure water shall be maintained for continuous flushing of the tool chamber until the gate is completely closed.
(6)   Replacement of cutting tools. Remove the guide screw and disassemble and assemble the screw, use the gourd to lift the cutter teeth and the tool holder out of the cutter cavity, and check the wear of the tool (teeth). If the tool tooth wear is serious, first remove the outer cover plate fixing bolt located at the end of the tool seat and remove the outer cover plate; then remove the inner cover plate fixed bolt and remove the inner cover plate; then remove the tool

tooth fixing bolt, remove the seriously worn cutter teeth, install the new tool teeth on the tool seat, tighten the tool fixing bolts, and install the inner and outer cover plates respectively. The installation program of the tool is basically the opposite of the disassembly and unloading program, which is not repeated here.

2.1.2.2 Classification by the Method of Rock Breaking

It generally can be divided into two categories: rolling tools and cutting tools. Rolling tools include single-edged hobs and double-edged rolls hobs, three-blade hobs, etc.; cutting tools include cutters, tooth cutters, scrapers, cutting knives, profiling knives, center fishtail knives, shell knives, etc. For the excavation of different strata, the shield cutters usually adopt different forms: when the excavation stratum is hard rock, the disc hob is used; when the stratum is soft rock, the tooth cutter is used; when the stratum is soft soil or broken soft rock, a cutter or a scraper can be used.

In hard rock excavation, a hob is used to break the rock. The characteristic of the hob breaking rock is that it relies on the rolling of the tool to produce impact crushing and shearing. The effect of crushing reaches the purpose of breaking the rock. The hob is mainly composed of a cutter ring, a cutter body (set), a cutter shaft, a bearing, a seal, an end cover and so on (Fig. 2.20).

The hobs can be divided into disc hobs, ball hobs and wedge hobs according to their shapes. According to the installation location, it can be divided into center hob, positive hob and contour hob. According to the material of the blade, it can be divided into all-steel hob and inlaid carbide hob. According to different structural forms, it can be divided into single-blade hob, double-edged hob, three-blade hob, four-blade

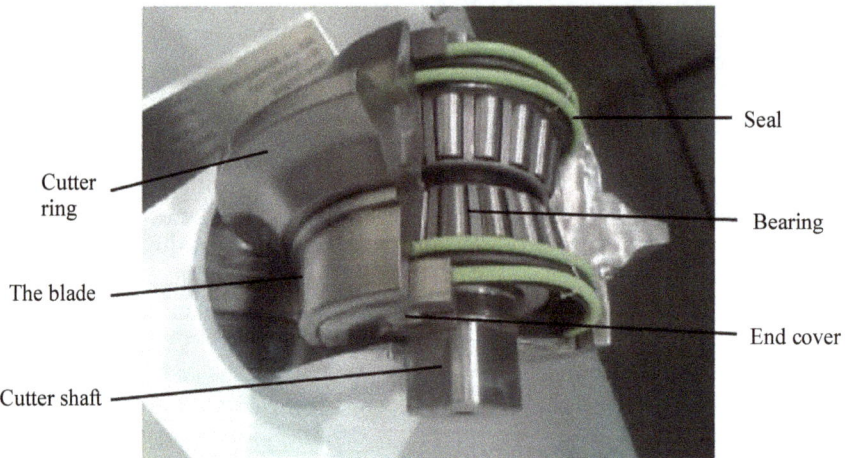

Fig. 2.20 Composition of the hob

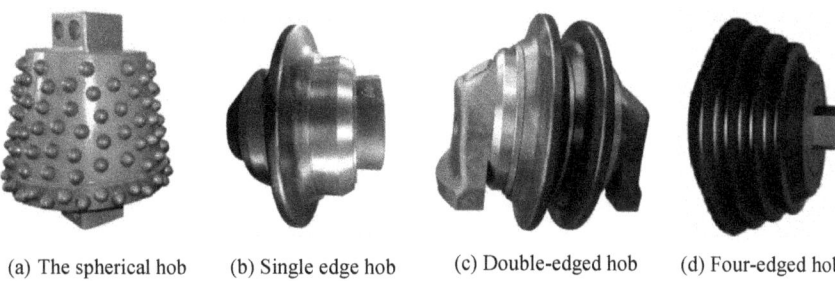

| (a) The spherical hob | (b) Single edge hob | (c) Double-edged hob | (d) Four-edged hob |

Fig. 2.21 Shield hob

profiling cutter, two-row hob, triple hob and eccentric hob. Some of the shield hobs are shown in Fig. 2.21.

Three-blade and four-blade hobs are only used in the center of the cutterhead to accommodate small radius rotations in the center of the cutterhead; double-edged hobs are usually used in soft rock such as mud rock and shale, which requires a small blade pitch, and can adopt a double-edged hob to adapt to geological requirements. When the rock mass is high, the blade pitch is too small, which leads to low tunneling efficiency. On the other hand, the double edge forced synchronous rotation will aggravate the wear under the condition that the excavation surface is not ideal leveling, and one of the edges will wear and the whole cutter will fail, so the double edge hob is not recommended in the middle of the hard rock.

2) Cutter

The cutter is a soft soil tool, which is installed on both sides of the opening slot of the cutter head and is used to cut the soil layer of the palm surface.

When the cutter disc rotates with the cutter, the soil body of the excavation surface generates axial shearing force and radial cutting force, and the cutter head and the cutting edge part are inserted into the stratum, and the soil body on which the excavated surface is continuously cut. The cutter is generally suitable for loose strata with particle size less than 40 mm, such as sand, pebble, clay and so on. The arrangement of the cutter on the cutter head is generally concentric and Archimedes spiral arrangement, and more is the Archimedes screw thread arrangement method.

Shield tool parameter design and material process selection are of great significance for shield tunneling. Cutter form and geometric parameters (Fig. 2.22).

The definition is as follows:

(1) Front edge surface: the surface through which the layer flows.
(2) Rear edge surface: the tool surface facing the new broken surface of soil and rock.
(3) Front angle γ: the angle between the front edge of the tool and the plane perpendicular to the cutting face through the main cutting edge.
(4) Rear angle δ: the angle between the rear edge surface of the tool and the cutting surface.

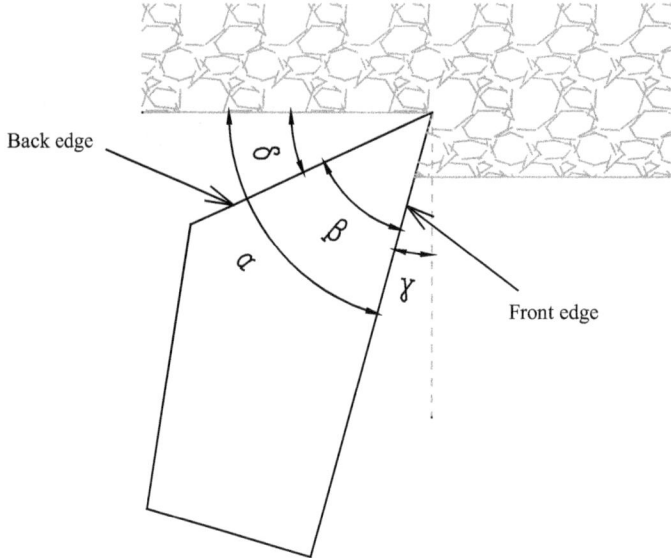

Fig. 2.22 Cutting tool geometry parameter

(5) Cutting angle α: the angle between the cutting surface and the cutting surface.
(6) The β angle of the cutting edge: the included angle between the front edge surface and the rear edge surface of the cutter.

Parameters and factors affecting cutting performance of cutting tools include: front angle γ, rear angle δ, blade cross-sectional dimension, alloy hardness, bending strength, alloy and blade body connection process (thermal insertion or medium, low-frequency welding).

3) Tooth cutter

When the shield is digging in the case of soft rock, there is no effective effect adhesion between the disc hob and the rock face, which will cause the hob to not roll and produce a string grinder. If the hob does not roll, it will lose the effective rock breaking function. Therefore, in the soft rock formation, part of the disc hob should be replaced with a replaceable cutter, and the rock cutter is used to break the rock, and the rock breaking trajectory is the same as that of the hob. The principle of breaking the rock of the shield cutter is similar to that of the cutter, and there are two cutting edges on the cutter. The rock can be broken when it is rotating forward or reverse. Part of tooth cutters used in the shield is shown in Fig. 2.23.

4) Scraper

The scraper is a tool for cutting soft rock and soil or scraping muck. Its main function is to clean rock debris and edge dregs. The cutting mechanism of the scraper is similar to that of the cutter, and the clods under cutting are scraped into the earth chamber.

2.1 Types and Arrangements of Shield Cutterhead and Tool

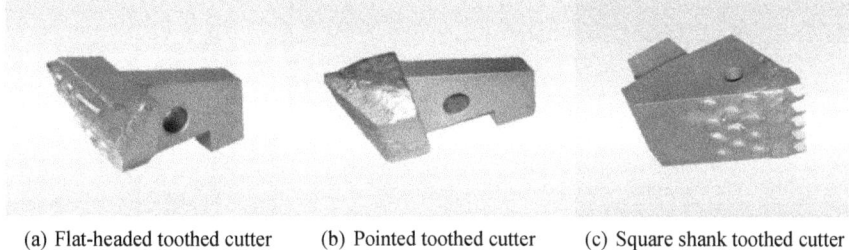

(a) Flat-headed toothed cutter   (b) Pointed toothed cutter   (c) Square shank toothed cutter

Fig. 2.23 Tooth cutter

According to the installation position, the scraper can be divided into a positive scraper and an edge scraper (peripheral scraper, gauge cutter). The downside is that it is subject to wear and tear. Part of the shield scraper is shown in Fig. 2.24.

5) Fishtail cutter

The fishtail cutter is installed at the center of the shield cutterhead. It is usually used for sand cobble formations or high-strength clay formations. It is used to advance the soil at the center of the cutter blade to improve the fluidity of the soil at the center and prevent mud cake. At the same time reduce the cutting resistance of other scrapers and reduce wear. The center cutter can be divided into a fishtail cutter and a cavel cutter, and the fishtail cutter can be divided into an integral type and a combined type. The fishtail cutter is shown in Fig. 2.25.

The size of the fishtail cutter is large, generally long 120–150 mm, high 400–500 mm, higher than the ordinary scraper 200–300 mm. The design and arrangement principle of the fishtail cutter is: The design is not in the same plane as the other tools, so the fishtail cutter cuts the soil body first and the original stratum; the root of the fishtail cutter is designed to be tapered, which increases the flipping motion of the soil cut by the fishtail cutter when the cutter head rotates, thus solving the cutting problem of the central part of the soil and improving the fluidity of the cutting soil.

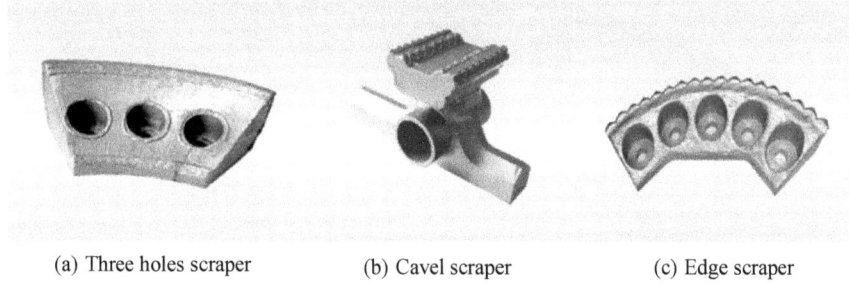

(a) Three holes scraper   (b) Cavel scraper   (c) Edge scraper

Fig. 2.24 Scraper

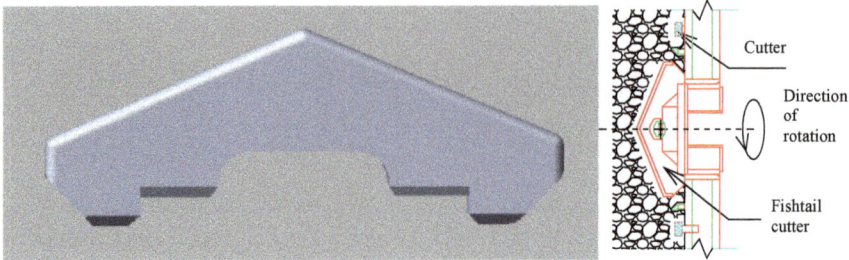

Fig. 2.25 Fishtail cutter

6) First cutter (Advanced cutter)

Arranged on the panel and spokes, between the cutting path of the scraper, the first cutter is usually 40–50 mm higher than the (ordinary) scraper, doesn't play a direct cutting role, contacts the formation early than the cutter, plowing soft rock and soil layer. The first cutter cuts the soil before cutting cutter cuts the soil, and cuts the soil to create good cutting conditions for the cutting cutter. Using a first cutter, generally will improve the fluidity of the cutting soil, greatly reduce the torque of the cutter, reduce the cutting resistance of the scraper, improve the cutting efficiency of the cutter, and reduce the wear of the cutter. The first cutter is divided into two types: replaceable and non-replaceable. The first cutter is shown in Fig. 2.26.

7) Shellfish cutter

It can be regarded as a peripheral cutter for the right-angled cutterhead. The outer diameter of the cutter is slightly larger than the outer diameter of the shield, reducing the propulsion resistance of the shield, preventing the cutter disc from being worn, and cleaning the soil at the bottom of the cutterhead to improve the efficiency of

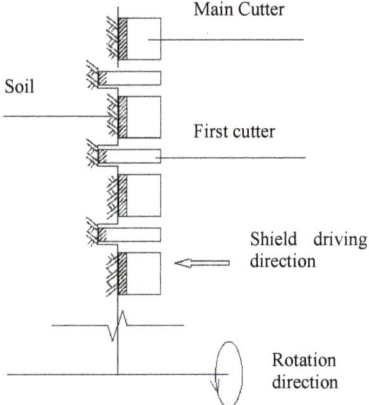

Fig. 2.26 First cutter

Fig. 2.27 Shell cutter

the tunneling. When the shield passes through the sand- pebble stratum, especially in the large-grained sand-pebble stratum, if a hob- shaped cutter is used, due to the loose body of soil debris, Under the condition of hob driving and extrusion, large deformation will be produced, and the cutting effect of hob will be greatly reduced, sometimes even loses the ability to cut and break. It is arranged on the front end face of the cutter disc coil by a rim bell cutter, which is specially used for cutting sand eggs stone. The difficult problem of cutting soil (sand and pebbles) by shield cutting can be solved by using circle shellfish knife (Fig. 2.27).

2.1.3 Arrangement of Shield Cutter Head

The arrangement of shield cutter head refers to the reasonable arrangement of the cutter head after the geological adaptability selection of the tool, which is one of the core contents of the shield cutter head design. This requires a variety of cuts for shield cutters in a limited space under various performance constraints let the tool to meet the needs of the shield tunneling project. In the case of complex shield conditions and varied geological conditions, the cutter head with dozens of times or even hundreds of times of random abrupt load, often causes vibration of the whole machine, damage of the tool, and even the disintegration of the cutterhead. This puts high requirements on the geological adaptability of the cutterhead and the reasonable arrangement of the cutter. Therefore, whether the tool arrangement is reasonable or not is not only related to the force of the cutter head, the life of the tool and the main bearing of the cutter head, but also affects the driving efficiency of the shield and the normal construction of the shield (Fig. 2.28).

2.1.3.1 Layout Principles of Shield Cutters

The tool layout should not only consider the type of tool, but also the motion characteristics and knots of different radii across the entire face of the cutter face.

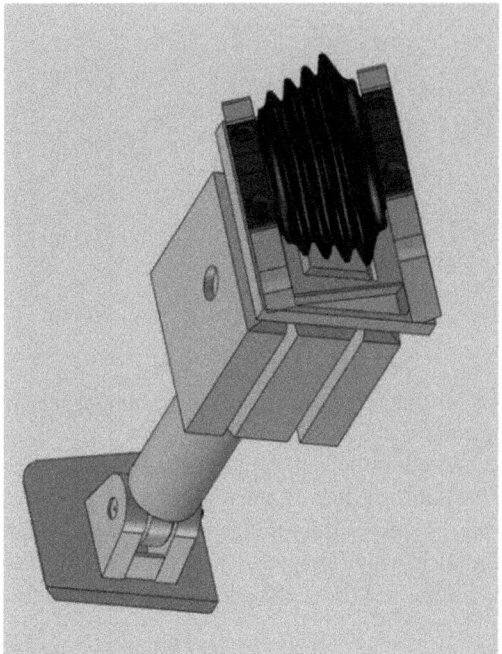 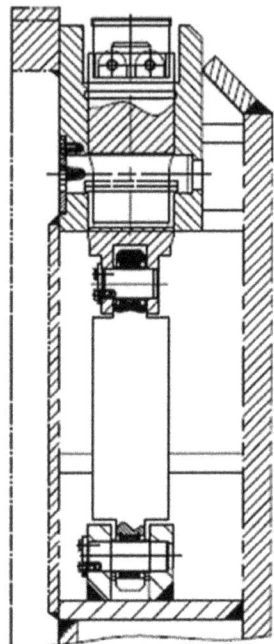

Fig. 2.28 Profiling cutter

Because the digging diameter of the shield is large, different digging diameters have different mining effects due to their different linear velocities, spatial structures and their interaction with the excavated materials. In the case of earth pressure balance shields, the cutters mainly are hobs and scrapers, and the general hob is 20–35 mm higher than the scraper, which plays the role of crushing the rock firstly, ake a crack on the rock or hard soil. The scraper can then scrape the broken rock and bring it to the opening. Due to geological formations, the soil quality of many soil layers is higher than that of rock or hard soil. The viscosity is relatively high and the strength is low. At this time, the use of the hob will often cause the phenomenon of the paste cutter, and the good condition is that it will cause the hob to fail to roll, resulting in excessive local wear; bad condition is that it causes the cutter head to rotate hard or die, damage main drive system. Therefore, for the more viscous formation, a tearing cutter is often used instead of the hob. The tearing cutter is generally made up of 4 vertical tooth, so that cutting force of the tearing cutter can instead the pressing force of the hob, and the structure is simple, and the cost is relatively low. The assembly of the tearing cutter is designed on the hob holder so that it can be easily realized the exchange of two tools on the same holder in mixed geology. When designing the tool height, the size of the cutter should be considered to be higher than the size of the cutter face. The appropriate amount can prevent the formation of the paste cutter or the mud cake. However, if the height of the tool is too high, the corresponding torque is increased and there is a danger of breaking the cutter. In short, to ensure the spindle's stability and excavation efficiency, each tool must be placed evenly on the entire excavation surface according to its own trajectory. In general, the arrangement of the tool on the cutter head generally follows the following principles:

(1) The principle of reasonable distance between knives. In engineering, the minimum distance between the two adjacent cutters in the radius direction is defined as the tool spacing, which is usually expressed as S. Reasonable tool spacing (mainly hob) arrangement should make the rock between adjacent hob can be completely broken by hob, that is to say, the crushing slot of adjacent hob can intersect. Because the size of the crushing groove is related to the hob size, rock properties, push force or cutting depth, the knife spacing is also related to the above factors. The choice of the tool spacing on the project is usually 40–120 mm, the harder and more complete the rock stone, the smaller the choice of cutter spacing should be. The number of tools placed on the same diameter cutter head should be more. On the contrary, the softer the rock, the more complete the development of the joint micro crack, the choice of the cutter spacing can be appropriately larger, and the number of tools can be less. For example, according to experience for a 17in hob, the maximum tool spacing in hard rock should generally not exceed 90 mm. Table 2.3 shows the requirements for the tool spacing of some formations.

In practice, the determination of the distance between the full-section rock-boring machine cutters is generally determined by two arrangements: First, the cutter spacing on the same road header is constant, and the blade strength is changed to adapt to the rock strength; The second is to change the blade spacing on the same tunneling machine by increasing or decreasing the number of cutters or changing the number of cutters on each disc-shaped cutter to adapt to the rock strength. Among them, the first arrangement is simple and highly usable, and is generally adopted.

(2) The principle of reasonable tool height. The height of shield tool arrangement refers to the distance between the tool and the cutter disk, which is related to the efficiency of shield tunneling. In soft soil layers with high viscosity and low hardness. The phenomenon of paste knife or mud cake will appear in the

Table 2.3 Requirements for tool spacing in some layers

Engineering geological condition	Uniaxial compressive strength	Distance between knives (mm)
Limestone formation	80	80
Granite strata	180	50
Limestone strata are dominated by granite	80–180	70
Soft rock is dominant and contains a small amount of hard rock	300	120

Table 2.4 Tool height in different geotechnical bodies (mm)

Engineering geology	Tool type	Tool arrangement height	Difference in height
Soft soil formation	Serrated knife	160	20
	Slicker	10	
Hardpan	Hob cutter	175	25–35
	Slicker	140	
Composite formation	Hob cutter	175	25–35
	Slicker	140	

excavation of too small tool height; and the excessive tool height will increase the corresponding torque, there is a danger of broken cutter.

In hard or composite formations, the shield is equipped with both breaking tools and cutting tools, but the heights of the two tools are also different. The height of the cutting tool can effectively scrape the soil layer into the earthen cabin and reduce the wear of the cutter head; the height of the rolling tool should meet the rock breaking requirements and reduce the grinding of the cutting tool. When the height difference between the hob cutter and the cutting tool is too small, the load on the main bearing of the cutter head will be increased, the wear of the cutter will be accelerated, and the service life of the cutter will be shortened. When the difference is too large, the cutter will cut too deep and cause cracking. Combined with the use of domestic shields, the relationship between tool height and rock and soil structure can be seen in Table 2.4.

(3) The principle of tool balanced wear. The ideal tool arrangement is to satisfy the condition of rock breaking (the tool broken rock ring should cover the excavation surface and do not leave the protruding rock ridge ring), and the wear amount of each tool position should be balanced as much as possible, that is, "the original rule of equal wear arrangement". However, due to the limitation of the maximum knife spacing, the smaller the rotation radius near the center of the cutter head, the smaller the tool wear. Therefore, only if the radius of rotation is greater than a certain value of the cutter head area, can we achieve "equal wear cloth knife".

(4) Principle of tool force balance. The combined force of the horizontal cutting force of each tool on the cutter head (perpendicular to the axis of the cutter head) is called the unbalanced force. The design imbalance should be as small as possible. In general, the ratio of unbalanced force to thrust should be less than 1%. The unbalanced force is too large, and the boring process is prone to the vibration of the cutter head, which is unfavorable to the bearing capacity of the cutter head spindle. Effective measures to reduce imbalance is to arrange the tool symmetrically as much as possible.

(5) Installation requirements. When installing hob, consider whether different components interact with each other.

2.1 Types and Arrangements of Shield Cutterhead and Tool

(6) The distribution of centroids. The center of the cutter head should coincide with the center of gravity of the hob as far as possible.
(7) The radial load is the smallest.
(8) The overturning moment is the smallest.

From a geometric point of view, the arrangement method of the tool on the cutter head mainly includes Archimedes spiral arrangement and concentric circles. In order to ensure full-section excavation, the Archimedes spiral layout method is mainly used. The cutters are distributed symmetrically on both sides of the spokes intersecting the spiral to meet the requirements of the forward and reverse directions of the shield, thus achieving layout, optimal design of structure and load. The Archimedes spiral is also called a constant velocity spiral. When a moving point moves along the ray, it moves linearly at a constant speed. When the ray is rotated at the same angular velocity as the center of the circle, the trajectory of this moving point is called the "Archimedean spiral" (Fig. 2.29), and the curve equation is:

$$\rho = \rho_0 + \alpha(\theta - \theta_0) \tag{2.1}$$

ρ—Polar diameter.
ρ_0—Initial value of polar diameter.
α—Constant.
θ—polar angle.
θ_0—polar initial value.

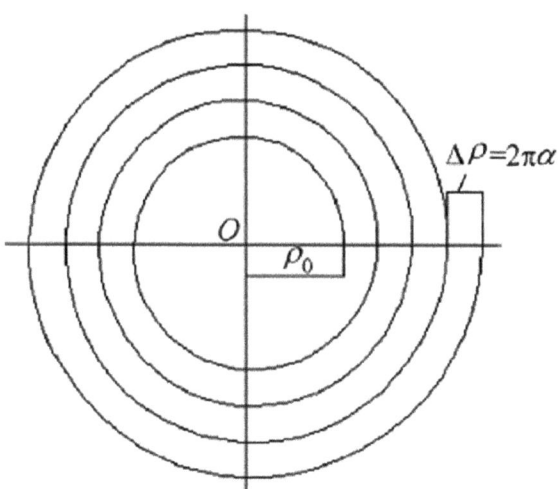

Fig. 2.29 Archimedes spiral diagram

2.1.3.2 Layout Principles of Hobs and Scrapers

The tool of the composite shield includes a hob, a cutter, a scraper, a first cutter, etc., and the arrangement of the cutter on the composite shield cutter head is generally divided. There are three areas, namely the central area, the front area and the edge area. The tools in the three areas play different roles in the tunneling process, and the tool layout features are also different, but they all follow the layout principle of the shield cutter. The following is a detailed description of the layout principle and content of the hob and the scraper.

1) Hob arrangement

The normal working speed of the cutter head is 0.9–1.2 r/min, and the closer the cutter head surface is to the center of the cutterhead, the lower the linear speed, and the center is basically. In the environment with poor material flow, in order to avoid clogging when excavating sticky materials, the opening ratio in the middle of the cutter must be large, and also avoid the wear caused by excessive distance of the excavated material from the center of the cutter surface to the structural opening. In addition, the geometry of the center of the cutter head is also very small, so only a single row of center hobs is provided at the center of the cutter head, and the pitch control is preferably about 84 mm. If a tearing cutter is used instead of a hob, set a center tearing cutter or a dovetail cutter, a dovetail cutter is usually designed to protrude a certain size of the cutter face, and it can effectively prevent the paste cutter when excavating the soft soil layer. The hob distribution in the radial direction of the cutter head is based on the principle of the track pitch of about 10 mm, and the position of each hob on the spokes is evenly distributed. If the excavated material has high viscosity and low hardness, and can appropriately increase the track pitch; if the excavated material has small viscosity and high hardness, track spacing can be appropriately reduced. The line speed on the periphery of the cutter head is the largest, the space structure is the most affluent, the movement is also the most unstable, and the hob distribution should be relatively dense to improve the excavation efficiency and ensure the stability of the main bearing. Therefore, generally, a hob is added to each of the webs, and the position of each of the hobs on the spokes and the webs is evenly distributed on the principle that the track pitch decreases with the increase of the radius. For the hob, only one cutter is assigned to each track, which can achieve the effect of breaking the soil.

2) Arrangement of the scraper

The scraper placed at the opening serves to remove the entire surface, and the cutting path of the single scraper should be a torus, the cutting path of all scrapers must cover the entire excavation face. For the arrangement of the scraper in the radial direction of the cutter head, a scraper of half the number of spokes should be arranged on each track, that is, if there are 6 spokes, 3 scrapers are arranged for each track. In order to make the cutting direction of the blade face is as close as possible to the radius of rotation. It is necessary to pay attention to the angle of assembly of the blade on the spokes. For example, a typical earth pressure balance shield with a diameter of 6.28 m is used. The blade angle is 25° below 120 mm radius, and the blade angle is 20° in the radius of 1200–1600 mm, the blade angle is 15° in the radius of 1600–2000 mm, the blade angle is 10° in the radius of 2000–2400 mm, the blade angle is 5° in the radius above 2400 mm.

2.1.3.3 Common Installation Location of Various Tools

Figure 2.30 shows the common installation positions of all kinds of cutting tools.

China has a vast territory and great differences in geological conditions. The appropriate cutter head structure and tool configuration scheme should be selected according to the specific conditions. The main results are as follows:

(1) It is suggested that spoke cutter head should be used in gravel composite strata represented by Beijing and Shenyang, and soft soil knives with high strength, such as shellfish knives, should be used.
(2) In the muddy strata represented by Shanghai and Tianjin, the panel cutter head or spoke cutter head can be selected, all of which are equipped with soft soil cutter panels, and the opening rate can be enlarged appropriately.
(3) The hard rock represented by Guangzhou strata and the composite strata with uneven soft and hard should be equipped with spoke plate cutter head, which can be used for rock breaking, or all of them can be replaced with soft soil knives, which can be used for soft rock excavation.
(4) The boulder stratum represented by Shenzhen can be said that there is no cutter head designed to cope with boulder, which can only be broken by auxiliary measures. The radial plate type composite cutter head can be used, equipped with a rolling cutter and a cutting tool, which is similar to that commonly used in Guangzhou.
(5) The pebble formation represented by Chengdu and Lanzhou can be equipped with spoke plate cutter head and tear knife with high strength. Hob plays little

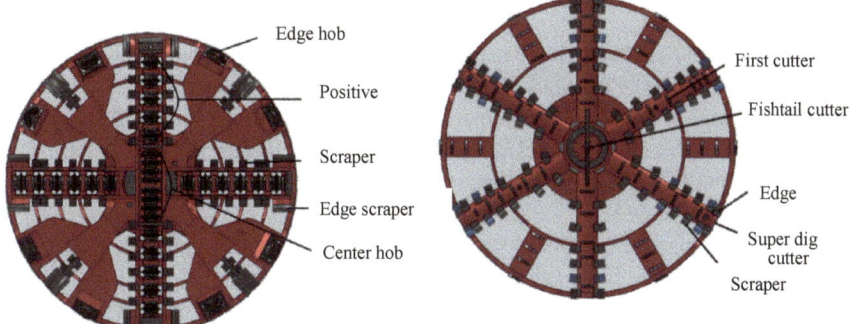

Fig. 2.30 Common installation positions for various types of tools

Table 2.5 Tool layout of Intercity Rail Transit Construction Project in Guangguan and Shenzhen of Pearl River Delta

Tool type	Central tool	Positive hob	Side hob	cutter	Overcut knife
Quantity (handle)	4	31	12	144	2
Height difference between tool and cutter head (mm)n	150	150	150	110	110–150
Blade spacing (mm)	89–110	89	89	290	40
Tool arrangement	Spiral arrangement			No	No

    role in this kind of formation, and the main function of cutting tool is stripping rather than breaking.

(6) The loess layer represented by Xi'an can be equipped with spoke type or surface plate cutter head, and the tool can be used as soft soil tool.

Taking the construction project of Haiguan deep intercity rail transit project of the Pearl River Delta as an example, the tool layout is described in detail.

The project site is a composite formation dominated by silty mudstone, silty clay and gravel sandstone. In the arrangement of cutting tools, not only the high strength excavation of hard rock, but also the influence of mudstone viscosity on shield excavation are considered. According to the uniaxial ultimate compressive strength of rock in the formation, it is determined that the hob spacing is 89 × 110 mm, and the spiral arrangement is adopted, and the cutting tools are arranged evenly at the edge of the six spoke plates with symmetrical cutter head and arranged in both directions. After the cutter head is put into use, the driving efficiency is high, and it is found that the tool wear tends to be uniform and there is no serious bias wear of the tool during the tool head inspection. The specific tool arrangement is given in Table 2.5.

2.2 Cutting and Wear of Tools

Serious tool wear problems have occurred in many complex shield projects at home and abroad, such as the Istanbul Tunnel in Turkey. For example, the Shield Tunnel of Istanbul Tunnel in Turkey, Green Heart Tunnel in the Netherlands, DTSS Tunnel in Singapore, Yangtze River Tunnel in Nanjing (Wei Seven Road), Beijing Underground Diameter Line, Wei Three River Tunnel in Nanjing, and the River-crossing Section of Wuhan Metro Line 8,

The occurrence of accidents such as severe wear of the cutter caused shield tunneling difficulty or forced shutdown has attracted the attention of the engineering and technical community. Clarify the cutting mechanism of different types of tools and the tool wear law helps to rationally arrange and select the number of tools,

2.2.1 Cutting Mechanism of Shield Tool

1) Cutting mechanism of the hob

The research on the cutting mechanism of the hob started from the coal mining in the mine. The fracture line of the coal seam after cutting is considered to be circular arc, and the cutting force of the rolling cutter is calculated based on the assumption. With the massive construction of the mountain tunnel, the machine for installing the hob has been applied more, and the research on the rock breaking mechanism of the hob is deeper. The way of cutting the rock by the hob is mainly squeeze damage, shear damage and tensile damage. The crushing of the hob is mainly caused by the path of the hob, Through the blade surface of the hob, the compressive stress is directly applied to the rock mass, and the surface of the rock mass is subjected to excessive compressive stress, which leads to failure, resulting in fine broken rock mass particles; and the shearing rock is mainly generated outside the trajectory line of the hob cutting edge, the destroy occurs because the shear stress of rock mass exceeds its shear strength. The most important rock breaking method of the hob is tensile failure, and the failure of the rock body is in accordance with the tension failure theory. By applying compressive stress to rock mass, rock mass is broken and cracks are produced in rock mass. Then, due to the extension and penetration of cracks, the rock mass produces excessive plastic deformation and final tensile failure, as shown in Fig. 2.31.

By considering three kinds of rock breaking modes of hob, such as tension, extrusion and shearing, combined with tool selection, cutting parameters, rock type and experimental data of linear cutting machine, etc. A semi-theoretical and semi-experimental CISM model was proposed by the Colorado University of Mining in the United States of America, it is used to calculate the force of the hob (Fig. 2.32),

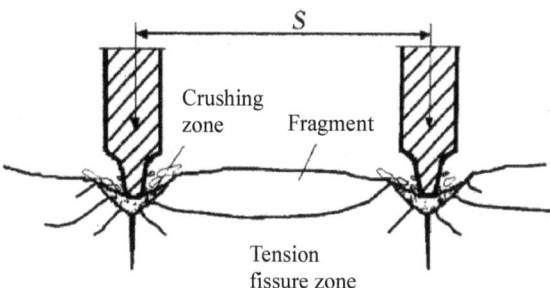

Fig. 2.31 Hob tension broken rock diagram

Fig. 2.32 Schematic diagram of hob force

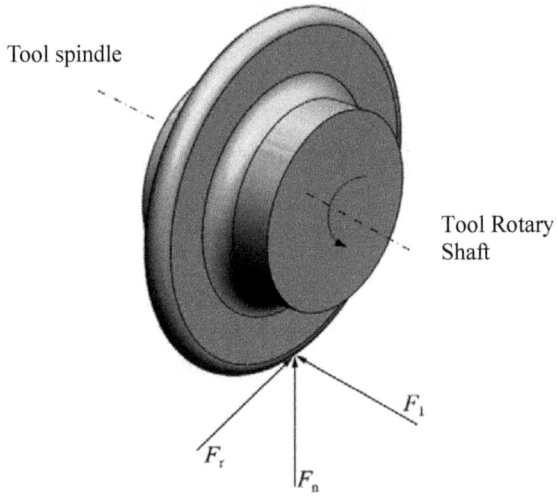

which provides the basis for the selection of the hob. But the model only considers the normal force and the rolling force of the tool, do not take into account the lateral force of the hob. Therefore, the hob force obtained based on the model is smaller than the experimental test value.

The estimated resultant force of hob in CSM model is as follows:

$$F_t = \int_0^\varphi TRp d\theta = \int_0^\varphi TRp_0 \left(\frac{\theta}{\varphi}\right)^\varphi d\theta = \frac{TRp_0}{1+\psi} \quad (2.2)$$

$\varphi = \arccos[(R-L)/R]$, $p = p_0(\theta/\varphi)^\varphi$.
F_t—The resultant force of hob.
Φ—Contact angle between hob and rock.
T—Hob tip width.
R—Hob radius.
P—Crushing zone pressure.
p_0—The basic pressure in the crushing zone directly below the hob is related to the strength of rock (rock compressive strength σ_c and rock shear strength σ_t), the size of hob (radius of hob R, spacing of tool tips S and one-time tunneling quantity p), and the shape of the blade (width of hob tips T, etc.), $p_0 = f(\sigma_c, \sigma_t, S, T, R, p)$.
ψ—Pressure distribution coefficient of tool tip, $\psi = -0.2$–0.2 (If the hob is V-shaped and sharp, $\psi = 0.2$; If the width of the rolling knife tip is larger, $\psi = -0.2$; General situation, $\psi = 0.1$).
θ—Blade angle.

However, the rock breaking mechanism of single-edged hob and double-edged hob is not exactly the same in the same rock. The single-edged hob presses the rock many times until the rock mass produces tension cracks and destroys, resulting in more regular cracks and larger slag blocks. For the double-edged hob, the rock is broken together through the cooperative action between the hob. The extrusion pressure of the hob on the rock mass is greater, the rock is easy to break and produce fine gravel and slag, and the fracture energy consumption is high, only the complete lenticular slag appears in the extremely hard rock. Therefore, priority should be given to single-edged hob in general rock mass to reduce the energy consumption of rock breaking, while double-edged hob should be equipped in extremely hard rock to strengthen the extrusion pressure of hob and improve the efficiency of rock breaking.

Different rock properties also have some influence on the cutting mechanism of hob. The discrete element method is used to simulate the hob cutting path. It is found that the joint characteristics of rock have obvious influence on the rock breaking of hob. The smaller the joint spacing is, the greater the penetration noise of hob is, and the cracks produced by rock are easy to develop along the structural plane of rock mass. Similarly, the rock confining pressure has a significant influence on the cutting performance of the rolling knife. The increase of the confining pressure prevents the development of the fracture to a certain extent.

2) Cutting mechanism of scraper

The early research on the mechanical model of scraper cutting mainly refers to the experience of cutting soil processes such as excavator bucket teeth and plough teeth in agriculture and forestry cultivation in earthwork. The basic cutting process of the scraper is: The stress and deformation of the excavated soil are caused by the cutting effect of the blade and the push and extrusion of the front surface of the scraper. Among them, the cutting effect of the cutting edge makes the stress of the cutting layer soil exceed the strength of the soil, and the cutting layer soil is separated along the direction of the cutting edge. The pushing action of the rake face makes the separated soil deform and separate from the parent to form soil debris. Soil debris enters the opening with the front of the scraper, so the tool has both cutting and loading functions.

The deformation of soil in cutting area can be divided into three deformation zones, as shown in Fig. 2.33.

Fig. 2.33 Soil mass deformation in cutting area

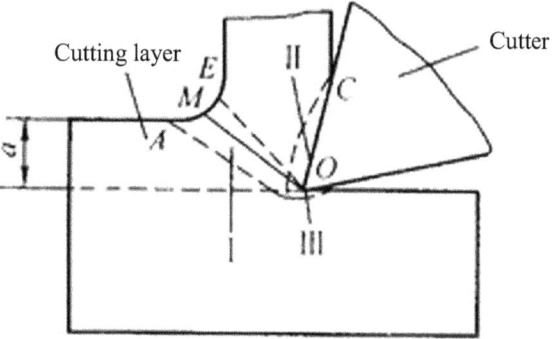

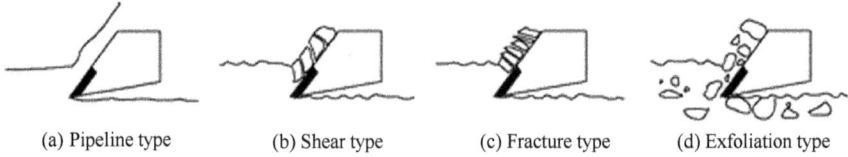

(a) Pipeline type   (b) Shear type   (c) Fracture type   (d) Exfoliation type

Fig. 2.34 Schematic diagram of soil flow model in scraper cutting

The first deformation region I: It refers to the deformation of the cutting layer near the front cutter face under the pressing action of the tool.

The second deformation region II: Refers to the extrusion friction deformation between the chip and the front blade during the outflow process.

The third deformation region III: Refers to the deformation zone produced in the face of the palm near the cutting edge. This deformation zone is mainly caused by extrusion and friction of the blunt circular part of the scraper and the rear knife facing the palm surface.

The most influential of the scraper cutting models is the flow model of Japanese scholars Shintaro Yano and Takeshi Fukuhiro. The soil excavated by cutting tool has different flow forms, which is mainly related to the cutting angle, speed, thickness and the properties of soil. There are four common flow patterns: flow type, shear type, fracture type and stripping type, as shown in Fig. 2.34. Flow cutting mainly occurs in silt formation, water-rich clay formation, silt formation and other low strength strata. With the movement of scraping knife, the soil produces continuous shear deformation from the blade; shear cutting mainly occurs in clay and silt layer with relatively high strength. During cutting, the soil mass is subject to compressive deformation, and then the cutting soil moves along the cutting edge surface until the soil mass is damaged by shear; the fracture cutting occurs in the sand formation, and the soil produces compression deformation in front of the cutting edge and maintains the stability for a certain period of time. With the cutting progress, the soil at the cutting edge produces cracks and destroys, and the slag soil is small in block; Spalling cutting occurs in common gravel and gravel strata, and scrapers play the role of peeling sand pebbles.

As the most commonly used force model in scraper cutting, Mckyes-Ali model divides soil failure into central failure zone and two half-moon failure zone, as shown in Fig. 2.35, and simplifies logarithmic helix failure line into straight line to derive the cutting force of the tool. On the basis of Mckyes-Ali model, most of the domestic research on scraper cutting model is based on the assumption that scraper cutting soil vertically or neglecting soil weight, so as to improve the calculation model. Guo Feng studied the interaction between tool and soil during rotary cutting, analyzed the rotating cutting process of cutting tool, obtained its cutting angle, and pointed out that the failure zone of rock is composed of central failure zone and one inner failure zone, and shear failure is on the outside of the center failure zone. Song Kezhi, Guan Huisheng analyzed the cutting process of the wedge cutter of shield tunneling machine. The cutting soil is taken as the isolated body for force analysis. The force

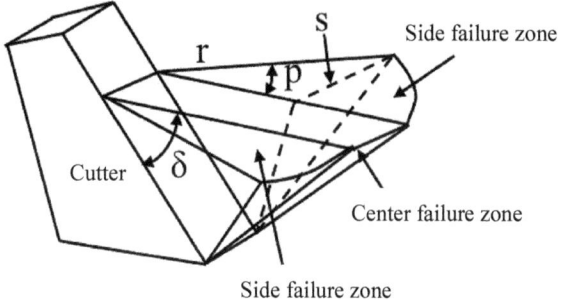

Fig. 2.35 Mckyes-Ali theory soil failure map

acting on the cutting soil mass includes the normal force, friction force, normal force, friction force and cohesive force of shear fracture surface, and neglecting the gravity and inertia force of the soil mass. According to the balance principle of force, a two-dimensional calculation model of cutting force is obtained. The force model of the force is shown in Fig. 2.36.

Song Kezhi and Guan Huisheng analyzed the cutting process of the wedged tool of shield roadheader, and took the cut soil as the isolator for stress analysis. The force acting on the cutting soil was the normal force, friction force, normal force, friction force and cohesion force of the contact surface between the tool and the soil, and ignored the gravity and inertia force of the soil. According to the balance principle of the force, the two-dimensional calculation model of the cutting force of the tool was obtained. The stress model is shown in Fig. 2.36.

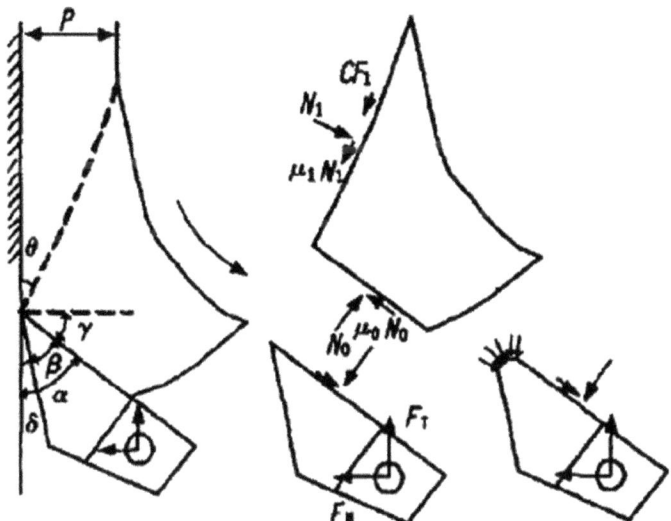

Fig. 2.36 Two-dimensional force model diagram of cutting tool of shield roadheader

Yu Ying et al. studied the total torque of cutting cutter head, and obtained the total torque of two-dimensional force model cutter head $T = T_1+T_2+T_3+T_4+T_5+T_6$. T is the torque generated by the cutting tool head, when the cutting tool is cutting in the soil, the soil in front of the cutting tool is extruded to the ultimate equilibrium state, and the pressure of the soil on the cutter surface is passive soil dynamic pressure, and the direction points to the cutter surface.

$$P_{11} = \delta_{\max}\frac{B_1}{2}h_1 = \left(\gamma H_o K_p + 2C\sqrt{K_p}\right)\frac{B_1}{2}h_1 \tag{2.3}$$

The friction between soil and cutting tool and the resistance of cutting soil on the side of cutting tool are as follows:

$$P_{12} = P_n\tan\varphi = \frac{3\gamma H_0 h_1 B_1}{\sin\left(45° + \frac{\varphi}{2}\right)}\tan\varphi \tag{2.4}$$

The total torque is as follows:

$$T_1 = \frac{D^2}{8}\frac{v_a}{N_c}q_u \tag{2.5}$$

γ—The severity of the soil (kN/m³);
H_0—Thickness of shield overlying soil (m);
B_1—Width of cutting surface of cutting knife (m);
h_1—Cutting depth of cutting tool (m);
v_a—Driving speed of shield machine (m/s);
N_c—Cutter head speed (r/min);
D—Cutter head diameter (m);
q_u—Unconfined compressive strength of soil (kPa).

Based on the cutting resistance model of Ivans wedge cutter cutting coal rock, and considering the influence of friction between cutting knife and rock and the action of vertical propulsion force, the force model of straight cutting soft rock cutting is established, as shown in Fig. 2.37. The horizontal cutting force of the cutter is:

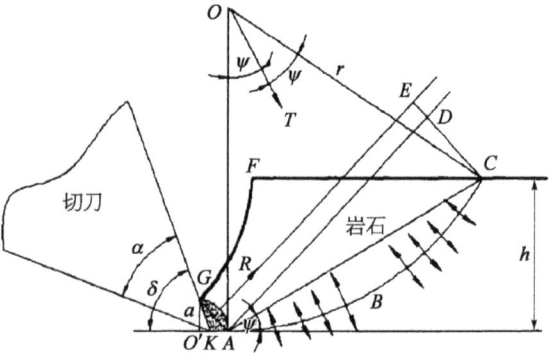

Fig. 2.37 Schematic diagram of force model for cutting soft rock with cutter

$$F_H = bh\sigma_c \frac{\sqrt{\frac{(\frac{\pi}{4}+\Psi)+\sigma_t}{\sigma_c}} - \cos(\frac{\pi}{4}+\Psi)}{\sin \Psi}$$
$$\times \left[\cos(\delta - 45° + \emptyset)(\mu \cos \sigma + \sin \sigma) + \sin(\sigma - 45° + \emptyset)(\mu \cos \sigma - \sin \sigma)\right]$$
$$+ 2\mu bh\sigma_c \tan \frac{\alpha}{2} \tag{2.6}$$

σ_c—Uniaxial compressive strength of rocks (kPa);
σ_t—Uniaxial tensile strength of rocks (kPa);
Ψ—Crushing angle of rock (°);
h—Tool cutting depth (m);
μ—Friction coefficient between cutting tool and rock;
α—Nose angle (°);
$\emptyset$—The friction angle of a rock (°);
b—The angle between the front edge of the tool and the soil layer (°).

2.2.2 Wear Types and Examples of Shield Cutting Tools

2.2.2.1 Shield Cutter Wear Type

The type of wear of the tool is mainly divided into normal wear and abnormal wear. Among them, the normal wear is due to the friction between the tool and the rock layer during the tunneling process of the tool, and the alloy and the weld area on the surface of the tool are continuously damaged by friction caused consumption situation.

Zhang Fengxiang, Zhu Hehua, Fu Deming, etc. proposed the calculation formula of the wear amount of the outer ring cutter of the shield cutter:

$$\delta = \frac{LK_n N\pi D}{v} \tag{2.7}$$

δ—the amount of tool wear (mm);
L—Driving distance (km);
K_n—Tool Comprehensive Wear coefficient when the tool is arranged with the same trajectory ($\times 10^{-3}$ mm/km);
D—Tool mining outer diameter (m);
N—Cutter head speed (r/h);
v—Propulsion speed (mm/min).

In the same cutting track, the relationship between the tool and only one knife is arranged, and the relationship between the wear coefficient is as follows:

$$K_n = K n^{-0.333} \tag{2.8}$$

Table 2.6 Empirical calculation value of tool wear

Tool number	Excavation distance L (km)	Cutter head speed N (r/min)	Advancing speed V (mm/min)	Number of tools with the same trajectory	Tool wear coefficient K_n (10^{-3} mm/km)	Tool mining outer diameter (m)	Tool wear calculation value δ (mm)
38	1.3	0.75	25	12	1.39	14.96	3.43
32				9	1.53	12.99	3.23
26				3	2.21	10.59	3.86
20				3	2.21	8.19	2.97

The wear of cutting tools is calculated by formula in Nanjing Yangtze River Tunnel (Weiqi Road) project. In the formula, the tool alloy is used in Japanese standard E3 cemented carbide (moderate toughness and hardness, suitable for gravel formation). The wear coefficient of mud and water shield in clay fine sand formation is 2.5×10^{-3} mm/km, in gravel layer, and the wear coefficient is 14×10^{-3} mm/km (wear coefficient is the empirical data corresponding to single knife in the same track), at present, the project is about 1100 m in silty clay and about 200 m in gravel layer. Calculate the weighted wear coefficient:

$$K_n = (2.5 \times 10^{-3} \times 11 + 14 \times 10^{-3} \times 2)/13 = 4.27 \times 10^{-3}\,\text{mm/km}$$

The calculation of tool wear at different cutting radii is shown in Table 2.6.

Based on tribological theory, normal wear can be divided into abrasive wear, adhesive wear, fatigue wear and diffusion wear. The main failure forms of tool wear are fatigue wear and abrasive wear. Contact fatigue wear includes surface initiation and surface initiation fatigue wear. The fatigue wear of the surface layer is caused by the stress concentration in the surface layer under cyclic load, and extends from the inside to the surface to form wear debris. The fatigue wear of surface eruption is due to the formation of cracks caused by stress concentration on the friction surface under cyclic load, and propagates into the surface layer, and then the bifurcation falls off to a certain depth to form a pit. Usually, the above two kinds of fatigue wear exist at the same time. Fatigue wear occurs in the process of tool cutting, and its main influencing factors are summarized into three aspects: load property, tool working state and material performance. The magnitude and properties of the load determine the macroscopic stress field of the friction pair, and it is the basic factor that determines the fatigue wear life. The experimental results show that: The contact fatigue life is directly determined by the surface contact stress. The cutting state of the tool directly affects the friction force, and the friction force is very small in the soft soil formation, but in the sand pebble or hard rock formation, the friction force on the contact surface will be significantly increased and the contact fatigue life will be reduced. The hardness of tool materials is related to the fatigue wear resistance. Usually, increasing the hardness of the materials can improve the fatigue wear resistance, but the hardness is too high and the brittleness of the materials

2.2 Cutting and Wear of Tools

increases, on the contrary, the contact fatigue life will be reduced. The fatigue wear resistance of the surface can be improved by reasonably increasing the hardness of the tool body material.

Abrasive wear is the phenomenon that the external hard particles or the hard protrusions or rough peaks on the grinding surface cause the surface materials to fall off during the friction process. The results of abrasive wear show that the wear amount is proportional to the normal load and wear distance, and inversely proportional to the hardness of the material. Because the hardness of tool material is higher than that of formation rock and soil, the abrasive wear in tool cutting process is mainly related to the shape of rock mass abrasive particle. In general, with the increase of abrasive size, the wear volume increases rapidly. When the abrasive size is more than 100 μm, the wear volume hardly changes with the increase of abrasive size. In addition, the load also significantly affects the degree of abrasive wear of various materials. The experimental results show that the wear amount of abrasive wear increases linearly with the increase of surface pressure. When the surface pressure increases to a certain value the wear amount becomes smooth with the increase of pressure. Therefore, the main factors affecting tool wear in cutting process include load properties and size, tool structure and material, and rock mass characteristics. The abnormal wear of the tool is due to the collision of the tool with the hard rock or the sand gravel layer and other mixed strata in the cutting process, and the abnormal cutting condition, which leads to the damage of the alloy block of the tool before it fully exerts its wear-resistant effect, thus making the cutting tool lose its cutting function. The abnormal wear of general cutting tools includes partial grinding of hob and scraper, crack of alloy teeth, bearing fracture of hob and so on. Compared with normal wear, abnormal wear has obvious sudden and irregular.

Some scholars also believe that the process of tool wear to damage can generally be divided into three stages. In the cutting process, the cutting edge of the cutting tool and the push and extrusion of the front tool surface make the cut soil layer deform, and the shear slip becomes the chip. Because of the friction between the front surface of the tool and the chip, and the friction between the back face of the tool and the surface of the cut layer, the tool is worn out in the process of cutting.

(1) The initial wear stage. When the shield tool is made, the front face and the back face are uneven. If you look at it with a magnifying glass, you can find that there are many sharp and small "convex peaks" on its surface. When subjected to cutting friction and chip impact, these "convex peaks" will be smoothed out quickly, this stage is very short, for the initial wear stage.

(2) Normal wear stage. After the initial wear of the shield tool, a wear band is formed on the rear tool surface, which increases the contact surface and reduces the pressure, so the wear increases slowly and evenly with the increase of time. With the increase of cutting time, the wear of tool surface will intensify.

(3) Rapid wear stage. In the later stage of normal wear of shield cutting tools, the wear of tool surface increases gradually, the tool wears sharply, and the complete wear of cutting edge can not continue to be used, and even leads to

the collapse of cutting edge. Tool wear enters this process, that is to say, tool wear is already abnormal.

2.2.2.2 Examples of Tool Wear of Shield Machine

Through a typical example of shield tool wear, this paper introduces all kinds of wear conditions of cutting tools.

1) Singapore DTSS Tunnel

DTSS Tunnel in Singapore uses two soil pressure balance shield machines with a diameter of 4.90 m, which mainly pass through three strata: hard rock, weathered rock and mixed strata of hard rock and soft soil. The cutter head is equipped with six double-blade rollers, 26 single-blade rollers, 12 scrapers and 4 disc-shaped peripheral scrapers. A telescopic device is arranged on the base of many rollers to extend out the hard rock formation to break the hard rock; The hob is retracted in soft soil, and the excavation is done with scraper, and the opening area of the knife disc is increased to improve the fluidity of residual soil. The shield can adapt to the change of rock and soil layer by changing the driving mode. However, in engineering practice, due to the high wear and corrosion of hard rock, the cutter head and tool have been worn and damaged to varying degrees, and the hob shown in Fig. 2.38 has undergone serious wear. By measuring the replaceable tool, the average life of the tool is less than 100m3/tool. In soft and hard uneven strata, the rotation of the tool collides with the rock block, which leads to the vibration of the cutter head, which is very disadvantageous to the stable operation of the shield machine and aggravates the wear of the tool and the cutter head.

By removing the double-edged hob in the center of the cutter head, the engineers can increase the opening rate of the cutter head and increase the number of scrapers accordingly to improve the mobility of slag soil, so that the larger rock particles can easily enter the soil cabin and reduce the secondary wear of the tool; at the same time, the slope of the screw conveyor is reduced, and a screw conveyor is added to

Fig. 2.38 Abnormal wear of hob

2.2 Cutting and Wear of Tools 49

increase the energy force and pressure control ability of the shield to discharge slag, so that the wear of the cutter head and the tool is greatly reduced, and finally the tunnel runs through smoothly.

2) Nanjing Yangtze River Tunnel (Wei Seven Road)

The Nanjing Yangtze River Tunnel (Wei Seven Road) adopts two mud-water shield machines with a diameter of 14.93 m. The shield section with a total length of 3020 m passes through strata containing silty clay formation, fine sand formation, gravel formation, round gravel formation and very few strong weathered mudstone, and some section strata are soft and hard, and the water permeability is strong. The cutter head adopts spoke plate structure and is equipped with 6 main spoke arms and 6 pair spoke arms. the cutting tools are composed of 189 scrapers (including 118 fixed knives and 71 atmospheric pressure replaceable knives), 16 leading knives, 6 pairs of circumferential scrapers, 6 pairs of small shovel knives and 2 copying knives. The shield machine adopts the replaceable tool equipment under atmospheric pressure for the first time, and the tool wear monitoring system is used on the tool. However, in sand and pebble formation, there are still a large number of tool wear, and finally because of the excessive torque of shield machine, it is forced to stop and repair. After stopping, the cutting tools are checked, and most of the blade bodies of the leading knife and the peripheral scraper are worn out. The height of a large number of leading knives and peripheral scrapers is less than the height of the front scraper, and it has completely failed, while a large number of front scrapers have the alloy block falling, as shown in Fig. 2.39.

The engineering technicians restored the normal tunneling of the shield by replacing the new tool. By replacing the replaceable tool and measuring the amount of wear, the wear factor of the new tool is significantly reduced, the number of cracks in the alloy block is reduced, and the tool change distance is increased.

3) Beijing Underground Railway Diameter Line

The Beijing underground railway diameter line tunnel adopts a mud-water shield with a diameter of 12.04 m, which mainly passes through the pebble, medium sand, coarse sand, pebble cement layer and other strata, and the pebble and pebble cement

(a) New cutter   (b) Severe abrasion of blade body   (c) Alloy teeth cracking

Fig. 2.39 Abnormal wear of scraper

(a) Scraper fracture (b) Hob eccentric wear

Fig. 2.40 Abnormal wear of tools

layer account for 95.21% of the whole process. Due to the relatively hard formation, the cutter head is equipped with 284 front scrapers (including heavy tearing knives), 10 central knives, 30 first cutters, and 16 peripheral scraping knives, 16 peripheral scraper protectors, 1 copying knife and 1 wear monitoring knife. After 67 m excavation, the shield machine enters the pebble cement layer with the maximum strength of 23 MPa, and the excavation parameters increase sharply until the main driven safe shaft breaks, and the shield machine is forced to open the cabin because of the failure of the driving shaft. Most of the leading knives on the whole cutter head have occurred the phenomenon of tooth collapse and partial grinding, and have been worn to the same height of the scraper, which has been completely invalid. Most of the alloy blocks of the front scraper on the cutter head crack, and the wear of the scraper in the center of the blade head is much greater than that of the blade outside the blade head. A large number of peripheral scrapers are worn and broken directly (Fig. 2.40a), resulting in direct contact between tool base and cutter head and rock and soil layer, resulting in unusual wear of tool base and cutter head.

Then, through the replacement of new scrapers and the addition of hob and other measures to restore the excavation. Both the new scraper and the heavy tear knife greatly increase the area of the alloy block, and weld the wear-resistant block in the surfacing area of the external scraper to increase its toughness, make the new tool more resistant to impact and wear, and more suitable for the hard formation; and rearrange the new cutting tools, replace the scraper at the center of the cutter head and 32 heavy tear knife with a double-edged hob, and add a single-edged hob to each plate, and set height difference of cutter, the hob is 25 mm higher than the heavy tear and the tear is 15 mm higher than the scraper. After the shield is re-excavated, the tool wear is still serious (Fig. 2.40b), but the abnormal wear tool is greatly reduced, the service life of the tool increases, and the one-time excavation distance increases. Because the pebble cementitious layer is too hard, the diameter line tunnel of Beijing Mass Transit Railway is connected after three knife changes before and after the project.

2.2 Cutting and Wear of Tools

4) Other shield engineering tool wear examples

In addition to Shanghai and individual tunnels passing through muddy clay strata, the large diameter mud-water shield projects that have been built and are under construction in China have almost encountered composite strata of upper soft and hard, upper soil and lower rock due to the fact that most of the large diameter mud and water shield projects have crossed rivers and seafloor. Chengdu, Beijing, Shenyang, Lanzhou and other urban subways have encountered large pebbles, gravel strata. In order to cope with this complex ground condition, most shield machines are equipped with scrapers for cutting soil layer and hob for extruding rock at the same time. Due to the excessive abrasion of scraper in the rock formation and the hysteresis of the roller blade in the soil layer, serious tool wear, abnormal wear and mechanical equipment damage occur in the process of shield tunneling. In view of the various wear phenomena of cutting tools, engineers have carried out a lot of analysis and experiments in this field, and accumulated a wealth of experience.

The No. 9 standard of Chengdu Metro Line 7 adopts earth pressure balance shield, which passes through dense pebble soil layer, medium dense sand pebble layer and medium coarse sand layer. The tool wear is very serious, so it is necessary to open the cabin and change the knife. The 12 standards of Guangfo Metro are driven by a compound earth pressure balance shield with a diameter of 6.260 m. There is much soft soil in the surface layer along the shield interval. The main strata through the tunnel are fully weathered to slightly weathered mudstone and argillaceous siltstone, with a maximum uniaxial compressive strength of 33.9 MPa. The hob has some phenomena, such as partial grinding, bearing damage, alloy cracking and so on. The excavation speed is 13 m/d. Every 2 weeks of excavation, it is necessary to check the knife and change the knife, with an average of 4.8 hob at a time.

The tunnel between Science and Technology Road Station of Xi'an Metro Line 3 and Taibei South Road Station is eastward westward. The left tunnel is 1017 m in length, the buried depth of interval tunnel is 9–14 m, and the buried depth of groundwater level is 9.7–12.3 m, it is mainly diving, and the aquifer is medium sand layer, coarse sand layer and silty clay layer. The moisture content is about 17%. From top to bottom, the interval strata are as follows: plain fill, loess soil, medium/fine sand, coarse sand, silty clay, coarse sand and so on. The mineral composition of sand layer is mainly quartz (44%) and plagioclase is 38%. The excavation diameter of the earth pressure balanced shield cutter is 6280 mm, the opening rate is 60%, the thickness of the disk body is 510 mm, a fish tail knife is installed in the center, the height is 400 mm, the total number of cutting knives is 36, and the number of cutters is 82. The number of edge scraping knives is 8, the height is 110 mm, and two copying knives are equipped, which can realize the overcut of 40 mm in radius direction. In addition, there are 12 diameter-keeping knives around the cutter head, 24 large circular environmental protection knives are inlaid and welded on the outer surface of the big ring, and 85 mm wear-resistant grid is surfaced on the outer surface of the cutter head. At the same time, 5 slag soil modifier injection entrances are uniformly arranged on the cutter head panel. When the shield is driven to about 480 rings (using bentonite mud and foam to improve the slag soil), the torque of shield heading cutter head increases and the excavation speed decreases, and to 504 ring, the torque reaches 5000 kN m, and the driving speed is in 10 mm/min; to 524 rings, the driving speed is only 4 mm/min, cutter head torque often exceeds the rating and trip. There is no mud cake in the soil cabin, and the tool wear is serious. The cutter head tool needs to be repaired before it can meet the remaining excavation task. From the shaft to the front side of the cutter head for tool cleaning and replacement, it can be obviously observed that the wear degree of the outer plate around the cutter head is the largest, and the wear degree of the center of the cutter head and the circumference of the cutter head is smaller, which is characterized by the greater the wear of the cutter head to the outer diameter, and the specific damage conditions are shown in the following aspects: The radial wear of the large circular ring of the cutter head is about 40 mm, with the local wear of 80 mm, 8 edge scrapers and tool holders, 16 diameter keeping knives have been worn out, and the surface of the cutter is about 40 mm in the radial direction, and the wear of the blade is about 40 mm in the radial direction; the cutter wear is 50×60 mm in the peripheral area of the cutter head and 10×30 mm near the center area of the cutter head; The outermost trajectory of the cutter head is all worn out, the front of the cutter head and near the surrounding part are worn 60×95 mm, and the first knife wear is 30×40 mm near the center area of the cutter head.

Through gravel clay soil and fully weathered granite, part of Shenzhen Metro Line 5 passes through gravel clay soil and fully weathered granite, locally entering medium weathered granite, Breeze granitic rock (bedrock), and granite weathered soil is uneven, and there is granite weathered residual body (solitary stone). In addition, there are many solitary rocks and nearly 80 m hard rock protrusions, the rock strength is 45.4–191.5 MPa, the size of solitoids is different, the thickness is 0.7–1.8 m, the maximum size of solitary rocks is 7.8 m $\times$ 2.5 m $\times$ 1.8 m (length $\times$ width $\times$ height), and the height of bedrock intrusive tunnel is 0–4 m. The thickness of soil cover at the top of tunnel is only 10.9–23.6 m. The selected shield is Herrick earth pressure shield, and the specific cutting tools are as follows: 4 mid-center double-edged hob, 31 single-edged hob, 64 scraper, 16 edge shovel and 1 overcut knife. Among them, the hob is designed as a 17in blade, the installation height is 175 mm, and the distance between the knives is 100 mm. The scraper is arranged in two rows in four directions of the cutter head, the height is 140 mm, and the difference between the height of the scraper and the hob height is 35 mm. In addition, the edge shovel is used to calibrate the excavation diameter of the shield. The overcut knife is used in the construction of shield curve and the adjustment of shield attitude. In the process of driving $175 \leq 240$ ring by changing knife for the second time, the parameters are basically normal, the thrust is about 1200 t, the torque of cutter head is 1.2–1.6MN·m, the excavation speed is easy to reach above 60 mm/min, and no abnormal unearthed and oil temperature is found. When the 192 ring is excavated, it is found that there is a complete double-edged hob ring in the slag soil, with only a little partial grinding. At the same time, it can be confirmed that a central hob has fallen into the earth cabin, but at this time it does not have the condition to check and replace the tool, so it can only continue to push forward. After the 240 ring is excavated, the driving parameters are gradually abnormal, and the oil leakage temperature of the cutter head is high, the thrust is large, the torque is large, and the driving speed is small. After entering the cabin with

Fig. 2.41 Hob wear of Shenzhen Metro Line 5

pressure, it is found that the wear degree of all the tools is very serious, and most of the tools have abnormal wear. Tool wear is shown in Fig. 2.41.

The serious wear of shield cutting tools will lead to the shield can not be excavated smoothly, forced to open the cabin to change knives, and the opening position is mostly in the center of the city, the buildings are dense, which is not conducive to the shield to maintain a balanced state. Therefore, it is particularly important to avoid the serious wear of shield cutting tools. The configuration and selection of shield tools have a great influence on the wear of shield tools. Generally speaking, semi-rational and semi-empirical methods are used to configure and select shield cutting tools. Firstly, the rock and soil layer of shield section is explored in detail, and the suitable tool configuration is selected according to the cutting mechanism of cutting tool, and the shield engineering of adjacent interval which has been constructed is investigated, and the tool configuration is optimized by comparing the wear coefficients of all kinds of tools. However, shield cutting tools still have serious wear because of the high ground erosion coefficient and the collision between cutting tools and rock mass. Through the calculation of the wear coefficient of the shield tool, the tool changing distance of the shield machine is preliminarily determined, and all kinds of tool wear monitoring methods are adopted to invert the tool change distance of the tool, and the smooth driving of the shield machine is maintained by updating and replacing the tool replacement tool at atmospheric pressure.

2.2.3 Analysis on the Causes of Tool Wear of Shield Machine

In most shield projects with complex geological conditions, tool wear occurs in different degrees. In view of all kinds of wear phenomena of cutting tools, combined with the formation of tool wear, the reasonable tool configuration and selection measures are carried out by summing up and analyzing the causes, so as to reduce the tool wear phenomenon and increase the effective cutting distance of cutting tools.

The normal wear phenomenon of the tool is the result of the normal use of the tool. The main reason for the wear is that the tool edge surface is in contact with rock and soil, sliding friction, so that the tool edge surface is constantly lost, and the main phenomenon is the uniform wear of the alloy block.

There are many kinds of abnormal wear of cutting tools, which are mainly due to improper tool configuration, improper tool selection, low financial embedding quality, insufficient tool assembly strength and so on.

The improper configuration of the tool means that the selection of the tool type is not suitable for the current formation. For example, in Guangfo subway, hob cuts soft soil layer and produces partial grinding. The improper selection of cutting tools is due to the fact that the section shape of cutting tools is not suitable for cutting strata, such as when Nanjing Yangtze River Tunnel (Wei Seven Road) passes through sand and pebble strata, the sharp blade of scrapers leads to a large number of fracture of alloy blocks of cutters. The causes of abnormal wear of the tool are analyzed as follows. The partial grinding of hob is due to the lack of friction force given to hob in soft soil layer, so that the hob does not rotate, so that only one side of hob cuts the formation, resulting in the arc oh the cutter ring being flattened and deflected, as shown in Fig. 2.42. After the partial grinding of the hob, the knife ring is ground into a plane, which leads to the smaller contact area between the palm surface and the hob, so that the friction force given to the hob is smaller, and the plane is more difficult to rotate than the circular arc surface, resulting in more serious abnormal wear. In addition, when shield is driven in soft and hard uneven strata, the knife ring of hob collides at the rock layer of soft and hard interface. Because the knife ring is brittle and hard alloy material, alloy cracking is easy to occur, which will lead to the fracture of bearing due to the excessive bearing moment. If the damaged hob is not replaced in time, the adjacent hob will bear more load, resulting in more serious damage.

The abnormal wear of scraper mainly occurs in hard rock and soft and hard strata. The collision between scraper and rock in palm surface will lead to tool damage such as collapse and fracture of alloy teeth, as shown in Fig. 2.43. In the process of excavation, alloy teeth also fall off due to excessive wear and tear of surfacing zone. Once the alloy teeth are damaged, the wear resistance of the scraper will be greatly reduced, and then the driving will lead to excessive wear or partial wear of

2.2 Cutting and Wear of Tools

Fig. 2.42 Partial grinding of hob and crack of knife ring

Fig. 2.43 Abnormal wear of scraper

the tool. In extreme cases, because the large rock can not enter the pressure chamber smoothly and stops at the scraping tool, the scraper repeatedly cuts the rock block, resulting in excessive wear or partial wear of the cutter body.

Another important reason for the serious wear of shield cutting tools is the error of shield operation. So far, there is no simple and reliable operation rules for the reference of construction personnel, so that man-made operation accidents are not uncommon. The improper operation mainly includes too large thrust, too large torque and improper rotation speed. In order to ensure the stable driving attitude of shield, the construction personnel need to set up greater thrust and torque in the hard rock formation or dense sand and pebble formation, However, when the thrust setting is too large or the formation lithology changes, the pressure set by the shield is larger than the formation pressure, the shield cutter head is too large in the formation, resulting in a large contact area between the cutter and the formation. At the same speed, the impact degree and friction area are increased, which makes the tool more likely to cause abnormal wear. When the torque of shield machine is too large, the stress of shield tool in its unit area increases, the reaction force and vibration degree of collision become larger, and the alloy block on the tool is more likely to break or fall off.

The shield machine has to stop and open the cabin to change the tool because of excessive or damaged tool wear. In engineering, the worn tool will be replaced, and the unsuitable tool type will also be improved. In the complex geological situation of Nanjing Yangtze River Tunnel, the effective part of the tool (alloy tooth) is worn very seriously in the gravel layer, which directly leads to the shield machine shutdown for 6 months. After the engineer stops and opens the cabin, the cutter on the cutter head is reformed, and in view of the problems existing in the size, material and welding process of the replaceable knife at atmospheric pressure, it is improved in order to improve the life of the replaceable knife at atmospheric pressure, and the driving behind is completed by replacing the replaceable knife at atmospheric pressure many times after the excavation. Combined with the abnormal situation of tool wear and the analysis of wear causes in this project, the normal pressure replaceable knife is improved from the following aspects:

(1) The sectional dimension of the carbide of the cutting edge is enlarged, the width of single alloy is widened, the geometric parameters of the blade section are improved, and the resistance to bending and impact resistance of the alloy are enhanced.
(2) The hardness of Roche HRA ≥ 86 and the bending strength ≥ 2400 MPa are guaranteed in the performance parameters of cemented carbide, which essentially improves the wear resistance and bending resistance of the blade.
(3) High frequency copper-silver composite welding (silver-based solder is used as far as possible) is adopted to prevent the gold alloy from falling off from the knife slot as a whole by using the high frequency copper-silver composite welding process between the blade cemented carbide and the knife body.
(4) The cutting tool front angle is changed to $10°$, and the rear angle is changed to $10°-15°$. The cutting ability is improved and the wear is reduced.

2.2 Cutting and Wear of Tools

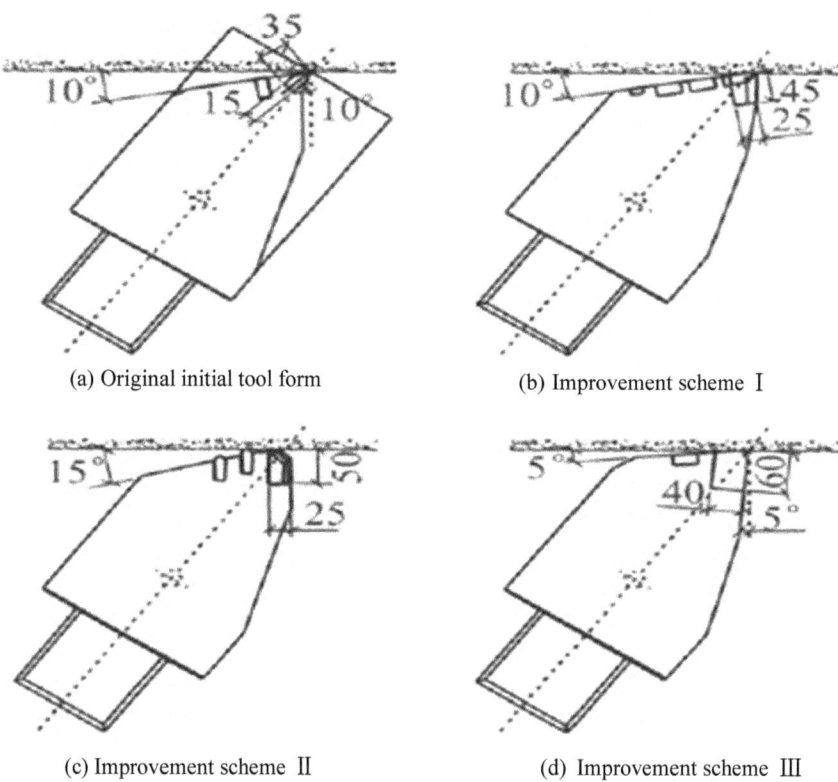

Fig. 2.44 Tool transformation plan

The specific scheme is shown in Fig. 2.44.

The structure of the improved scheme I is shown in Fig. 2.44b. The cross section size of the main alloy tooth is increased to 45 mm × 25 mm, the width is 48 mm, the original tool is changed from 11 small alloy teeth to 5 large alloy teeth, and the anti-wear alloy on the back of the knife is also optimized. The structure of the improved scheme II is shown in Fig. 2.44c. The cross section size of the main alloy tooth is increased to 50 mm × 25 mm, and the section size of the main alloy tooth is increased from 11 small alloy teeth to 5 large alloy teeth, the width of the alloy teeth on both sides is 52 mm, the width of the master alloy teeth is 40 mm, the rear angle is increased to 15°, and two vertical small alloys are set on the back, the first one is flat with the main alloy, the main alloy is protected from impact when reversed, and the second way prevents the back wear of the knife.

The structure of the improved scheme III is shown in Fig. 2.44d. The cross section size of the main alloy tooth is increased to 60 mm × 40 mm, the width is 55 mm, and the alloy tooth is changed from 11 small alloy teeth to 4 large alloy teeth. The alloy teeth are only inverted at the head R = 10 mm fillet, and one antiwear alloy is arranged on the back of the knife, which is flat and has a net distance from the main alloy teeth to 10 mm.

The wear coefficient of the new tool is obviously reduced, and the number of cracks in the alloy block is reduced, which increases the tool changing distance. Through the transformation and renewal of the cutting tools, the tunnel runs through smoothly.

2.2.4 Tool Wear Monitoring Method for Shield Machine

In the process of tunneling, all kinds of wear will inevitably occur, and the wear condition of the tool can be monitored more effectively by various wear monitoring methods, which provides an effective reference for changing the tool and maintaining the effective cutting effect of the tool. The common methods to judge the wear and damage of cutters are as follows:

(1) Adding odour additive. By adding odour additive to hob bearing lubricating oil, whether the bearing of hob is broken or not is judged, but this method has no obvious effect on the normal wear and partial wear of hob bearing.
(2) Analysis of excavation parameters. With the wear of the tool, under the condition of constant thrust, the excavation speed will generally decrease and the torsional moment will increase, according to which the wear of the tool can be roughly estimated. However, the driving speed and torque are affected by many factors, including thrust change, rotational speed change, formation change, soil cabin pressure change and so on, so it is often difficult to judge directly.
(3) Shape analysis of rock slag. Generally, the slag produced by the new knife is large in size, mostly in the shape of piece, and the angle is clear. After tool wear, the degree of slag becomes smaller, the edge and angle wear, and the powder increases. In addition, you can also pay attention to the presence of metal blocks in the slag, the cracked knife ring will often be discharged with the slag.
(4) Wear monitoring device. At present, the most commonly used monitoring methods are electrified and ultrasonic, and their devices indication are shown as Fig. 2.45. The so-called electrified monitoring method refers to the fact that when the tool wear reaches a limited amount of wear, the electrified wire will be worn off, so the circuit will be cut off and alarm will be issued. However, the measurement accuracy of the electrified monitoring method is poor, and it can not be continuously monitored. By adding single chip microcomputer to the electrified wear monitoring device, the electrified monitoring device can continuously monitor the tool wear, which makes the electrified monitoring more widely used. Ultrasonic monitoring is to continuously send ultrasonic waves to cutting tools through monitoring instruments to achieve the purpose of monitoring. The ultrasonic monitoring device has high accuracy and can

2.2 Cutting and Wear of Tools

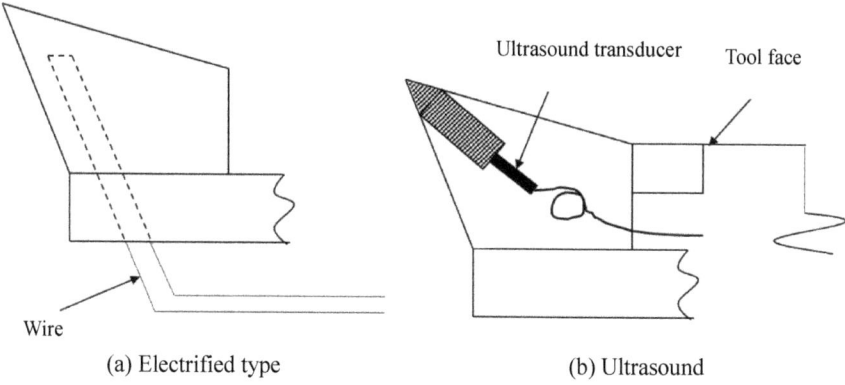

Fig. 2.45 Schematic diagram of energized and ultrasonic monitoring methods

obtain the wear value of the tool without interruption, but the ultrasonic monitoring instrument is complex in installation, expensive in price, easy to damage the probe, and less used in shield cutting tools.

The wear detection device of the tool is installed in the middle railway tunnel, and the detection point is connected with a pressure sensor by setting the detection point in the tool edge area, and the wear of the tool is reflected by the pressure change. When the detection point finds that the pressure of the tool is less than 60 MPa, it is determined that the tool is damaged.

Tool wear monitoring system has been applied to shield construction, and its application is helpful to monitor the real-time situation of tool wear. However, due to the complex manufacture and high cost of the tool installed with the monitoring device, only a small number of partial cutting tools of the shield machine in the project have been installed with the wear monitoring system. For the engineering with tool monitoring system installed, the technicians often rely too much on the wear early warning of the system, but reduce the vigilance to the abnormal situation in the process of shield excavation parameters and slag discharge. For example, in the Nanjing Changjiang Tunnel Project, although some cutting tools have installed electrified monitoring devices, because of the excessive sand and pebble content in the formation and excessive wear and erosion, the monitoring system has not issued an alarm and will be damaged. Technicians also failed to detect the serious wear and tear of the tool in time, which eventually led to a large number of damage to the cutter head and forced to carry out long downtime maintenance. Construction technicians found that the tool monitoring device installed in the tool is easy to be destroyed with the tool, and the monitoring effect is not good.

Harek has developed a Disc Cutter Rotation Monitoring (DCRM) system that includes hob rolling monitoring devices, including sensors, antennas and self-provided power sources, and the antenna of the monitoring device is wirelessly sent to the receiver, as shown in Fig. 2.46.

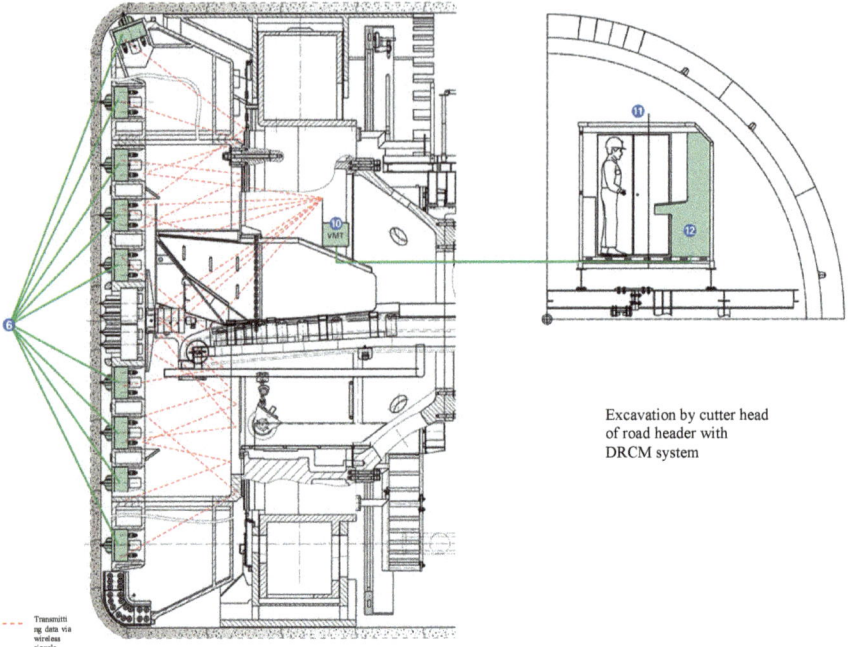

Excavation by cutter head of road header with DRCM system

Fig. 2.46 Harek DCRM system

It is a tool wear monitoring technology for hob. By installing pulse generator in hob hub, the rolling condition and temperature of hob are continuously monitored, and the rolling trajectory is displayed on the controller, which is convenient to monitor the rolling cutting situation in time. The rolling monitoring system of DCRM hob takes up very little space, and the installation and maintenance of components is convenient. The operator of roadheader can pass through the hob with damaged or blocked positioning and evaluate and analyze the measured data, and make individual adjustments to the hob, optimizing the maintenance interval of cutter head. The Mobiddick system developed by Harek Company has been well used in the Tuen Mun-Chek Kok Cross Harbour Tunnel Project in Hong Kong.

Sun Zhihong et al. installed a tool monitoring and sensing device on the hob holder, and monitored the running condition of the hob by detecting the distance between the cutting edge and the monitoring point and the rotation of the hob, and classifying the working condition of the hob into the following four types:

① Normal wear and tear. The sensing distance is judged by the distance sensor.
② Cutter ring partial grinding. When the cutter head is in normal operation, the rotation of the hob can be judged by the hob rotation speed.
③ The cutter ring is cracked. Small cracks cannot be judged, but as the cracking condition increases, the distance sensor will appear the pulsating signal during work, at the same time considering the possibility of cutter ring eccentricity.

2.2 Cutting and Wear of Tools

④ The cutter ring is displaced or peeled off. It is similar to normal wear and tear in severe cases.

Wu Quanli et al. installed a photographic device in the shield bubble compartment, and its conductors are connected from the bubble chamber through the hollow threaded pipe to ensure the sealing property of the whole photographic device. The wear of the tool can be directly observed in the shield control room by means of the photographic device.

(5) Digital photogrammetry technology. The shield cutter is calibrated in the three-dimensional control field, and the camera is installed on the cutter head, and the cutter head tool is modeled in the computer to visualize the tool wear. However, the digital photogrammetry technology has not been really applied to the engineering practice.

2.3 Shield Opening Technology

In the course of shield excavation, excessive wear or damage of the tool, such as excessive wear or damage, the shield needs to be stopped, replaced or repaired and other work, in order to restore the normal construction of the shield. And tool replacement and other jobs, first, it needs to establish a stable working space, and then open the cabin into the vicinity of the work surface can be implemented. Therefore, before the shield tool replacement and other operations, carry out the shield open cabin operation, it can also be said that the shield opening cabin technology is part of the shield tool change rafter technology.

2.3.1 Brief Introduction of Shield Tunnel Opening

At present, most of the shield is closed chest shield, which is usually equipped with pressure chamber. The core of shield tunnel technology is to balance the earth pressure and water pressure on the excavation surface by applying appropriate mud pressure or soil pressure in the pressure chamber to ensure the stability of the excavation surface. The so-called "open cabin" refers to the construction behavior of opening the shield pressure chamber. Under the condition of releasing the pressure in the cabin or maintaining a certain pressure, the technical and technical personnel enter the pressure chamber to carry out tool replacement, cutter head maintenance or equipment maintenance and other operations. It is generally believed that the pressure chamber is the heart of the shield, and the opening of the shield is equivalent to a heart operation, the risk of which can be seen. Figure 2.47 shows the basic structure of the mud-water pressurized shield made by Harek Company of Germany, which is used in the Nanjing Yangtze River Tunnel. The mud tank and the pressure tank together form the pressure chamber of the shield. Opening refers to the behavior of

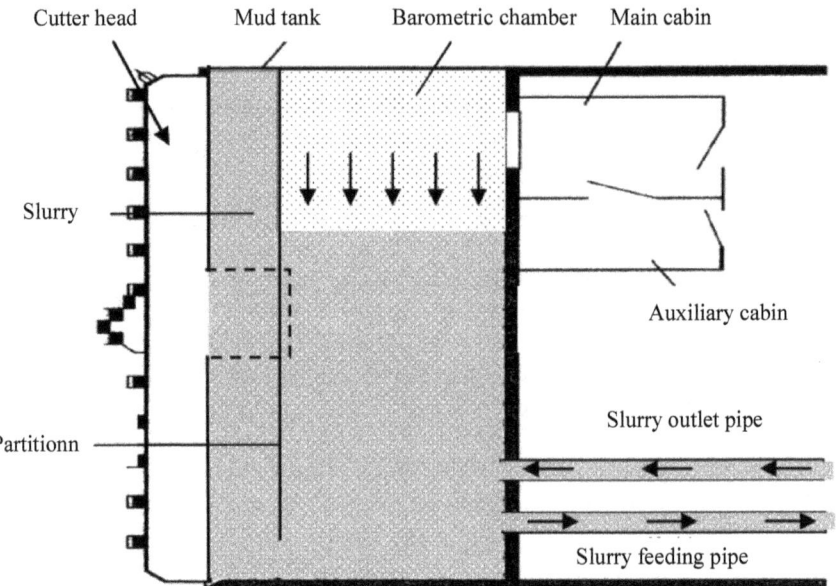

Fig. 2.47 Simple diagram of mud water shield structure of Nanjing Yangtze River tunnel

technicians entering the pneumatic chamber or mud cabin through the brake, through the auxiliary cabin, and into the main cabin to carry out related operations.

At present, the implementation of cabin opening operation is the main hand section for shield tool replacement, cutter head maintenance, equipment troubleshooting and obstacle cleaning. Because the shield is forced to stop in a deep buried depth, complex hydrogeological conditions or adverse environment such as the bottom of the river, there is a great safety risk under the action of soil pressure and water pressure of the formation. The improper construction measures and construction management during the opening of the cabin will often lead to the collapse of the excavation surface, groundwater breakdown and other accidents, such as personal safety when it is bad, and even the scrapping of the whole tunnel. For example, the collapse of excavated surface occurred when the shield was opened by air method in domestic shield construction, when the pressure gas is used to maintain the stability of the excavated surface in foreign countries, the workers in the tunnel have decompression symptoms, such as eardrum rupture, hearing impairment and so on, due to improper operation. Therefore, how to open the cabin safely has become a key technical issue of great concern to the shield tunnel industry in China.

Shield opening will encounter a series of technical problems, such as the stability of excavation surface, the safety of personnel under pressure conditions, the welding operation under pressure bars and so on. However, the most basic and core problem is the stability guarantee of excavating surface, that is, to construct a more stable open space maintenance environment. Up to now, there are few reports on the construction examples of cabin opening in foreign countries, but many shield construction projects have been carried out in China. Although there are many examples of successful cabin opening at home and abroad, most of them are organized by engineering technicians according to experience, which is biased to the construction process flow, and there is no theoretical study on the stability guarantee of the excavation surface during the opening. At the same time, because most construction enterprises take technical confidentiality measures to the core technology of cabin opening, the solutions to the problems such as the stability of excavation surface, the influence on the surrounding environment, the safety of tunnel structure, the safety of personnel and so on, which may be encountered in the opening construction, have not been widely disclosed, and the corresponding technologies and related theories have not been deeply analyzed and systematically combed. The shield opening technique, which is very risky, is in urgent need of a systematic, comprehensive and standardized technical method through the study of the theoretical problems, the summary of the process and methods, and the formation of a systematic, comprehensive and standardized technical method. Therefore, the collection and arrangement of shield opening examples, the analysis of theoretical problems and the clarity of technical problems have become an urgent task for the development of shield tunnel technology in our country.

The purpose of shield opening is generally inspection, maintenance and special construction. The inspection is generally when the shield construction can not advance normally when the special condition is encountered, the state of the excavating surface, the cutting tool, the cutter head and the accessories in the pressure chamber are checked to see if the excavating surface encounters special geology, whether the cutter head tool is abnormal, whether the attachment is damaged, and so on. Maintenance refers to excessive wear and tear of cutter head or failure of equipment such as shredder, and have to enter the pressure chamber to repair or replace it (Fig. 2.48). The special construction is that in the excavation, there are isolated stones, tree roots, metal and other foreign bodies, or abnormal conditions such as the knife head has a serious cake, blockage in pressure chamber, or shield ground docking and disassembly, and so on, need to manually enter the pressure chamber to remove the operation.

From the working procedure, the shield opening operation generally includes the analysis of the reasons for shield shutdown before opening, the selection of opening scheme, the stable treatment of excavating surface, the maintenance in the cabin, the construction technology in the cabin and the safety guarantee of the incoming personnel, and so on. The problems encountered in the process of cabin opening involve many disciplines, such as geological engineering, geotechnical engineering, tunnel structure, mechanical engineering, medicine and so on.

2.3.2 Common Opening Technology of Shield Tunnel

In the current shield tunnel construction, there are three common shield opening methods: After strengthening the soil body in front of the cutter head of the shield,

Fig. 2.48 Opening welding maintenance cutter head operation

the pressure of the pressure chamber is released, the door of the shield is opened, and the operator enters the cabin under atmospheric pressure; After strengthening the soil in front of the cutter plate and stabilizing the excavation surface, the shaft is excavated from the ground down to the front of the cutter head, and the technicians work in the shaft, and the pressure in the pressure chamber can also be released; The mud pressure or soil pressure in the pressure chamber is replaced by the pressure of compressed gas. According to the principle of pressure gas construction, the excavated surface is supported by air pressure, and the operator enters the pressure chamber under a certain pressure environment to carry on the operation.

There are three main types of gases needed to open the cabin: 5% of the Helium–oxygen mixture, the gas composition mainly includes 5% oxygen, 75% Helium and 20% nitrogen. This type of mixed gas is the most important use in the saturated shield opening technology, and it is the main gas breathed by the personnel in the whole opening process. 10% Helium–oxygen mixture, the gas composition mainly includes 10% oxygen, 55% Helium and 35% nitrogen; Medical oxygen. 10% Helium–oxygen mixture and medical oxygen are mainly used to regulate the oxygen content in life chamber and shuttle chamber and the treatment of decompression sickness.

The process flow for opening the cabin is as follows:

(1) Before the shuttle cabin and the living cabin are docked, the shuttle cabin should be washed with Helium and oxygen mixture gas to drain the air, and

2.3 Shield Opening Technology

the gas composition and content in the shuttle cabin shall be determined by the gas analysis system of the shuttle cabin. After meeting the requirements, it can be docked with the living cabin.

(2) The shuttle cabin is docked and fixed with the life cabin, and the pressure of the shuttle cabin is the same as that of the life cabin, and the staff enter the shuttle cabin.

(3) Lift the air conditioning system and shuttle compartment to the transport vehicle, connect the shuttle cabin and the air conditioning system and the hybrid gas delivery system, and the air conditioning system supplies electricity through the generator. Before the shuttle cabin is transported, the safety officer shall check the smooth condition of the road and drain it in front of the transport vehicle, as shown in Fig. 2.49a–c.

(4) The mud water and air of shield mud tank are replaced at the same time during the transportation process of shuttle cabin.

(5) The shuttle cabin will be transported by the transport vehicle to the end of the shield No. 1 trolley port component. The shuttle cabin is rotated to the docking platform through the transport track and assembly machine, as shown in Fig. 2.49d, e.

(6) The shuttle cabin is in butt joint with the human brake cabin. And the adjusting and assembling machine is used for aligning the traveling wheel of the shuttle cabin to the rail of the docking platform. Adjust the concentric degree between the shuttle cabin and the center of the man gate cabin, dock and check the sealing condition between the shuttle cabin and the man gate cabin. The pressure of the pressurized brake chamber is the same as that of the shuttle cabin, and the staff enter the man gate cabin, the pressure of the pressurized man gate cabin is the same as that of the bubble tank, and the staff enters the mud tank to work, as shown in Fig. 2.49f.

(7) Enter the mud tank work. During the opening operation, the communication inside and outside the cabin is ensured by radio, and the gas supply system for the breathing of the staff is in good condition.

(8) After each shift of opening operation, the staff return to the shuttle space, the shuttle cabin is undocked with the man gate cabin, the shuttle cabin is transported through the shuttle cabin to the ground and docked with the living room, and the staff return to the life space.

The first two shield opening methods generally need certain site conditions for formation reinforcement. When the third shield opening method can not maintain air pressure stability in sand and other high permeable strata, it needs to increase the gas closure performance of the auxiliary construction to complete. According to the state of maintaining pressure in the pressure chamber, the shield opening technology can be divided into two kinds: normal pressure opening and pressure opening. The normal pressure open cabin refers to a method for operating personnel to enter the pressure cabin or to carry out the operation in front of the pressure cabin after the pressure of the pressure cabin is released into normal pressure, as described in the previous two shield opening methods. Open chamber with pressure refers to the

(a) Transport vehicles transport shuttle cabins to tunnels

(b) Hoisting shuttle cabin

(c) The shuttle module is in orbit

(d) The shuttle compartment moves horizontally and is connected with the jib of the assembler

(e) Rotary hoisting of shuttle cabin by assembler

(f) Docking between shuttle cabin and human lock cabin

Fig. 2.49 Saturated pressure opening process

2.3 Shield Opening Technology

method of operating personnel into the pressure chamber under the condition of maintaining pressure, such as the third method of shield opening.

For pressure opening technology, the risk of atmospheric pressure opening technology is much smaller and will generally be preferred. Most of the normal pressure open compartments need to strengthen the soil around the excavation surface first. When the shield layer has high self-stability, the infiltration of ground water is small, and the excavation surface can stand on its own and stable surrounding rock environment, the reinforcement and precipitation treatment measures of the excavated surface can also be saved. However, when there are structures above the tunnel, busy roads, and at the bottom of rivers and lakes, due to the complexity of the ground quality, the maintenance position of cutting tools is often unpredictable, and it is often difficult to have the conditions for reinforcement construction from the surface. At the same time, the construction period of strata reinforcement is long and the cost is high. These problems restrict the applicability of atmospheric pressure cabin opening technology.

Compared with the pressure entry tank, the normal pressure opening operation of the pre-strengthened soil must first have certain construction conditions, and there should be enough space on the surface for the construction operation. For example, cross the river tunnel, the undersea tunnel should not use this kind of pre-reinforcement square method, there are some urban subway tunnels, because there are buildings, roads and other important facilities on the surface, there is not enough space for pre-reinforcement. But the safety of the normal-pressure open cabin is good compared with the press-in cabin, and the health of the construction personnel is not adversely affected during the step-up and the step-down process. At the same time, the welding safety is good, the construction is convenient, the difficulty is low, and the repair accuracy is high under atmospheric pressure environment. In the process of shield construction, the opening method should be selected on the basis of comprehensive consideration of various related factors.

When shield only needs short-term inspection, priority will be given to the method of opening the cabin with pressure. In general, under the environment of higher mud pressure and gas pressure, the professional diver enters the pressure cabin, and the condition of the shield and the excavation face is checked in the mud or in the high-pressure constant-gauge compressed gas, and the access time is short, which belongs to the inspection strip pressure-opening cabin (Fig. 2.50).

In that case of shield machine designed with a normal pressure replaceable knife, the tool wear condition inspection can be completed by checking the normal pressure replaceable knife in the radial arm of the cutterhead, so that the cumbersome open cabin operation is omitted, and the required time is less and more secure. At present, this design is widely used in large river shield construction, such as Nanjing Yangtze River Tunnel (Weiqi Road), Nanjing Metro Line 10 Crossing Tunnel, Wuhan Metro Line 8 Crossing Tunnel and so on, which play an important role in construction.

When it is necessary to carry out maintenance or replacement of the cutter head, fault removal and other operations, in particular, it is necessary to carry out long-time operation such as crushing and welding operation, while the construction environment and the limit of the construction period cannot be carried out at normal pressure,

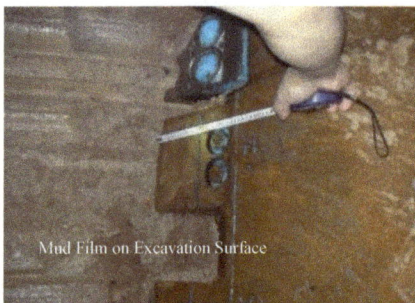 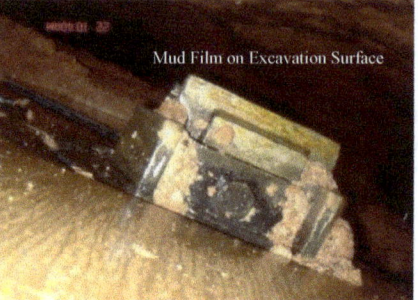

Fig. 2.50 Inspection of tool wear in cabin opening under pressure

and the maintenance of the maintenance pressure is often used. This kind of operation usually takes up to several hours at a time, and needs to be supported by compressed gas to form a local relatively stable working space, and the technicians enter the pressure chamber under the pressure environment to carry on the operation. When the air permeability of the local layer is large, the auxiliary construction for improving the air permeability of the formation is to be carried out first. At present, the most commonly used method to reduce the air permeability of soil is to inject mud into the front of the excavation surface, so that the mud film with good air tightness can be formed in front of the excavation surface. This method can be suitable for the opening of earth pressure balance shield and mud and water pressure shield, and has little influence on the surrounding environment, low cost and short construction period. Before the project, it has been successfully applied in Guangzhou Metro, Chengdu Metro, Nanjing Yangtze River Tunnel and so on. It is difficult to open the cabin with pressure, and there is a great risk, especially when the bottom of the river is open with high water pressure and high permeability, how to form the mud film with good air tightness is still an important problem in the shield opening technology.

2.3.3 Brief Introduction of Typical Shield Tunnel Opening Examples at Home and Abroad

1) Brief introduction of opening of typical shield tunnel engineering abroad.

Martin and Bapple (2007) report an example of the opening of the fourth tunnel of the Elbe River at the bottom of the river in Germany. The tunnel is constructed with a mud-water pressure shield with a diameter of 14.2 m. The shield is a double-blade disk structure (with a central knife disk with a diameter of 2.5 m) and is equipped with a normal-pressure replaceable tool on the five spokes. The five spokes on the cutter head are designed into a hollow state, and the operator can check the wear condition of the normal pressure replaceable cutter inside the five main spokes of the main cutter disc under the normal pressure condition, and perform the replacement operation of the partial cutter. When the bottom of Yibei River passes through the Quaternary moraine layer composed of sand, marl and drifting stone, the tool wear is serious, and the normal excavation can no longer be maintained by replacing the replaceable tool at atmospheric pressure. The central tool must be replaced and forced to open the cabin with repairable belt. Because the opening place is located at the bottom of Yibei River, the head height is 42 m, mainly for high permeability sand and drifting stone and other strata, so the technicians enter the cabin under the pressure of 0.4–0.45 MPa to carry on the maintenance operation. The whole process took 6 weeks, the number of people entering the cabin reached 2738, the total operation time was 10,920 h, the highest inlet pressure was 0.45 MPa, and the incoming pressure was more than 0.36 MPa. This example is particularly concerned with the health problems of the personnel entering the cabin, and it is reported that 21 persons have symptoms of decompression disease, and all occur under the condition of pressure less than 0.36 MPa, which indicates that the necessary decompression measures are very important for health. However, this report does not involve the guarantee method of excavation surface stability under air pressure support. It is estimated that the closed gas action of mud film should be used to maintain the stability of excavation surface under the premise of mud infiltration to form mud film. Figure 2.51 shows the technician replacing the tool under the action of 0.35 MPa pressure.

Martin and Bapple (2007) also report an example of pressure opening in Weser tunnel in Germany. The tunnel is constructed by a mud-water pressurized shield with a straight diameter of 11.71 m, and the strata are mainly glacier sediments. This example belongs to the maintenance belt pressure opening, which is due to the wear of cutter head and the failure of stone crusher. The stop position is 40 m below the bottom of the sea floor (from the arch roof of the tunnel), and the formation is moraine gravel. In this operation, the repair of cutter head is carried out under the

Fig. 2.51 Replacement of cutting tools in the fourth tunnel of Yibeihe under the condition of 0.35 MPa

condition of greater than 0.45 MPa pressure, while the maintenance of stone crusher (located in the lower part of shield pressure chamber) is carried out by professional divers in bentonite mud greater than 0.5 MPa pressure. The total working time under pressure is about 5,000 h, and the number of persons entering the cabin is 14,000 person-times, among which the working pressure of 600 h is greater than 0.36 MPa. Similarly, the health problems of the entry staff are also described in this example. A total of 15 cases of decompression diseases were reported, all of which occurred below 0.36 MPa, and the method of ensuring the stability of the excavated surface under air pressure was not introduced.

During the construction of soft and low plastic clay, silt and fine sand in the red line subway of St. Petersburg, Russia, the maintenance zone pressure opening was also carried out because of the serious wear and tear of the cutter head tool. The tunnel vault of the stop position is 65 m from the ground, the formation is low-plasticity clay, and the static water pressure is 0.56 MPa. The opening is carried out after the pressure chamber is replaced with air pressure. When the air pressure is 0.55 MPa, the personnel wear breathing mask to enter the cabin. The time for decompression is about 5 h, and the operation time is within 4.5 h. The compressed air pressure of this example is basically the same as the water pressure, and it is in the state that the water pressure is balanced but the pressure of the soil is not balanced. The stability of the excavation face makes use of the strength of the soil body itself, but there is no introduction to the analysis of the stability of the excavation face under the compression condition in the literature.

Heijboer etc. in The Westerschelde Tunnel: Approaching Limits in introduced the example that the Dutch Westerschelde Tunnel took the initiative to open tank. This is an ultra-long two-line shield tunnel with a total length of 6600 m. Two mud-water pressurized shielded tunnels with a diameter of 11.34 m are constructed in the same direction. As the construction distance is long, in order to prevent excessive wear of cutting tools, the shield can pass through sand strata with buried depth of over 60 m, high pressure and large permeability coefficient, construction organization set to carry out active open cabin maintenance before entering the sand formation. This example belongs to the maintenance with pressure opening, the reason for the opening is to actively carry out the overhaul of cutter head cutting tools. The strata at the opening place are clay strata with permeability coefficient 10.6 cm/s. The arch roof of the tunnel is 45 m away from the ground, and the water pressure is 0.45 MPa. The opening of the cabin is carried out after the mud pressure is replaced by the air pressure method, and the air pressure during the opening is 0.45 MPa pressure to carry out the maintenance operation. The decompression time after entering the cabin is generally 2 h. During operation, the stability of the excavation surface is monitored by monitoring the change of air pressure in the excavation chamber.

2) A brief introduction to the opening example of Shield Tunnel Engineering in China.

During the construction of subway in China, there are also reports on the examples and construction methods of open-cabin inspection or overhaul, such as Chengdu Metro, Guangzhou Metro, Beijing Metro, Shenzhen Metro, Kunming Metro, etc. The technicians of the engineering unit have made many attempts and practices in this field, and accumulated many successful experience. For example, many shield sections of Chengdu Metro Line 1 have adopted a repair with pressure to open the cabin. The shield section of Chengdu Metro Line 1 is 6.28 m, most of which are constructed by earth pressure balanced shield, and pass through a large number of strata composed of sand pebbles, alluvial clay and silt. The gravel content larger than 5 cm in the strata is above 50%, and the maximum drifting stone diameter can reach more than 20 cm. Because of the high quartz content, more drifting stone, large particle size, high strength and fast wear of excavation tool, it is necessary to change the knife frequently. According to the construction statistics, the tool inspection and replacement shall be carried out every 130–200 m of the excavation.

Because the shutdown position is mostly located in the busy area with dense ground buildings, the ground traffic is busy, and the foundation cannot be pre-reinforced, so the cabin opening method is adopted for cabin opening and repair. Most of the opening positions are located in sand and pebble strata. The buried depth of tunnel overlying soil (arch roof position) is 12 m 20 m, and the groundwater level is generally 4 m 8 m. firstly, the mud in the pressure cabin is replaced by the mud when the cabin is opened, and the mud is formed on the excavation surface, and then the mud is replaced by the compressed air by a compressed air method (generally adopting a high-viscosity bentonite slurry), Through the action of the mud membrane with good air-tight performance formed on the excavation surface, the pressure of the ground water and a part of the earth pressure are balanced by the air pressure force, and the stability of the excavation surface is maintained. Generally speaking, the operator enters the pressure chamber to replace the knife head in the pressure ring of 0.1–0.2 MPa, and most of the opening can be completed under the condition of stable air pressure and stable excavation surface. Therefore, when the condition of formation reinforcement is not available in the shutdown place, the earth pressure balance shield can carry out the opening and maintenance equipment by adding mud injection equipment and using the method of mud blocking excavation surface combined with air pressure support. The method is also reported in the construction of Guangzhou Metro and Shenzhen Metro.

When the shield stop has good ground reinforcement conditions, most of them will use precipitation, plain concrete + mixing pile or chemical reinforcement methods to strengthen the strata, carry out normal pressure open cabin inspection or maintenance equipment, such as Guangzhou Metro Line 4 university city special line Lentou-university city shield section, Chengdu Metro Line 1 phase 1 interval and so on. In fact, there are still a large number of examples of cabin opening that have not been officially reported. In the exchanges of many colleagues and various engineering argumentation, we have learned that many shield tunnel construction with open cabin has been encountered, such as the partial interval of Chengdu Metro Line 2 passing through the moderately weathered medium weathered strata with good self-stability, and the strong weathered mudstone strata, because of the high viscosity of the strong weathered mudstone and the serious problem of the knife head knot cake, which makes the torque of the shield tunneling increase dramatically. The construction efficiency was greatly reduced and the mud cake on the cutter head had to be removed. Before opening the cabin, the well point around the excavation surface is dewatering, and then the soil in the pressure chamber is emptied. The staff enters the pressure chamber through the brake under atmospheric pressure, and uses the shovel to remove the mud cake attached to the cutter head. Because of the good self-stability of the formation, the incoming personnel generally do not use special protection device. This is an open cabin example which makes use of the strength and impermeability of surrounding rock.

In the construction of large-scale cross-river tunnel, there have also been many examples of successful cabin opening, such as Nanjing Yangtze River Tunnel (latitude 7), Nanjing Weishuan River crossing tunnel, South-to-North Water transfer Middle Line Crossing Yellow River Tunnel and other typical large-diameter underwater shield tunnel projects, which will be introduced in detail in the following chapters. In addition, Wuhan Yangtze River Tunnel, Beijing Railway Underground diameter Line, Shiyang Tunnel and so on have carried out various ways of open maintenance or open inspection, and there are few officially reported examples of open cabin. This paper summarizes various examples of shield tunnel opening, and forms a systematic shield opening and maintenance technology, which is of great significance to the development and safety construction of shield tunnel technology in our country.

Based on the examples at home and abroad, it can be considered that it is necessary to release or replace the stress in the pressure chamber, that is, to change the stress state in the pressure chamber, whether it is open at atmospheric pressure or with pressure.

Therefore, it is the key to ensure that the stability of the excavation surface is the success of the shield opening in the open-cabin operation condition. In most examples of open cabin with pressure, air pressure is used to replace mud pressure or soil pressure. Whether air pressure can ensure the stability of excavated surface is the key to open cabin technology with pressure. If the stability of excavation surface is classified into the category of personnel safety and security measures during open cabin operation, some foreign examples of shield opening cabin pay more attention to the health problems of operators, that is, the health problems caused by pressure and decompression process. Therefore, there are at least three key problems involved in the process of shield opening, that is, the stability of excavation surface, the closed gas effect and mechanism of mud film, and the protection of personnel safety and health.

2.3.4 Summary and Prospect of Shield Opening Technology

(1) Frequent open-cabin is one of the more and more prominent problems in the construction of shield construction in China. The core of the open-cabin technology is the stability analysis and stabilization measures of the excavation face, whether the normal-pressure open-cabin or the under-pressure open-cabin method is adopted, and the problem should be solved first.

(2) Opening at atmospheric pressure refers to the way in which operators enter the cabin to carry out maintenance work under atmospheric pressure. According to the strength and stability of the soil around the excavation surface, it is necessary to judge whether dewatering and reinforcement are needed. The principle of stability of excavation surface is similar to that of soil slope stability and foundation pit stability.

(3) When the shield is pressed to open the cabin, the mud must form a dense mud film on the excavation surface, which makes the formation have a good closed gas effect, which is the premise of carrying out the pressure opening. The gas closure and mechanism of mud film is the key scientific problem, and it is the theoretical basis for the establishment of open cabin technology with pressure.

(4) Shield opening is a systematic project, which should be considered comprehensively from the aspects of the stability of excavating surface, the maintenance of equipment and equipment, the safety barrier of personnel, the cost and so on, and the appropriate scheme should be worked out according to the characteristics of the project. The theoretical system related to shield opening technology also needs to be constructed urgently, and the development of new technology is very urgent. These two points are of great significance to promote the development of shield technology in our country.

2.3.5 Tool Replacement Technology of Shield Machine

According to the pressure state of shield tool changing, the shield tool replacement technology can be divided into atmospheric pressure tool exchange technology and pressure tool exchange technology. According to the shield opening technology and the structure of the cutter head, the tool replacement technology can be divided into conventional cutter head atmospheric pressure tool changing technology, atmospheric pressure replaceable cutter head design atmospheric pressure tool changing technology, knife changing technology with pressure and so on.

The conventional cutter head normal pressure tool changing technology refers to the release of the cabin pressure after strengthening the soil in front of the shield machine cutter head, and the operator enters the shield cutter head in front of the shield cutter head under normal pressure to replace the tool; or dig the shaft down from the ground to the front of the cutter head, and the technician carries on the tool replacement operation from the shaft to the front of the cutter head. The opening of the yellow tunnel project in the middle line of the South-to-North Water.

Transfer Project is an example of a typical press-change tool in China (Fig. 2.52). The shield of the project is stopped in the flat yellow river beach, the water level of

Fig. 2.52 Replacement of cutting tools under atmospheric pressure in the middle line of South-to-North Water transfer Project through the Yellow River Tunnel

2.3 Shield Opening Technology

the site is 4 m below the natural ground, and the top soil of the tunnel vault is 29 m (in which the upper part 28 m is medium coarse sand stratum and the lower part 1 m is a loam formation), And finally, the technician enters the pressure cabin through the shield man gate, and the cutter head cutter is repaired under normal pressure.

The atmospheric pressure changing tool technology designed by the atmospheric pressure replaceable cutter head means that for the shield machine with the atmospheric pressure replaceable tool, the technician can replace this part of the specially designed tool at the atmospheric pressure through the space in the cutter head spoke arm. Nanjing Yangtze River Tunnel (Wei Seven Road), Yangzhou thin Xihu Tunnel, Nanjing MetroLine 10 Crossing Tunnel, Wuhan Metro Line 8 Crossing Tunnel are equipped with this kind of atmospheric pressure replaceable tool on the shield of the Yangtze River Tunnel (Wei Seven Road), Yangzhou thin Xihu Tunnel, Nanjing Metro Line 10 Crossing Tunnel and Wuhan Metro Line 8 Crossing Tunnel. The Nanjing Yangtze River Tunnel (Wei Seven Road) project has experienced serious tool wear on the right side of the shield, and the left shield has learned its experience. It is through this design that many replaceable tools have been replaced at atmospheric pressure to ensure the smooth passage of the composite formation without causing serious damage to the cutter head. A total of more than 700 replaceable knives under atmospheric pressure have been replaced in the construction of Nanjing Yangtze Tunnel, which can be said to contribute to the smooth and beneficial propulsion of Nanjing Yangtze River Tunnel.

The technology of changing knife with pressure refers to the technology of replacing the mud pressure in the pressure chamber with the pressure of compressed air, balancing the water and earth pressure on the excavation surface with air pressure, and the operator enters the front of the cutter head through the brake under the air pressure environment to change the knife. In case of encountering high-permeability stratum such as sandy soil, the auxiliary construction to improve the air-tightness of the formation shall be carried out first to ensure the stability of the excavation surface under compressed air. Nanjing Yangtze River Tunnel (Wei Seven Road), which has been opened to traffic, is a typical example of conventional compressed air pressure opening in China. The shield machine is shut down about 50 m below the river surface, in which the overlying soil is about 22.5 m, about 1/4 in the upper part of the section is ⑧ fine sand, the permeability coefficient is 6×10^{-3} cm/s, about 3/4 in the lower part is ⑩ gravel, and the permeability coefficient is about 3×10^{-2} cm/s. The air pressure support with pressure tool changing technology is used to replace the tool. Because of the high permeability of silt and gravel formation, the mud in the pressure chamber is adjusted before opening the chamber to form a mud film with good gas closure on the excavation surface (Fig. 2.53), Then the slurry level was reduced within 3.0 m of the upper arch of the shield, replaced as the compressed air

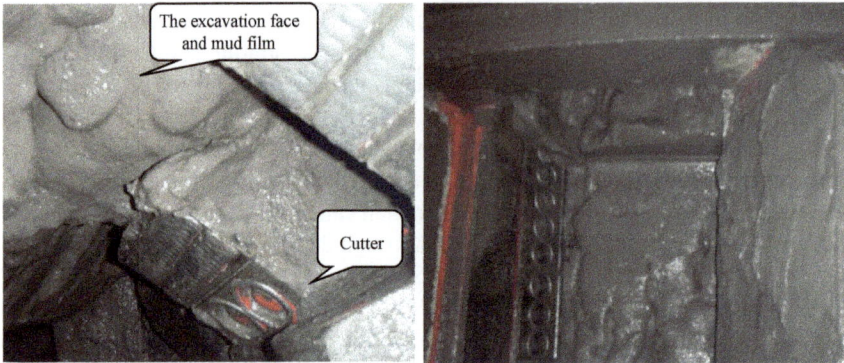

Fig. 2.53 Operation of pressure support with pressure changing cutter in Nanjing Yangtze River Tunnel (Weiqi Road)

with 0.6 MPa, the staff performed tool replacement and repair in 3.0 m high pressure gas. The open cabin maintenance of Yangzhou thin West Lake Tunnel Project is also the conventional compressed air belt pressure opening mode.

Chapter 3
Conventional Cutterhead and Tool-Changing Under Normal Pressure Technology

Conventional cutterhead and tool-changing under normal pressure technology refers to the technology that the workers safely enter the front of the cutterhead under normal pressure, and repair or replace the severely worn tool on the cutterhead. The technology to achieve this mainly through strengthening stabilization of the formation and releasing pressure in the shield pressure cabin, the construction workers then change tools under normal pressure.

In the shield tunneling process, if you encounter the following two conditions, you can use conventional cutterhead and tool-changing under normal pressure technology.

When the shield encounters a hard rock stratum or other upper soft and lower hard composite stratum, a large number of tools on the cutterhead are severely worn or abnormally worn. When the shield cannot continue to dig, the cutterhead and tools need to be replaced after the stratum is strengthened. Although this method is more complicated than opening cabin and changing tool under normal pressure technology, it can solve the problem of most worn tools at one time, and repair or replace the cutterhead at the same time.

There are no replaceable tool devices in the shield. Some old-style shields are not equipped with worn tool-monitoring and tool-changing device. When the shield tools are worn out and cannot continue to dig, this technology must be used.

Conventional cutterhead and tool-changing under normal pressure technology is mainly divided into two types according to whether the reinforcement layer is formed or not: the self-stabilizing stratum tool-changing technology and the reinforcement stratum tool-changing technology.

3.1 Self-stabilizing Stratum Tool-Changing Technology

The self-stabilizing formation tool-changing technology refers to the technology that when the layer has good self-stability, after the shield cutterhead and tools wear, there is no need to strengthen the stratum, and the construction workers directly enter the front of cutterhead from the shield human gate to change tools. This method is suitable for environments with good geological conditions, such as surrounding rock with high strength and few cracks.

When the shield is in the environment with a high self-stability, small groundwater penetration, and self-supporting excavation surface, the normal pressure opening cabin can save the steps of reinforcement and precipitation on the excavation surface. The self-stabilizing stratum tool-changing technology has higher requirements on the stratum, but it has the advantages of fast construction speed, high safety and low cost. The main construction process of this technology is shown in Fig. 3.1.

For the well-stabilized stratum, the layer properties of the rock and soil in front of the excavation surface must be confirmed before the shield is opened to prevent accidents due to geological exploration difference and other conditions. After ensuring the shield stopping interval, the space in front of cutterhead need to be left through the retraction function of the shield cutterhead to meet the construction workers' demand for the operation space.

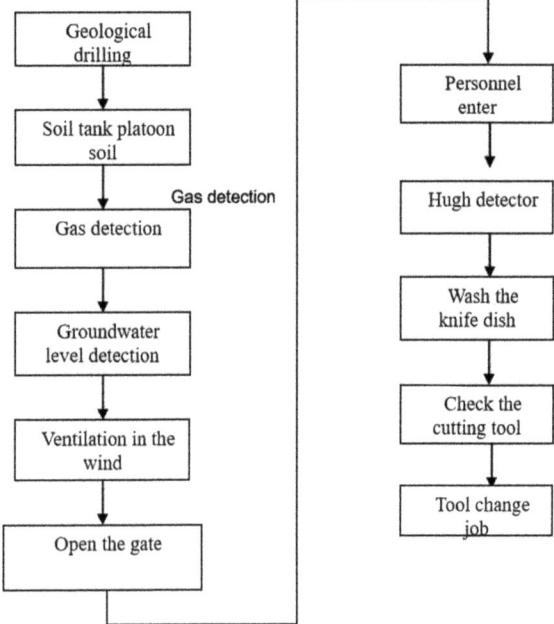

Fig. 3.1 Flow chart of tool changing technology for self-stabilized formation

3.1 Self-stabilizing Stratum Tool-Changing Technology

The gas detector is used to detect the gas in the soil cabin while opening, and it is forbidden to use the fire near the soil cabin and use the lighting without explosion-proof function before the test. Once the abnormal gas is detected, the personnel near the soil cabin and human gate should be evacuated immediately, and forced ventilation to soil cabin is needed to reduce the abnormal gas concentration to a safe state before entering the cabin. Uninterrupted ventilation and gas detection are needed after entering the cabin. Because the space for the self-stabilizing stratum to change tools is small, and the tool-changing operation needs to be cut, it is necessary to pay attention to the protection of the construction workers. The oxygen and acetylene bottles must be kept at a safe distance of no less than 5 m, and the fire damper and anti-collision rubber must be installed. If the tunnel face suddenly emits a large amount of water, we should immediately withdraw all personnel, close the soil cabin gate, and observe the soil cabin pressure changes from the operation room and measure and monitor the surface settlement.

The following describes the specific process of tool-changing through the example of self-stabilizing stratum tool-changing under normal pressure in the Kunming subway.

The first phase of the Kunming Rail Transit Project (B) is located in the underground section from Xiaowangjiaying Station to Chenggong North Station. The stratum mainly intersecting the fully weathered argillaceous siltstone layer, clay layer, silty clay layer and powder layer. The project uses the earth-pressure-balanced shield made by Herrenknecht, Germany. The shield has an outer diameter of 6.480 m and an inner diameter of 5.5 m.

According to the descriptions of the tunnel geological report, it is decided to carry out the tool-changing work at the 750 ring. The stratum is strong medium weathered siltstone layer. The top of the shield cutterhead is the strong weathered ⑦2-1 zone, whose bearing capacity is 350 kPa, while the cutterhead and below is the weak weathered ⑦2-2 zone, whose bearing capacity is 400 kPa. The rock mass has high hardness and strong self-stability, and the natural water content is low, which is suitable for changing the tools under open conditions.

Before opening the shield, push it about 300 mm without adding foam and water. And cut the soil that has been infiltrated by water and foam before opening, so that the excavation face in front after shutdown has good stability. Then drain the soil in the cabin as far as possible by the spiral soil remover to release the pressure in the soil cabin. At the same time, observe the pressure changes in the soil cabin while removing the soil to judge the soil self-stability. When there is no problem in ensuring the soil stability, proceed to the next step. Open the two ball valves above the soil bulkhead, insert the steel bars with a diameter of about 15 mm into the soil cabin, judge the solidified soil according to the insertion condition, and use the toxic gas detector extending into the soil cabin to detect the content of toxic gases and make a record. If the toxic gas content exceeds the standard content, add compressed air to the soil cabin for dilution, and emit it through the ball valves. At the same time, ensure the operation of the ventilator and let fresh air in from outside the tunnel. Then check the water level of the ball valves on the soil bulkhead from top to bottom in the same way, and ensure that the lower ball valve keeps the drainage smoothly during the whole tool-changing process. When the toxic gas content is within the allowable range and the water level in the soil cabin is only 1 m or less, the human gate can be entered to open the cabin door. When opening the door, firstly loosen all the hatch bolts 2–3 teeth, and observe whether there is muddy water infiltrating into the human gate from the joint gap. If there is no muddy water infiltrating, continue to loosen the remaining 3–4 teeth of the bolts and shake the door frame to make it move. If there is still no muddy water, then unscrew all the bolts and open the door. The toxic gas detector is used to detect the toxic gas content in the soil compartment. When it is not exceeded, the doorway soil is removed, the illumination is introduced, introduce the lighting, and observe the soil and muddy water infiltration in the soil cabin. Then the personnel enters the cabin and one person is guarded at the door. After the workers enter the soil cabin, firstly, scrape the soil attached to the bottom of it, and then rinse the knife seat with water to expose all the steel plate and tools and bolts on the tool holder. Clean the tools so that they don't wrap the soil for tool-changing. Select the optimum tool-changing position and weld the lifting lugs of the tools in the soil cabin to facilitate horizontal transport of the tools while changing. During and after the welding process, introduce the compressed air into the soil cabin, so that the smoke generated during the process can be quickly emitted. The content of toxic gases should be detected again. If the content doesn't exceed the standard, change tools. The operators turn the cutterhead to the tool-changing position and enters the soil cabin to remove and change tools. At the same time, the coordinator transports the tools into the human door compartment for preparation and convey the removed tools out of the hole.

3.2 Reinforcement Stratum Tool-Changing Technology

Reinforcement stratum tool-changing technology strengthen the stratum by using various types of methods. The construction workers enter the front of the shield cutterhead from the ground or the shield's human gate for tool-changing work after the soil in front of the shield excavation face is stabilized. This method is suitable for dangerous stratums with poor geological conditions or high groundwater levels.

Except for a few self-stabilizing stratums, most of the stratums where the cabin under normal pressure is located are very complicated. It is necessary to reinforce the soil around the excavation surface firstly. Therefore, the conventional cutterhead and tool-changing under normal pressure technology mainly refers to the reinforcement stratum tool-changing technology (Fig. 3.2). The advantages of the tool-changing technology are: Since the excavation surface has used the appropriate stratum reinforcement technology, the tool-changing operation under normal pressure is relatively safe; since the stratum has been reinforced and under normal pressure, the worker can carry out the repair work safely for a long time; when the most part of the cutterhead is damaged and the tools are worn extensively, the conventional cutterhead and tool-changing under normal pressure technology is lower in cost and

3.2 Reinforcement Stratum Tool-Changing Technology

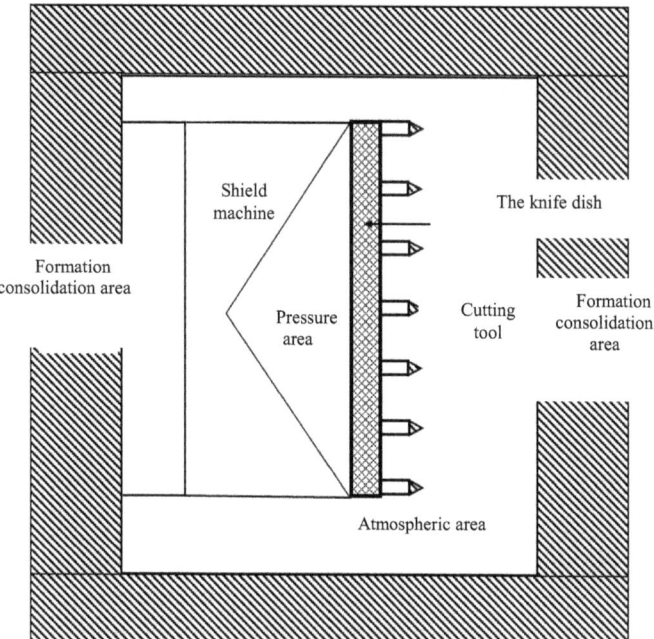

Fig. 3.2 Schematic diagram of tool change for stratum reinforcement

less risky than pressure opening technology; the construction worker change tools under normal pressure, which won't lead to decompression sickness.

The disadvantages are: The timing of the tool-changing is determined according to the shield tunneling parameters, the tool wear and the stratum conditions, and there are many factors to be considered; this method needs to strengthen the ground layer from the surface, which has a great influence on the surface environment; the tool cannot be changed at any time, which can be replaced at a specific location; the time from the tool wear to the completion is long, which affects the construction period.

Soil reinforcement technology is an auxiliary technique in the shield construction. Commonly, there are rotary spray reinforcement, deep mixing reinforcement, freezing reinforcement and underground continuous wall reinforcement.

Rotary spray reinforcement and deep mixing reinforcement are generally used for soft soil layers. In shield construction, it is common for the end reinforcement and connecting passage surface reinforcement to improve the soil stability and bearing capacity to block the water and replace the bad soil, so that we can ensure the safety of the soil mass during the excavation from the beginning to the connecting passage. In the Guangzhou metro granite and gneiss mixed weathering stratum and Chengdu rich water sand stratum, deep mixing piles, triple pipe spinning method and other methods are widely used in the stratum reinforcement while opening cabin under normal pressure, which plays a good impact on maintaining excavation face. The freezing method is often used for the connecting passage construction in the shield section, and for the end reinforcement construction under special circumstances. The freezing method is applicable to Shanghai Metro, Nanjing Metro saturated sand and soft clay stratum. The construction process is long has high cost.

Rotary spray, agitation and freezing are used to improve the self-stability, bearing capacity and water-blocking of the soil in the poor stratum, and improve the shield safety from the beginning to the reach of the terminal well and the connecting passage. When choosing the reinforcement scheme of the shield tunnel for opening cabin and changing tools under normal pressure, the economical and reasonable reinforcement method should be selected according to the requirements of the reinforcement, engineering requirements, geological conditions and construction conditions. The composition of the stratum, the characteristics of the every layers, groundwater and other factors should also be taken into consideration to ensure the quality of reinforcement.

3.2.1 Rotary Jet Reinforcement Technology

Rotary jet reinforcement technology refers to the technology using a rotary jet drilling machine to send a grouting pipe with a special nozzle to a predetermined depth, and then slowly lift the drill pipe, rotating it. At the same time, the nozzle sprays liquid with a certain pressure and the gas jet stream to mix and solidify the slurry stream with the soil layer to form a consolidated pile with a certain strength reinforcing the stratum.

The rotary jet reinforcement technology is mainly suitable for the Quaternary alluvium, residual layer and artificial fill. It can reinforce the sand soil, cohesive soil, loess and silt. However, if the gravel with excessively large diameter and content or humus with a large amount of fiber, it is slightly inferior in quality, and sometimes even less effective than static pressure grouting. If the groundwater flow has large rate, the sprayed slurry cannot be condensed around the filling pipe. And the karst zone without filler, the permafrost and the foundation with severe corrosion to the cement are not suitable for the rotary jet grouting method.

The rotary jet reinforcement method is a ground reinforcement method based on the chemical grouting method using the principle of jet cutting. The rotary jet grouting method directly destroys the soil by high-pressure jet, so that the slurry and the soil mass are self-mixed into a uniform rotary jet slurry and form a induration reinforcement stratum. This method is applicable to both sand and clay, and has the advantages of lower cost, faster construction speed, high induration strength and high reliability. It uses a high-speed jet to forcibly destroy the soil to form a induration. Generally, there is no problem about irrigating in the covering layer. At the same time, since the high-speed jet is restricted to the soil broken range, the slurry is not easily lost, so that we can ensure the expected reinforcement range and control the shape of the consolidation; it can be applied in any section of the borehole or sprayed at the bottom or middle of the hole. In addition, it is also possible to spray in the

3.2 Reinforcement Stratum Tool-Changing Technology

horizontal direction and oblique direction. The rotary spray reinforcement method usually adopts cement slurry, which does not cause environmental and groundwater pollution, and has good durability and low construction noise. The single pipe and the two methods of pipe construction are relatively simple.

1) Features of rotary jet reinforcement

(1) Wide application range

① It can be used to reinforce the foundation of new construction, and can also deal with the foundation during construction process to improve the foundation strength, reduce or rectify the settlement and uneven settlement of the building.
② The deep foundation pit side retains soil or water to protect adjacent buildings and protect underground construction.
③ The bottom of the foundation pit is reinforced to prevent piping and bulging.
④ Reinforcement and waterproof curtain of the dam body.
⑤ Reinforcement on the slope and at the top of the tunnel.

(2) Simple construction. In the construction of rotary jetting, only a small hole (usually 50–108 mm) is drilled into the soil layer, which can form induration with a diameter of 0.4–3.0 m or even 5.0 m by swirling in the soil. Conjuncts. The rotary jet reinforcement method can be flexibly formed, and the diameter of the pile can be arbitrarily changed within the depth of the hole, and can also be used in the case where the obstacle is occasionally encountered in the stratum.

(3) The strength of the consolidated pile is high. The unconfined compressive strength of the jet grouting pile by using cement slurry in clay can reach 5–10 MPa, and is higher than that in sand, generally up to 10–20 MPa. In addition, the jet grouting pile has the advantages of good durability, wide material source, low price, little slurry loss, simple equipment and no pollution.

2) Classification of the rotary jet reinforcement method

The classification of this method is shown in Fig. 3.3. The rotary jet reinforcement method can be divided into vertical rotary jet and horizontal rotary jet according

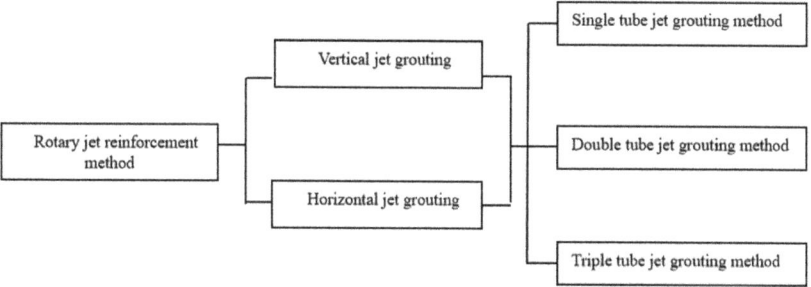

Fig. 3.3 Classification of rotary jet grouting

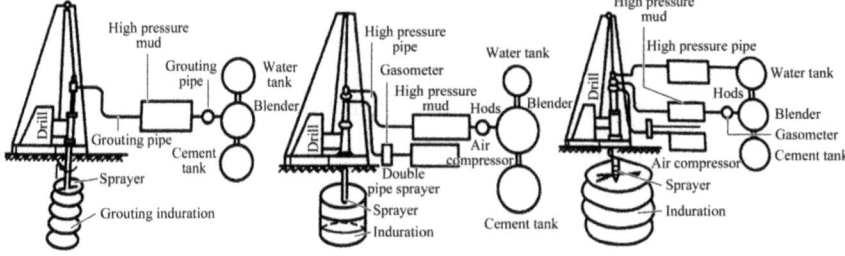

Fig. 3.4 Schematic diagram of high-pressure rotary grouting

to the direction of drilling and pile stratum; according to the number of injection pipelines, it can be divided into single-pipe rotary jet grouting method and double-pipe rotary jet grouting and triple-pipe rotary jet grouting, respectively referred to as single-tube method, double-tube method and triple-tube method for short.

(1) Single-pipe jet grouting method.

The method uses the single pipeline and the high-pressure slurry (about 20 MPa) to jet and cut the soil mass. The single-tube rotary jet reinforcement has good quality, fast construction speed and low cost, but the diameter of the induration is small, which is shown in Fig. 3.4a.

(2) Double-pipe jet grouting method.

This method adds compressed air based on the single-tube method and use double grouting tubes in a dual channel. Coaxial double nozzle are arranged on the bottom side of the tube, and the high-pressure slurry is ejected from the inner nozzle at a high pressure about 20 MPa. Waht's more, compressed air about 0.7 MPa is sprayed around the periphery of the jet. The induration significantly increases in diameter in the soil as shown in Fig. 3.4b.

(3) Triple-pipe jet grouting method.

A triple-root grouting pipe that separately transports water, gas, and slurry respectively is used. The high-pressure water jet and the surrounding airflow are coaxially jet-punched to destroy the soil, and a cylindrical air jet is added around the nozzle of the high-pressure water jet to make coaxial injection of water and gas, which can reduce the friction in the water jet and surrounding medium. To avoid premature atomization of the water jet, and to enhance the jetting water cutting ability, the nozzle is lifted while rotating, and a large negative pressure zone is formed in the foundation. When the intersices is filled with the injected slurry, the induration in large diameter is formed in the foundation to reinforce it. When the three-tube method grouts, the water pressure is 20 MPa, the gas pressure is 0.7 MPa, and the slurry pressure is 2–5 MPa, as shown in Fig. 3.4c.

3) Form of high-pressure jet flow

The basic form of high-pressure jet flow is related to the type of rotary jet grouting process. The types of rotary jet grouting are divided into single-tube rotary jet method, double-tube rotary jet method and triple-tube rotary jet method.

(1) Jet flow of a single-tube. When the single-tube reelingly jets ement slurry under high pressure, which can be studied by using the method of a continuous jet flow pattern by a high-speed water stream in still air. The jet flow of the single-tube rotary jet is composed of three regions, that is, an initial region in which the outlet pressure is maintained, a main region in which the turbulent flow is developed, and a final region in which the jet flow becomes discontinuous.

In the initial region, the velocity at the nozzle exit is even and the axial dynamic pressure is constant. As the distance from the nozzle increases, the portion that remains evenly distributed becomes narrower and narrower until the velocity on the section at a certain position is no longer even. The portion where the velocity remains uniform is called the jet core (E section), and the transition portion where the axial dynamic pressure is reduced is called the migration zone. The length of the initial zone is an important parameter of the jet flow, from which we can judge the destruction of the soil and the agitation effect.

The main area is after the initial area. In this region, there are several features: the width of the jet continuously increases, the axial dynamic pressure is rapidly weakened, the flow velocity is further reduced, the diffusivity is constant, and the jet flow diffusion width is proportional to the square root of the distance. When swirling in the soil, the jet and the soil are mainly stirred and mixed in the area.

In the final area, the jet energy attenuates greatly, the width is large, the end is atomized, and finally mixed with the air and dissipated in the atmosphere. The effect of the jetting on the pile is not significant. The effective injection jet length is the sum of the length of the initial zone and the length of the main zone. The longer the effective jet length, the larger the range of the agitated soil and the larger the diameter of the duration.

(2) Jet flow of a triple pipe. When the jet grouting is started, the pressure of the high-pressure jet is attenuated sharply, and even when the jetting pressure is extremely high, it is often hard to achieve the desired effect of the crushed soil. To this end, a triple-tube rotary jet technique that simultaneously jets a high-speed water stream outside the high-pressure jet to expand the effective jet length is created.

4) Rotary jet pile mechanism

(1) Destructive effect of high-pressure jet on soil mass. The role of high-pressure jets in destroying soil mass is multi-faceted, including jet flow pressure, jet pulsation load, water hammer impact force, cavitation phenomenon, extrusion force and air flow agitation. According to the momentum theorem, when the jet is injected in the air, its destructive force is:

$$F = \rho Q v_m \tag{3.1}$$

F—destructive force (N);
ρ—the density of the spray medium (kg/m³);
Q—flow (m³/s);
v_m—the average velocity of the jet (m/s).

$$Q = v_m A \tag{3.2}$$

A—nozzle sectional area (m²).
Equation 3.1 can be expressed as:

$$F = \rho A v_m^2 \tag{3.3}$$

It can be known from the formula 3.3 that when the jet medium density and the nozzle cross-sectional area are invariable, in order to get greater destructive force, it is necessary to increase the average flow rate, that is, to increase the injection pressure. Generally, the working pressure of the high-pressure pump needs to be above 20 MPa, so that the jet has sufficient energy impact to damage the soil mass. However, simply relying on increasing the injection pressure to improve the jet cutting effect is a waste of energy, and is not the best way to obtain a larger pile diameter. From formula 3.1, the factor determining the jet cutting effects is the impulse rather than speed. Therefore, the main parameters (nozzle diameter, pressure, shot volume and lifting speed) should be considered comprehensively to achieve the best results.

(2) During the jet grouting time, the high-pressure jet flows slowly rotates to make the soil mass around the grouting hole cut and destroyed. After destroying the surrounding soil mass, a part of the fine soil particles are replaced by the jet slurry, and are carried to the surface with the liquid flow (commonly known as mud spillover). The remaining soil particles are mixed with the slurry. Under the joint influence of the rotary pressure, centrifugal force and gravity, the soil particles are regularly arranged in the cross section according to quality and mass. The small particles are mostly in the middle part, most of which move to the outer side or the edge part to form the main body of the slurry. With the process of mixing, compressing and infiltrating and After a certain period of time, the main body of the slurry solidifies into a cement-soil network structure consolidated body with high strength and small permeability coefficient, that is, a jet pile. When the soil quality is different, the structure of the jet grouting pile formed in the cross section is slightly different, and the soil that has not been cut around is compacted and compressed. In the sand soil, a part of the slurry penetrates outside the compression layer to form a permeation layer. The cement content and strength in each part of the jet grouting pile are different. The general cement content is 30–50%, and the central part strength is low, which the edge part strength is high.

3.2 Reinforcement Stratum Tool-Changing Technology

5) Slurry material

Cement is the basic material of high pressure rotary jet grouting. Cement type slurry can be divided into the following types:

(1) Ordinary slurry. It generally uses ordinary Portland cement without any admixture. The water-cement ratio is generally (0.8:1)–(1.5:1), and the compressive strength (28d) of the induration can reach 1.0–20 MPa, which is suitable for projects without special requirements.
(2) Rapidly setting and early-strength type. It is suitable for projects with high groundwater level or requiring early load. Add calcium chloride, triethanolamine and other quick-setting early strength agents to the cement slurry. The compressive strength of the induration incorporating 2% soil was 1.6 MPa, and it was 2.4 MPa after incorporating 4% calcium chloride.
(3) High-strength type. The average compressive strength of the jet induration is under 20 MPa. You can choose a cement with a high strength grade, or a compounding agent composed of a high-performance diffusing agent and an inorganic salt. When incorporating 2–4% water glass into the cement slurry, the impermeability is significantly improved. For the purpose of impermeability in projects, it is better to use "flexible material". The cement slurry can be mixed with 10–50% bentonite (percentage made up of cement mass). If there is only impermeability and no anti-freezing requirements, volcanic ash cement can be used.

6) Vertical rotary jet process

The vertical swirling process is shown in Fig. 3.5. When the vertical jet is sprayed, the rotary jet grouting process has single-tube, double-tube and triple-tube rotary jet grouting, and there are differences in the slurry materials and types and quantities injected into the formation, but the construction steps are basically the same. They are all composed of drilling emplacement, drilling, intubation (lower grouting pipe), rotary jet operation, flushing and other processes.

(1) Drilling emplacement. It is the first step of the rotary grouting reinforcement construction work. The drilling emplacement is placed in the designed hole. After the drilling rig is in place, the drilling machine is horizontally corrected so that the drill rod axis is vertically aligned with the hole center. The deviation between the drilling position and the design position shall should not exceed 50 mm.
(2) Drilling hole. The purpose of drilling hole is to insert the jet grouting tube into the formation and reach a predetermined depth. The drilling method is based on the formation conditions, reinforcement depth and equipment, and the depth can reach more than 30 m. It is suitable for sand and cohesive soil with a standard value less than 40. When encountering a relatively hard formation, or when constructing a double-pipe and a triple-pipe, it is advisable to use a geological drill forming a hole.

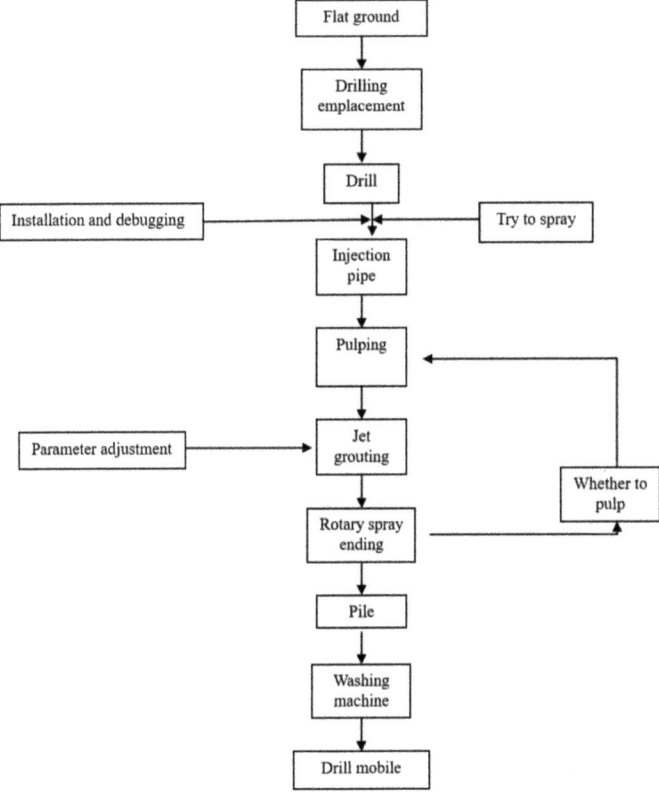

Fig. 3.5 High pressure jet grouting process flow

(3) Intubation. When using the rotary jet drilling machine, the two processes of the lower grouting pipe and the drilling are combined into one process. The drilling and the lower grouting pipe simultaneously complete the process. When using a geological drill, after the drilling is completed the core tube needs to be pulled out, replaced with a nozzle, and inserted into the borehole to reach a predetermined depth. In the process of intubation, it is advisable to inject the water while spilling at the side to prevent the sediment from clogging the nozzle, but the water pressure should generally not exceed 1 MPa to prevent the hole wall from collapsing.

(4) Jet grouting. After the nozzle is inserted into the borehole and reaches a predetermined depth, the slurry is immediately agitated according to the design mix ratio. Rotate from bottom to top, lift the nozzle while rotating, and check whether the parameters such as initial setting time, grouting flow, air volume, pressure and rotation lifting speed meet the design requirements, and record them at any time. After the nozzle is lifted to the design elevation, the rotary jetting operation ends.

3.2 Reinforcement Stratum Tool-Changing Technology

(5) Flush the machine. After the rotary jet grouting construction ends, the grouting pipe and other equipment should be rinsed clean, and the cement slurry should not remain in the pipe or inside the machine. While flushing, the pipette of the grouting pump is moved into the water tank, spray on the ground, and the slurry in the mud pump and the grouting pipe is completely eliminated by the high-speed water flow.

(6) Move the equipment. Move the rig and other equipment to the new hole position to prepare for the rotary injection of the next hole.

7) Horizontal rotary jetting process

Horizontal rotary jetting refers to a construction method in which a basic horizontal drill hole is arranged in a soil layer, and a grouting pipe is inserted horizontally or slightly upwards to start a rotary jet grouting. The injection pressure of the horizontal jet is mainly determined by the diameter of the jet grouting pile and the engineering geological conditions of the soil layer, generally around 60 MPa.

Compared with vertical rotary jetting, horizontal jetting is mainly characterized about horizontal construction, including horizontal drilling and horizontal grouting. Due to the different pile conditions, the horizontal jetting process requires higher than the vertical swirling. Compared with vertical rotary jetting, the horizontal rotary jetting process has the following features: in the range of reinforcement depth, the drill pipe is removed as much as possible in the middle process; the drill pipe is required to be used as a grouting pipe; The pipe pull-out speed should be kept as uniform as possible, which jet from the central of it to the surroundings; when the diameter of the jet spray is required to be large, the process of double spray (slurry sprays after water spray) can be used, or the spray pressure can be increased, or the pull-out speed can be appropriately slowed to make it reach the predetermined one. Requirement; when the jet slurry is lost too much, the slurry needs to be replenished. The slurry can be recharged at the orifice. When the slurry shrinks too much, the slurry should be replenished in time to avoid the depression of the jet grouting pile; when the length of the horizontal jet is too long and one drill pipe cannot be drilled and swirled at a time, it can also be segmented and swirled, and one drill pipe is still used in each segment.

8) Construction examples of jet grouting pile.

The first phase tunnel section of Chengdu Metro Line 1 is mainly passed through the strong permeable pebble with abundant water content and sufficient supply. The buried depth of the tunnel structure is 1.0–14.0 m. The pebbles in the tunnel crossing the formation account for 55–80%, and the particle size is mainly 30–70 mm, and the partial particle size is partly 80–120 mm. Containing a small amount of large-diameter boulders, the general content is 5–10%, and the maximum diameter of boulders is 650 mm. The filling material is mainly sand and medium sand, whose content is 10–35%. The cobble aquifer in this section has strong permeability, good water-richness, rich water content, and the permeability coefficient is generally 20 m/d. Groundwater level burial depth is generally 4.1–8.75 m. The shield method is used in the water-rich sand and gravel stratum, which puts higher requirements on the type, combination, wear resistance and replacement method of the tools. The Chengdu metro stratum is dominated by sand and gravel. The framework effect is better after precipitation, which can maintain the stability of the soil mass. However, in the bottom of the tunnel or the sand and clay layer weathered by mudstone, the precipitation effect is general, and other strata need to be assisted on the basis of precipitation.

In the right line YDK14 + 077.9, the shield is mainly located in silty clay layer and sand pebble layer. Before the shield tunneling to the predetermined tool change mileage, the excavation parameters are abnormal. It is estimated that the tools wear seriously causes the cutterhead wear. In order to repair the damaged cutterhead, it is decided to use the combination of manual digging pile and high-pressure jet grouting pile to open cabin under normal pressure.7 artificial digging piles with a diameter of 1200 mm were constructed directly above the shield ring, and the depth of the hole was 50 cm to the shield shell; 4 rows of high-pressure jet grouting piles were constructed in front of the cutterhead with a diameter of 450 mm, which was arranged in a plum-shaped shape. There is 30 roots in total. In order to avoid the squeezing of the cutter slurry pasting cutterhead and tools, so that it is difficult to clear the slag in the cabin, the construction depth of the first row of jet grouting piles is controlled to the depth of the contour of the shield. The construction depth of the other three rows of jet grouting piles is 20.5 m. In view of the geological characteristics of drilling holes in Chengdu sand and gravel stratum, it is difficult to drill into holes, the steel casings are followed up, and the construction process of the steel casing is sprayed out after the spinning core is pulled down. Figure 3.6 shows the specific arrangement of the jet grouting pile during construction.

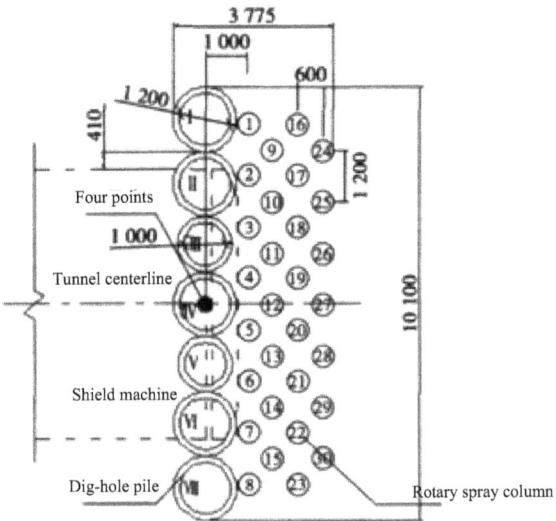

Fig. 3.6 Specific layout of rotary spraying reinforcement

3.2 Reinforcement Stratum Tool-Changing Technology

The starting distance of the left line of the Shenzhen Metro Line 1 is 525, the starting distance is SK25 + 008.700, and the ending mileage is SK25 + 733.674. The length is 724.974 m. This section has a plane turning radius of 400 m and a turning angle of 90°. The tunnel mainly passes through clay and sandy cohesive soil mass. In the section of SK25 + 232.02 + 452.71, there are rock formations in the tunnel area, in which SK25 + 280.58 + 406.326 section is medium weathered granite. Through the slag discharge situation and ground drilling analysis, there is a large amount of sea mud in the tunnel area, not sandy cohesive soil. And it is not fully weathered rock, where the degree of weathered rock is low, and through the strength of the unearthed stone detection, the axial compressive strength reached 161 MPa, far exceeding the designed rock breaking strength of the tools by 120 MPa. The intrusion tunnel range is about 4 m, which is a typical "soft and hard" formation.

According to the geological section map, the original plan was successfully passed under the premise of low cutter speed and small thrust. However, when drilling to SK25 + 257.52, the cutter wheel rotation torque and the cutter wheel speed jumps whose range is 12–20 MPa, and the sound was loud. Based on past experience, it is known that the tool has been damaged, and it is necessary to open the cabin to replace the tools with severe wear. This opening is carried out by using the stratum reinforcement method for atmospheric pressure changing. Therefore, it is necessary to formulate a special and reasonable stratum reinforcement scheme. After several consultations with technical experts, the grouting pile reinforcement method is used to reinforce the stratum, so as to achieve normal pressure opening to change the knife. 600 mm@500 mm jet grouting piles are used in the construction, the reinforcement range is 10 m × 4.5 m, and the reinforcement depth is 12 m upwards. The elevation of the tunnel floor is −14.2 m and the elevation of the ground is +5.8 m. 42.5 ordinary Portland cement is used, the cement blending amount is 150 kg/m^3, the slurry water-cement ratio is 1:1, the lifting speed is 8–15 cm/min, the rotating speed is 5–10 r/min; the jetting pile The hole position deviation is <50 mm, and the verticality of the pile is <1.5%. When the same pile body needs several injections, the overlap of the upper and lower pile bodies is larger than 200 mm. The specific layout of the jet grouting pile reinforcement is shown in Fig. 3.7.

The construction process of the rotary jet reinforcement method used in this project is shown in Fig. 3.8.

The precautions for construction of jet grouting piles are as follows:

(1) Before the construction, check the design hole position of high-pressure jet grouting according to the site environment and the location of underground buried materials. And the slurry draining tank and the mud pool need to be dug. During the construction, the discarded slurry will be introduced into or discharged into the mud pool, and the sediment will be transported to the off-site storage or disposal.

(2) The drilling rig is placed horizontally, and the drill pipe is vertical whose inclination is not more than 1.5%. Check the high-pressure equipment and piping system before construction, and the pressure and flow rate need to meet

Fig. 3.7 Spouting reinforcement plan

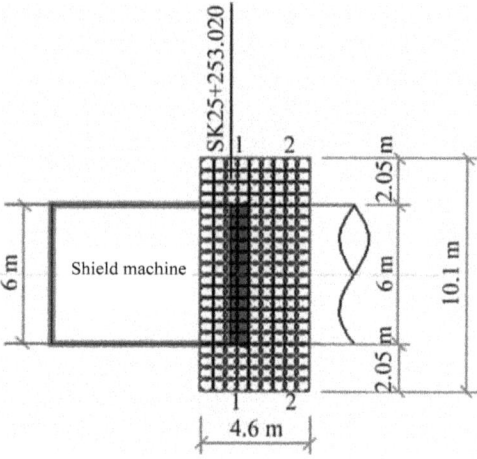

Fig. 3.8 Spray spinning reinforce process

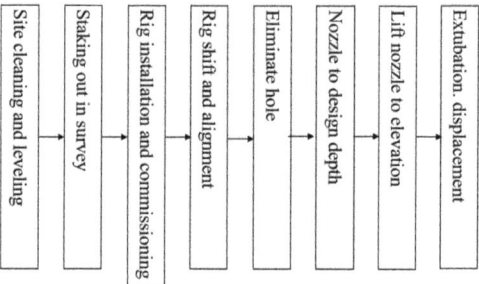

the design requirements. Clean the debris in the grouting pipe and the nozzle, and make sure sealing ring of the grouting pipe joint is good.

(3) Test piles are taken before formal construction to confirm reasonable parameters such as water pressure, lifting speed, slurry ratio and pressure.

(4) The continuity of the pile body is ensured during the rotary jetting process. If it is stopped for some reasons, the length of the second rotary jetting pile must be larger than 20 cm.

(5) If there is a large amount of slurry in the construction, stop immediately and take measures.

(6) The deviation between the drilling position and the design position is not more than 50 mm. The actual hole position, hole depth and underground obstacles, caves, water inrush, water leakage and inconsistency that discrepancies the engineering geological reports in each borehole are recorded in detail.

(7) After the high-pressure jet grouting is completed, the grouting pipe is quickly pulled out, and the grouting pipe and the grouting pump are thoroughly cleaned to prevent solidification and blockage. In order to prevent the solidification and shrinkage of the slurry from affecting the elevation of the pile top, if necessary,

measures such as slurry recharge or secondary grouting need to be used in the original hole position.

3.2.2 Deep Mixing and Reinforcement Technology

The deep mixing and reinforcement method refers to the technology that transporting the cement, lime and other materials as the main agent of the curing agent into the ground through drilling, relying on the deep mixer to force the soft soil and the curing agent (slurry or powder) to be stirred on the ground. A series of physical and chemical reactions between the curing agent and the soft soil form deep mixing piles to improve the physical and mechanical properties of the soft soil. The deep mixing and reinforcement process is shown in Fig. 3.9.

① Deep mixer is in place.
② Stir and sink. Start the motor and relax the hoist according to the soil conditions, so that the mixing head will mix and sink from top to bottom until reaching the design depth.
③ Grouting agitation lift. After the cement slurry reaches the mixing head, lift the mixing head according to the calculated speed. Grouting, while stirring and lifting, the cement slurry and the original ground soil are thoroughly mixed until the height of the piles reach the top, and then turn off the pump.
④ Repeat the stirring and sink. Once again, the mixer is allowed to sink to the design level while agitating.
⑤ Repeat the agitation (no grouting) and lift. Lift to the natural ground with agitation, turn off the mixer, and complete the work of one pile.

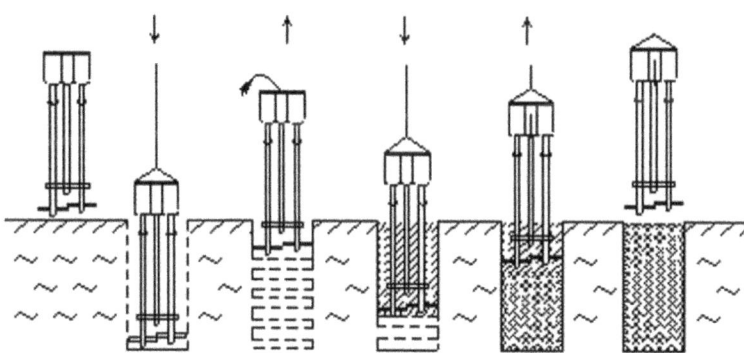

Fig. 3.9 Schematic diagram of deep stirring reinforcement process

1) Application range

The deep mixing and reinforcement method is suitable for foundation reinforcement of saturated soft clay, silty sub-clay, new-filling soil, marsh soil, sedimentary silt and other soil layers, which is used to improve the bearing capacity of soft soil foundation, reduce settlement and also improve the stability of the slope. It is mainly used in the following aspects:

(1) Stabilizing the soil outside the shaft of the tunnel for the reinforcement for the start and end of the shield into the hole.
(2) As the foundation of buildings or structures.
(3) Strengthening the slope of the deep foundation pit to prevent the slope of the foundation pit from landslide.
(4) Strengthening roads or bridges and culverts.
(5) As an underground cut-off wall, preventing groundwater from penetrating.

2) Features of deep mixing and reinforcement method

(1) There is basically no squeezing effect, and the disturbance to the surrounding foundation is small.
(2) According to the requirements of different soil quality and engineering design, the curing agent and formula can be reasonably selected, and the application is more flexible.
(3) The construction has no vibration, no noise, and low pollution. It can be constructed in urban areas and in densely populated areas.
(4) After the soil has been reinforced, the gravity is basically unchanged, so there is no additional settlement.
(5) The structure of the reinforcement piles is flexible and diverse, and it can be selected in the form of block, column, wall or grid according to what the engineering needs.

3) Reinforcement mechanism

The mechanism of strengthening the formation by deep mixing is related to the curing agent and type. The curing agent is composed of cement, lime, gypsum, slag and the like. After the cement and soil are stirred, a series of physical and chemical reactions take place to strengthen the soil. These physicochemical reactions are different from the hardening mechanism of concrete. The hardening of concrete is mainly carried out by hydrolysis and hydration, and the setting speed is faster. But in cemented soil, the amount of cement is small, generally accounting for 7–15% of the weight of the reinforced soil. And the hydrolysis and hydration reaction of the cement is carried out under the encirclement of the soil particles with certain activity, so that the soil particles will chemically react with the cement hydrate at any time to form a new compound. The role of clay particles and cement hydrates mainly includes ion exchange and agglomeration, hard coagulation and carbonation. Since the chemical reaction takes place between the cement hydrate and the soil particles,

3.2 Reinforcement Stratum Tool-Changing Technology

the time completing the chemical reaction is inevitably long. Therefore, the hardening speed of the cement-reinforced soil is slow and complicated, and the strength of the reinforcing soil also increases slowly.

4) Construction points

(1) All underground obstacles must be identified and removed before starting the machine. The parts that must be backfilled need to be backfilled in batches to ensure the quality of the piles.
(2) The pile rails and sleepers of the pile driver shall not sink, and the vertical deviation of the pile machine don't exceed 1%.
(3) The cement must use 32.5 grade Portland cement. The cement blending ratio should be 8%-16%. Different types of admixtures can be added according to different geological conditions and construction period requirements.
(4) The amount of grouting and the speed of lifting must be strictly controlled to prevent the occurrence of sandwich layers or broken pulp.
(5) The speed of the two lifting heads should be controlled at 2.5–3 m/min. The outlet pressure of the grouting pump should be controlled at 0.4–0.6 MPa.
(6) The following matters should be noted in the construction of piles and pile lapping:

   ① The time between the pile and pile lapping should be no more than 24 h.
   ② If it exceeds 24 h, pay attention to increase the amount of grouting during the construction of the second pile, which can increase by 20% and slow down the lifting speed at the same time.
   ③ If the interval between the two piles is too long, the second pile cannot be lapped, adopt partial pile or grouting measures under design approval.

(7) Pre-stirring and sinking of pile driver should ensure that the structure of the undisturbed soil is fully broken according to the original soil conditions, so that it is suitable for even mixing with the cement slurry.
(8) Adopt standard water tank, and strictly control the water-cement ratio. The mixing time of cement slurry is not less than 2–3 min, and pour the slurry into the collecting tank, then continue to stir to prevent the cement from segregating. The grouting process should continue, not be interruptible.
(9) Each shift must be make a test block (three pieces). After 28 days, measure the unconfined compressive strength and the number it if reaching the design requirements

5) Quality standards

In addition to referring to relevant national standards, the following points should also be noted:

(1) The vertical deviation of piles shall not exceed 1%, and the deviation of pile placement shall not exceed 50 mm.
(2) The lapping pile should be evenly stirred and the surface should be dense and smooth.

(3) The pile top elevation and pile depth should meet the design requirements.
(4) There is no abnormality in the amount of cement slurry.

6) Specific design

This section discusses in detail how to determine the extent of the ground-based reinforcement area. Table 3.1 is the three methods and details of the deep mixing pile. A schematic diagram of the three methods is shown in Figs. 3.10–3.12.

(1) Reinforcement thickness. Considering the open part of the reinforced solid as a freely supported disc, based on the damage test of the simply supported beam, calculate the thickness of the external force (active earth pressure + water pressure) that can resist the back side as shown in Fig. 3.13.

$$t = F_s \sqrt{\frac{3W \cdot r^2}{\sigma_t}} \qquad (3.4)$$

t—requires thickness (m);
F_s—safety factor, 1.2 in soil mechanics, 2.65 in non-seismic test of concrete structure;

Table 3.1 Detailed forms of three construction methods of deep j stirring pile

Classification	Type A	Type B	Type C
Application condition	① Ground stabilization ② Watertight requirement ③ The basal reinforcement	① Ground stabilization ② Watertight is not required ③ The basal reinforcement ④ The preset ground beam in the middle	① Defective part of retaining soil ② As waterproof curtain
Basic interval	① As standard, for lap configurations ② Horizontal interval $l_1 = \frac{\sqrt{3}}{2}D$ ③ Longitudinal interval $l_1 = \frac{3}{4}D$	① As a standard, the contacts are configured ② Horizontal interval $l_1 = D$ ③ Longitudinal interval $l_1 = \frac{\sqrt{3}}{2}D$	① As standard, for lap configurations ② Effective thickness $t_0 = 2\sqrt{\left(\frac{D}{2}\right)^2 + \left(\frac{l}{2}\right)^2}$ ③ Splicing length $h = D - \sqrt{D^2 + t_0^2}$ $h \geq 0.2m$
Application	① Sealing up ② Sudden welling ③ Sand boiling	① Resistant soil uplift ② Grade beam	① Sealing up ② Temporary wall material
Perpendicularity	Precision: 1/250 $l' = l - \frac{Z}{250}$	Precision: 1/250 $l' = l - \frac{Z}{250}$	Precision: 1/250 $l' = l - \frac{Z}{250}$

Note D is the effective diameter of the pile and Z is the depth

3.2 Reinforcement Stratum Tool-Changing Technology

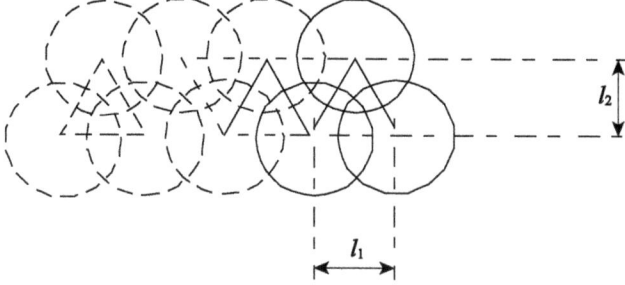

Fig. 3.10 Type A foundation reinforcement

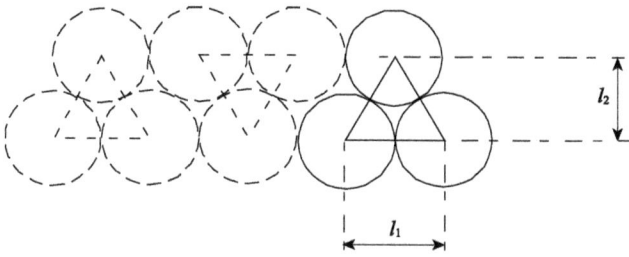

Fig. 3.11 Type B foundation reinforcement

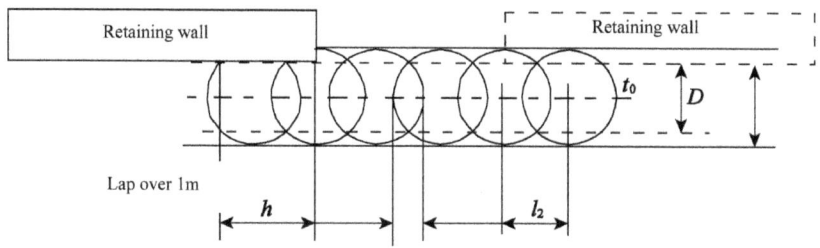

Fig. 3.12 Type C foundation reinforcement

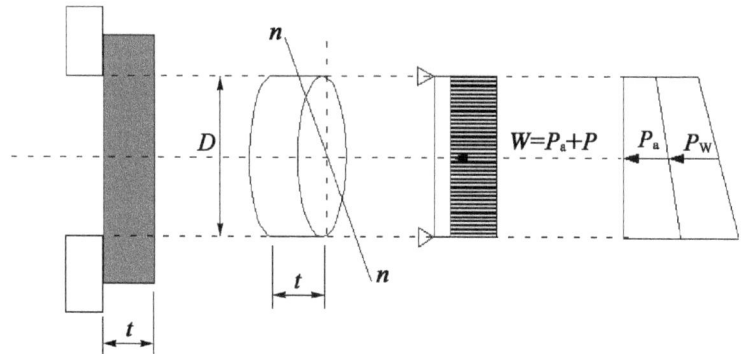

Fig. 3.13 Calculation model of reinforcement thickness

Fig. 3.14 Upper reinforcement area

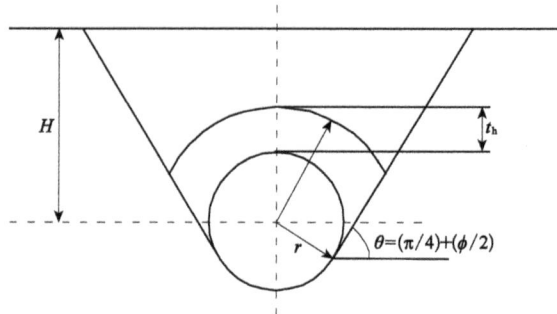

W—External force (mainly including active earth pressure and water pressure) (kN/m²);
r—the radius of the shield excavation (*D*/2) (m);
σ_t—the bending tensile strength (kN/m²) of the surrounding rock after reinforcement.

(2) Reinforcement range.

① The upper part requires thickness. The thickness of the upper part of the shield is studied based on the additional stress generated by the surrounding tunnel, as shown in Fig. 3.14.

$$t_h = F_s \left\{ \exp\left[\frac{(H-r)\gamma_t}{2C} + \lambda_n r \right] - r \right\}; \quad (t_h < 2.5, \ t_h = 2.5\,\text{m}) \quad (3.5)$$

H—the depth to the center of the tunnel (m);
r—the radius of the shield excavation (*D*/2) (m);
γ_t—the unit volume weight of the surrounding rock after reinforcement (kN/m³);
C—the cohesion force of the surrounding rock after reinforcement (kPa);
F_s—Safety factor, 1.2 in soil mechanics, 2.65 for non-seismic verification in concrete structures.

② The thickness is required on the side. The side portion needs to have a thickness that is in the upper limit of the intersection of the required thickness and the collapse angle ($\theta = \pi/4 + \pi/2$), as shown in Fig. 3.15.

$$\beta = \cos^{-1} \frac{r}{r + t_h} - \left(\frac{\pi}{4} - \frac{\phi}{2} \right) \quad (3.6)$$

$$t_x = (r + t_h) \cdot \cos\beta - r \quad (t_x < 2.0\,\text{m}, \ t_x = 2.0\,\text{m})h \quad (3.7)$$

r—the shield excavation radius (*D*/2) (m);
t_h—the upper reinforcement thickness (m);
ϕ—the internal friction angle of the surrounding rock (°);

3.2 Reinforcement Stratum Tool-Changing Technology

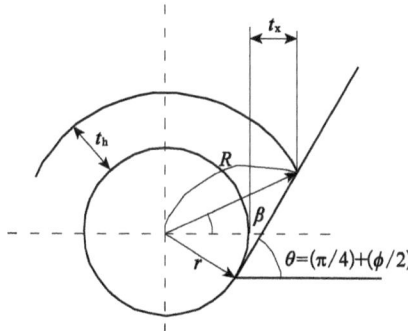

Fig. 3.15 Side reinforcement range

t_x—side reinforcement hardness (m).

③ The thickness is required at the bottom. Consider water repellency to ensure minimum thickness. Note: The minimum thickness in the reinforcement range, taking into account the actual use value to date, is 2.5 m on the top, 2.0 m on the side, 1.5 m at the bottom as the standard value. Wherein, the thickness of the lower portion should be more than 60% of the thickness of the upper portion.

The construction case of deep mixing piles is based on the Guangdong Foshan Rail Transit Project and the Shenzhen Metro Line 1 and Qianhai-Xin'an Station. The application of the rotary jet reinforcement technology in practical engineering is specifically described. The shield section is located in the first phase underground of the rail transit project in Foshan, Guangdong. The section tunnel mainly passes through fine sand, fully weathered mudstone, strongly weathered mudstone and moderately weathered mudstone, and partially passes through silty soil. The radius of the plane curve in the left and right tunnels is 600 m, and the distance between the left and right lines is 12.0–16.5 m. Two composite earth pressure balance shields are successively launched from the end shield tunnel of Dengzhou Station. At the same time, the left and right tunnels are excavated in the interval, and the middle wind shaft is adopted by means of passing the station. There are two types of groundwater in this section: quaternary pore water and bedrock fissure water. Among them, the communication channel is adjacent to a large waterway, which has good water-richness and strong water permeability, and is a medium-strong permeable stratum. water mainly appears in the coarse sand stratum in the Quaternary Upper Pleistocene-Geolian alluvial stratum. Because the upper part is covered with the fourth layer of silty clay, silty soil and other water-repellent layers, the lower part is also a strong weathered sandy mudstone water-repellent layer, which results in a certain pressure bearing property. The other is the bedrock fissure water that mainly occurs in the strong weathered and moderately weathered fissures of the bedrock. The groundwater depth varies with the undulation of the foundation rock surface, generally is 0.5–4.6 m. Due to lithology and fracture development, the degree of water richness and permeability are not the same, generally poor. Since the upper weathered rock and residual soil in the upper part of the strong weathering zone are mainly soil-based and its water permeability is poor whose water-blocking water is in a certain degree to the relative water-blocking effect, the bedrock fissure water has confined water characteristics. When the right-hand shield tunnels into the 370 rings (that is, about 555 m), the shield begins to break down, so the tunneling is slow, and it is necessary to open the cabin for inspection. And replace the damaged and worn tools according to the situation. The corresponding location is in the highway auxiliary road, and the stratum condition is that the overburden soil thickness is 27.603 m. The cave stratum is mainly composed of silty soil, medium coarse sand and strong weathered mudstone and sandstone. Considering that the soil layer in this area is relatively weak, so it is difficult to achieve the normal pressure to open the cabin, and the risk on the open cabin is relatively high. Considering that there is no building above the position where the tool change is needed, and it is located near the communication channel, so it needs to reinforce the stratum. By comprehensively considering, the technical plan for strengthening the stratum with deep mixing piles is confirmed. To carry out the tool inspection and replacement in a planned manner to realize the normal pressure tool-changing.

The specific reinforcement scheme is as follows: the reinforcement depth 4 m above the channel to 0.5 m into the strong weathered formation uses the deep mixing pile of 50 mm@600 mm. After reinforcement, the unconfined compressive strength should be greater than 1.0 MPa, and the permeability coefficient should be less than 1.0×10^{-5} cm/s. The reinforcement arrangement of the deep mixing pile is shown in Fig. 3.16. In Shenzhen Metro Line 1 Qianhai Bay-Xin'an Station section of the main line shield tunnel section (right line), the range mileage is SK26 + 338.300–SK26 + 678.085, and the line length is 339.785 m. The tunnel crosses the stratum with gravel clay and sandy clay. There are localized rock face bulges and underlying micro-weathered granite. After geological reconnaissance, the surface of about 50 m raised bedrock is a reclamation area, without any important buildings or pipelines. When the shield tunneled smoothly to the 118th ring, a small amount of granite particles began to appear in the muck, with particle sizes ranging from 5 to 25 mm. Occasionally, basketball-sized stones can be seen. The project department immediately adjusted

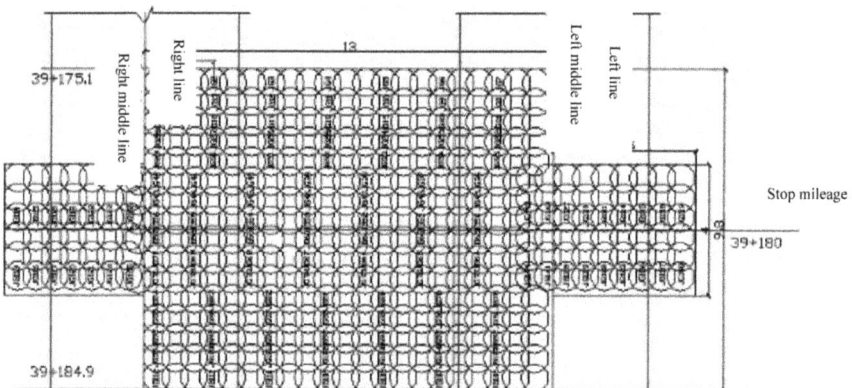

Fig. 3.16 Left and right line reinforcement location range

3.2 Reinforcement Stratum Tool-Changing Technology

the excavation parameters until the 128th ring, and there was no abnormal changes in the excavation parameters. After the tunneling to the 129th ring, the excavation parameters began to fluctuate greatly.

The thrust increases from 900 to 1100t, and the cutter torque increases from 1100 to 2100 kN/m, but the tunneling speed is only 16 mm/min. After the one meter shovel of the 130th ring, that is, when the shield cutter head is at SK26 + 539.600, the thrust and cutter torque suddenly increase to 1800t and 3200t, but the tunneling speed is only 2–5 mm/min. According to the geological re-extraction data, the length of the interval SK26 + 510.000–SK26 + 560.000 has a "protrusion bedrock" section with a length of about 50 m. The lower bedrock is very hard and complete, and the uniaxial compressive strength of the rock is 100–120 MPa. The thickness of the tunnel cover is about 12 m. The body of the hole is 2.2–3.7 m. It has gravel clay soil and sandy clay soil with large void ratio, good hydrophilicity, being easy to disturb and break, and to disintegrate in water. Silt, miscellaneous fill. The lower part of the hole is 2.3–3.8 m, which has hard micro-weathered granite. It is judged that the shield peripheral tools have been seriously worn at this time. According to the analysis of the results about the core sampling of the foundation in the relevant construction units, the cement mixing piles have good reinforcement effect under the geological conditions of this site, and the cost is lower and the construction speed is faster. Considering the construction progress and project quality, 550 mm mixing pile is adopted to reinforce the stratum, and 150 mm of interlocking pile is adopted. The piles length are about 15 m. About 6 m in the lower part is cement mixing piles (solid piles), and 9 m is the upper empty piles. In order to ensure the water stop effect when opening the cabin, rows of solid piles to the ground at the shield ring is designed. The specific arrangement of the deep mixing pile is shown in Fig. 3.17.

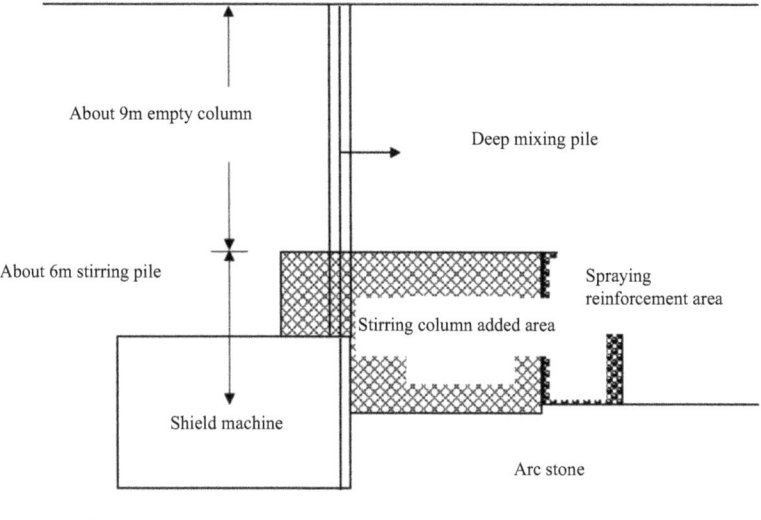

Fig. 3.17 Schematic diagram of stirring pile reinforcement area section

3.2.3 Freezing Method Construction Technology

The freezing method (Fig. 3.18) use artificial freezing method to reduce the temperature of the soil below 0 °C, so that the soil is consolidated to form a better stable soil. At the same time, it acts as a water barrier in the frozen soil. The construction of the structure is carried out within the scope, and the method of freezing and grouting is used after the completion of the construction. The freezing method is often used for pipelines, buildings, etc. on the surface, which has increasing cost and poor geological conditions in the ground reinforcement construction, and is located in the unfavorable geological conditions such as sand layer and rich groundwater.

The freezing method is a type of foundation consolidation method in which a freezing pipe is buried at a pitch of 60–100 cm in a predetermined freezing zone, and a cooled brine (aqueous solution of calcium chloride) is circulated in freezing pipes by using freezing devices to freeze and solidify the foundation.

1) Characteristics of freezing method construction

(1) Use the principle of heat conduction, which is suitable for the foundation from the clay soil to the gravel layer.
(2) The frozen soil sand is quite strong and has a high water stopping performance.
(3) There is no pollution to the foundation and groundwater.
(4) It is easy to freeze management and predict the scope of freezing.

2) Installations of refrigeration equipment, brine pump, cooling water pump and its piping system at the freezing station (Fig. 3.19) carry out the "Construction and Acceptance Specifications for Installation of Refrigeration Equipment and Air Separation Equipment" (GB50274-1998), "General specifications

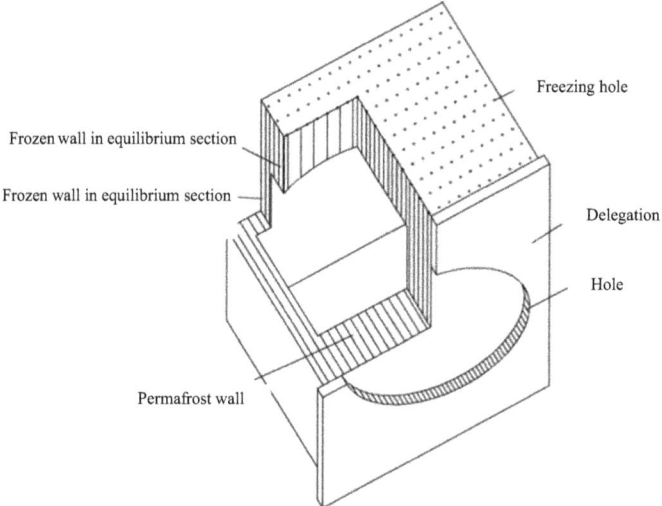

Fig. 3.18 Schematic diagram of freezing method reinforcement

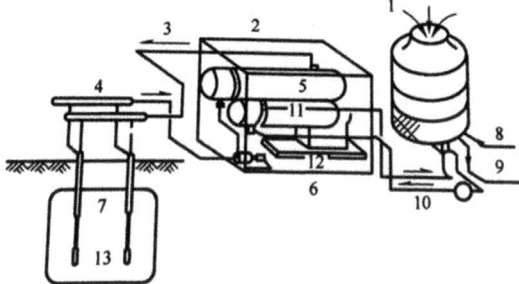

Fig. 3.19 Schematic diagram of freezing station. 1—cooling tower; 2—refrigeration unit; 3—pipe; 4—a water collecting pipe; 5—a cooler; 6—a water pump; 7—freezing tube; 8—water supply; 9—a drainage; 10—cooling water; 11—a condenser; 12—a compressor; 13—permafrost

for construction and acceptance of mechanical equipment installation works" (GB50231-2009) and the "Code for Construction and Acceptance of Industrial Metal Pipeline Engineering" (GB50235-1997). Installation and commissioning of the distribution system shall be carried out in accordance with the "Code for Construction and Acceptance of Electrical Installations for Panels, Cabinets and Secondary Circuits" (GB50171-2012). After the frozen section excavation and masonry construction are completed, the freezer can be shut down and the refrigeration station equipment can be removed.

3) Design points of stratum freezing

(1) The frozen layer reinforcement shall ensure the safety of earthwork excavation and structural construction within the designed time, and the surrounding environment and buildings (structures) shall not be damaged. The formation freezing design should include the following contents: comparison and selection of frozen wall structure schemes; deformation check of bearing capacity and frozen walls (except for type I frozen walls); freezing hole layout design; frozen wall formation check; freezing refrigeration system design; Monitoring and protection requirements; analysis of possible impacts on the surrounding environment and buildings; monitoring and protection requirements for impacts on the surrounding environment and buildings.

(2) During the formation of the frozen wall, precipitation measures should not be taken in the permeable sand layer in the frozen wall or within 200 m outside the frozen wall. If precipitation construction is necessary, the freezing design should fully consider the adverse effects of precipitation.

(3) The temperature of the low brine should be determined according to the average temperature of the frozen wall, the formation environment and the climatic conditions. When the average temperature of the frozen wall is low, the ground temperature is high, and the temperature is low, the lower brine temperature is taken.

Control the brine temperature according to the following requirements: actively freeze the 7d brine to below −18 °C, and actively freeze the 15d brine temperature to below −24 °C (if design the minimum brine temperature is above −24 °C, use the minimum brine design temperature) and during excavation process the brine temperature should drop below the design minimum brine temperature. Maintenance freeze can be carried out after support in the construction, but the temperature of the frozen brine should not be higher than −22 °C.

(4) During the excavation process, the salt water temperature may appropriately increase, but not higher than −25 °C, under the condition that the average temperature and thickness of the frozen wall meet the design requirements and the frozen wall is determined to be safe.

(5) When excavation, the temperature difference of the returning brine should not exceed 2 °C.

(6) The internal force and deformation calculation of the frozen wall can be considered the conditions provided with internal support, but the structural form, bearing capacity and construction sequence of the inner support must be clearly designed. When designing the inner support, the free time of the frozen wall is not more than 24 h.

(7) The freezing hole arrangement parameters include the freezing hole forming hole control spacing, the freezing hole opening spacing, the freezing hole position, the freezing hole depth and the freezing hole deflection precision requirement. The frozen wall formation parameters include the frozen wall intersection time, the expected frozen wall expansion thickness, and the average temperature of the frozen wall.

(8) The control hole spacing of the freezing holes shall be determined according to the freezing period requirements, the designed brine temperature and the freezing average temperature, but no more than the frozen wall design thickness. When multiple rows of frozen holes are densely arranged, Internal freezing hole into the hole control spacing can be 1.2 times the edge hole. The freezing hole deflection accuracy is selected according to Table 3.2.

Table 3.2 Freezing hole deflection accuracy requirements

Freezing hole type	Horizontal or inclined freezing hole			Vertical freezing hole	
Freezing hole depth H (m)	≤10	10–30	30–60	≤40	40–100
Maximum deflection of freezing hole R_p (mm)	150	150–350	350–600	150–250	250–400

(9) The refrigeration system of the refrigeration station consists of a refrigerant circulation system, a refrigerant circulation system, and a cooling water circulation system. The solution of the refrigerant circulation system is an aqueous solution of calcium chloride, which should meet the following requirements: the freezing point of the aqueous solution of calcium chloride should be lower than the designed brine temperature of 8–10 °C; the relative density of the aqueous calcium chloride solution should not be higher than 1.27.

4) Freezing hole and freezing tube

(1) Freezing hole.

① The opening pitch of the near-horizontal freezing hole, the slope of the drilling hole, the spacing of the adjacent drilling holes and the drilling depth should meet the design requirements.
② When constructing an aquifer, it is necessary to use a secondary opening method to open the hole and install an orifice sealing device to prevent water from rushing into the hole during drilling through the tunnel segments and into the hole.
③ When starting to drill into the freezing hole, it should be lightly pressed. Gradually pressurize when drilling. When pressing, observe the oil pressure gauge indicating the drilling pressure to ensure that the oil pressure gauge cannot exceed the allowable value and keep the medium speed drilling.
④ The drilling depth should not be shorter than the design depth.
⑤ The drilling rig can only be dismantled after all drilling holes have been tested and qualified.

(2) Freezing pipe.

① Freezing pipe must use seamless steel pipe.
② The wall thickness and outer diameter of the freezing tube: the wall thickness of the freezing tube is not less than 5 mm, and the outer diameter of the freezing tube is 89–127 mm.
③ The freezing pipe can be connected by being wired and welded, or welded with a pipe clamp; when welding with pipe clamp, the material of the pipe clamp should be the same as that of the frozen pipe.
④ Freezing pipe material and connection method: The freezing pipe is made of No. 20 low carbon steel seamless steel pipe, and the connection mode is divided into two types: wire-bonding connection and inner-ring butt welding connection.
⑤ The depth of the freezing pipe should not be less than the design depth. Each section of the freezing pipe placed under each freezing hole should have length, pipe diameter and number record. It is strictly forbidden to place any impurities in the pipe.

⑥ After the freezing pipe is drilled into the borehole, pressure test must be carried out. The test pressure shall be twice the sum of the pressure difference between the brine column and the clean water column outside the pipe and the working pressure of the brine pump. The pressure drop shall not exceed 0.05 MPa after 30 min of test pressure, and if the pressure shall remain unchanged after 15 min, it can be seen as qualified.

⑦ Leakage tube treatment: The freezing tube with unsatisfactory sealing must be processed to meet the sealing requirements. The preferred solution is to place the small diameter freezing tubes, followed by the plugging method, and finally the filling the freezing hole method.

(3) Acceptance. For the actual hole position, the depth of the freezing pipe, the temperature measuring hole, the water temperature hole, the final measuring result, and the pressure test data of the freezing pipe, the relevant department shall conduct the acceptance. When there is doubt about the result of the final measurement of the frozen hole, the third party will repeat the retest.

5) Frozen wall detection and judgment

(1) Temperature measurement hole detection.

① After the temperature measuring components in the temperature measuring tube are set, the nozzle should be protected to prevent the temperature measuring components and cables from being damaged.

② The accuracy of the temperature measuring component should reach ± 0.5 °C.

③ From the freezing station to the formation of the frozen wall, it should be observed every 8–24 h; during the completion of the excavation (opening the door) to the structure of the frozen wall, it should be observed every 4–12 h; after the structure is finished It should be measured every 1–3d, and the monitoring points can be reduced appropriately. When there are special requirements, the observations should meet the design requirements. All observations should reserve the original record and be signed by the observer.

(2) Pressure relief observation holes and other observation holes.

① Before the freezing station is operated, it is necessary to understand the conditions of the formation. The pressure measurement of the pressure relief hole should be done, which should be consistent with the original data. If the abnormality is found, the cause must be identified for disposal.

② The freezing station should be observed every 24–48 h before the operation. After the pressure starts to rise, it should be measured every 6–24 h. All measurements should have an original record.

③ After the formation of the frozen wall, the pressure of the pressure relief hole should be at least 0.1 MPa higher than the ground pressure.

3.2 Reinforcement Stratum Tool-Changing Technology

(3) Judgment of the frozen wall formation.

① The pressure of the pressure relief hole in the aquifer should be higher than the corresponding ground pressure 0.1 MPa, and it can be maintained after two or more pressure relief, except when there is no pressure relief hole or the frozen wall is not closed.
② Calculate the thickness of the frozen wall, the average temperature of the frozen wall, and the temperature on the excavation boundary according to the temperature measured by the temperature measurement hole, which should meet the design requirements.
③ Freezing the brine and the temperature difference of the loop should be gradually reduced and stabilized.
④ When there is no pressure relief hole or the frozen wall is not closed, the salt water flow rate of each freezing hole and the temperature of the detour and return brine should be monitored and meet the design requirements to ensure the normal formation of the frozen wall. If necessary, longitudinally measure each frozen hole to ensure the safety of the frozen wall.

6) Excavation and construction

When excavating and constructing the freezing method for reinforcement construction, in order to keep abreast of the construction conditions, a video surveillance system and a communication system must be installed on the site.

(1) The conditions that should be met when the communication channel is excavated.

① Water temperature hole is not related to external water force.
② The freezing time reaches the design value, the brine temperature reaches the designed minimum salt water temperature; the freezer brine circulation. The absolute value of temperature difference in the freezing brine circulation and loop should not be higher than 1.5 °C.
③ According to the temperature measurement results of the temperature measurement hole, the average temperature and thickness of the frozen soil curtain reach the design value.
④ The pressure of the pressure relief hole rises more than 7d, or the water is no longer flowing after opening the pressure relief valve.
⑤ The distance from the outer side of the freezing hole is 400 mm. The temperature of the frozen soil at 5 cm in the curtain is less than -5 °C. The freezing width at the interface between the frozen curtain and the tunnel segment is not less than 1.2 m. The position of the hole is selected at a larger hole spacing or where there is an abnormality in freezing.
⑥ Open the pressure relief hole to confirm that no muddy water is flowing out.
⑦ The protective door has been installed to confirm that the protective door is normally opened and closed, and the gas supply pipe is connected.
⑧ Complete the tunnel support reinforcement.

⑨ It is necessary to formulate a sound emergency plan and prepare emergency materials and equipment such as cement, water glass and liquid nitrogen; there should be a complete inspection report of mechanical equipment such as refrigerators and power supplies, and be approved by the supervisor; a sound structural monitoring program and the surrounding environment monitoring program should be arranged and implemented in place; a video monitoring system should be deployed and a reliable communication system should be provided.

⑩ There should be a trial excavation notice document issued by the supervision department.

(2) Test digging.

① Three holes are evenly distributed in the edge of the excavation area and in the temporary support area. The hole diameter is 30–40 mm and the depth is 500–800 mm (excluding the thickness of the outer wall) to check the soil freezing condition. If the temperature of the soil in the three holes is below 0 °C (that is, icing), it is considered that the freezing effect meets the excavation requirements; if the temperature of one hole is above 0 °C (that is, there is no ice), the freezing effect is not considered to meet the excavation requirements, so it is not possible to excavate and seal the holes. And continue to actively freeze until the requirements for the hole are reached.

② On both sides of the excavation area, the outer edge of the effective frozen wall and the middle of the zero-degree line are designed to inspect the soil freezing. The principle of judging the hole and the principle of freezing the soil are the same as above.

③ Open a trial window in the middle of the excavation area, and the window size should be no more than 500 mm × 500 mm. Drill 500–600 mm deep into the inside. Check the soil condition. If the soil is dry and self-supporting, it is considered to have reached the excavation condition; otherwise, continue to dig deep to 1000 mm. If the result is still not good, the soil freezing quality in the excavation area does not meet the excavation requirements. Backfill and close the trial and digging window, continue to actively freeze until the next test is completed.

④ Carried out the excavation according to above order. If each article meets the requirements, it is considered as a trial digging. If one does not meet the requirements, the test digging fails and the test is stopped.

(3) Formal excavation. The excavation section can be excavated by full section and step method.

(4) Construction of permanent structural reinforced concrete. Strictly in accordance with the "Construction Quality Acceptance Code for Concrete Structure Engineering" (GB50204-2015).

(5) Filling the grout behind the wall.

① Filling grouting design and technical requirements are as follows: a grouting pipe is reserved between the interface between the support layer and the frozen wall and between the sprayed concrete and the permanent structure. The back of the wall is filled and grouted when the permanent structure reaches a certain strength to reduce the deformation of the surrounding soil. Generally, concrete can be filled and grouted after one week of pouring.

② Strictly control the grouting pressure and grouting volume not to exceed the design range. Combined with the monitoring data, according to the principle of less than multiple injections, the stratum settlement will be gradually controlled to be stable.

③ The joint thread at the end of the grouting pipe should be inspected intact, the valve is sealed reliably, and it can be closed in time when the hole is sprayed.

④ Monitor the tunnel convergence deformation during grouting, and ensure that the tunnel and surface deformation under the grouting pressure are within the design allowable range.

7) Engineering monitoring

(1) Monitoring contents: surface subsidence monitoring; tunnel deformation monitoring; tunnel settlement monitoring; tunnel horizontal and vertical convergence deformation monitoring; ground building settlement monitoring; underground pipeline settlement monitoring; channel freezing wall, bracket convergence deformation monitoring; channel earth pressure and support force monitoring.

(2) The arrangement and measurement method of the measuring points should comply with the relevant provisions of the "Code for the Measurement of Underground Railways and Light Rail Transit Engineering", "City Measurement Specifications" and "Engineering Measurement Specifications".

3.2.4 Shaft Reinforcement Technology

1) The basic structure of the shaft is generally composed of the well wall, the blade foot, the inner partition wall, the well hole groove, the bottom plate, the top cover, etc., as shown in Fig. 3.20.

(1) Well wall: The outer wall of the shaft is the main part of the shaft. It should have sufficient strength to withstand the load during the sinking and using process. It also requires sufficient weight to make the shaft under its own weight to sink smoothly.

(2) Blade foot: The lower end of the well wall is generally made into a blade-shaped blade foot, and its function is to reduce the sinking resistance.

(3) Partition wall: It is installed in the shaft, and its main function is to increase the rigidity of the shaft during the sinking process. At the same time, it can divide the whole shaft into multiple construction wells (take the well), so that the excavation and sinking can be carried out in a more balanced way, and it is also convenient to correct the deviation of the shaft.

Fig. 3.20 Shaft

- (4) Groove: It is placed on the inner side of the well wall above the cutting edge. The function is to make a better connection between the bottom concrete and the bottom wall and the well wall to transmit the base reaction force.
- (5) Back cover: When the shaft sinks to the design elevation, after technical inspection and well bottom cleaning and leveling, the bottom can be sealed to prevent groundwater from penetrating into the well.
- (6) Top cover: The reinforced concrete roof is poured on the top of the well, and the pier can be built after the top cover reaches the design strength.

2) The characteristics of the shaft is that the building or building foundation located at a certain depth of, first makes a shaft at the surface. Then by continuously excavating from the well under the protection of the well wall, the shaft gradually sinks under the effect of self-weight. After reaching the predetermined design elevation, the bottom is sealed to construct the internal structure. The technology is relatively stable and reliable, the amount of excavation is small, the impact on adjacent buildings is relatively small, the foundation of the shaft is buried deep, the stability is good, and the load can be supported.

(1) Advantages.

① It has deep buried depth strong overall, good stability, large bearing area, and can withstand large vertical and horizontal loads.
② Shaft is not only the foundation, but also the retaining and retaining structure during construction. It is not necessary to install pit wall support or sheet pile surrounding wall during the sinking process, which simplifies the construction.
③ Shaft construction has less impact on adjacent buildings.

(2) Disadvantages.

① The construction period is longer.
② Construction technology requirements are high.
③ Sanding is likely to occur during construction, which makes it difficult to tilt or sink the shaft.

3.2 Reinforcement Stratum Tool-Changing Technology

2) Shaft production

(1) Shafts shall be made on weak foundations, sand, gravel or gravel cushions shall be used, compacted by a tamping machine to make them dense, and the thickness shall be determined according to calculation.
(2) When the local base soil is good, it should be finished once in a sub-section, and then sink; for a high ($\geq$12 m) shaft, firstly dig 3–4 m earthwork, sink once in the foundation pit or segmented making and sinking to reduce the free height of the shaft, increase stability and prevent tilting.
(3) Shaft preparation should adopt the method of setting wooden cushions or brick pedestals under the blade feet. The size and spacing should be determined according to the load calculation. When installing steel cutting feet, make sure that the outside is perpendicular to the ground so that it acts as a cutting guide.
(4) The pouring of the shaft and the concrete of the shaft should be segmented, symmetrical and continuous, to prevent the occurrence of inclination and crack. In the first section, when the first concrete strength level reaches 70%, the second section can be poured.
(5) The concrete of the poured body should be dense and the outer surface can be smooth. When there is waterproof requirements, the template through-wall bolts shall be welded with water-stop ring in the middle; the horizontal seam of the barrel shall be provided with a convex joint or a steel water stop belt, and the wall part of the protruding cylinder shall be shovel after the form removal to be benifical to waterproof and sinking.

4) Shaft sinking

(1) Before sinking, the appearance of the wellbore should be inspected, the concrete strength and impermeability grade should be checked, and the ultimate bearing capacity should be calculated according to the survey report. The segmental frictional resistance of the shaft sinking and the subsidence coefficient of the segmentation should be calculated ($\geq$1.15–1.25), as a basis for judging whether each stage can sink or not, whether there is a sudden sinking, determining the sinking method, and taking measures.
(2) Before sinking, the bolsters (brick pedestals) under the blade feet should be removed (demolition) in sections, groups, sequentially, symmetrically and synchronously. After each slat is pulled out, immediately fill with sand, pebbles, or gravel at the foot of the blade.
(3) Manual or pneumatic tools are used for excavation of small shafts; large shafts are used for excavation in small wells. The excavation shall be carried out in a layered, symmetrical and uniform manner. Generally, it will gradually be dug from the middle of the shaft to the surrounding area. Each layer is 0.4–0.5 m high, and a soil embankment with a width of 0.5–1.5 m is retained around the edge of the blade. Then along the wall of the shaft, the soil layer is completely, symmetrically and uniformly be thinned every 2–3 m in the direction of the blade foot. Each time it is cut 5–10 cm, when the soil layer cannot be crushed by the blade, the shaft will be broken. Under the action

of self-weight, the vertical squeezing soil sinks so as not to cause excessive tilting. The height difference of each tank should be within 50 cm.

(4) In the process of excavation and sinking, the foreman, surveyor and excavation workers should cooperate closely to strengthen observation and correct deviation in time.

(5) The method of sinking and sinking is adopted for the sinking of the shaft. The common method is to set up the drainage of the open ditch and the collecting well, and dig a drainage ditch in the shaft 2–3 m away from the cutting edge, and set up 3–4 collecting wells. It is 1.0–1.5 m lower than the bottom of the excavation face, and the depth of the ditch and the bottom of the well is deepened with the excavation of the shaft. A centrifugal pump is installed on the well wall or a submersible pump is installed in the well to discharge the groundwater out of the well. If the local quality conditions are poor and there is sand flow, a light well point, a jet well point or a deep well point can be set around the outside of the shaft to reduce the groundwater level, or a combination of well point and open channel drainage can be used for precipitation.

(6) The method of sinking the shaft is to project the horizontal line around the outer wall of the shaft, and the vertical axis is marked in 4 or aliquots in the wellbore. Each suspension line is dropped one by one, and is controlled by the lower target. Observation time, three times per shift, once 2 h close to the design elevation. Keep an eye on the analytical observations. When the line falls 50 mm from the vertical or the height difference is 100 mm, it should be corrected immediately. During the excavation process, the excavation can be corrected by adjusting the excavation elevation or labor.

(7) When the wall of the cylinder sinks, the outer soil will sag, and a gap will be formed between the outer wall and the wall of the cylinder. Generally, the outer side of the dry cylinder wall is filled with sand, which is not less than 30 cm high, and is poured into the gap with the sinking to reduce the sinking frictional resistance and the dredging work in the future. In the rainy season, a water barrier should be made on the outside of the sand filling to prevent rainwater from entering the gap, and prevent the phenomenon that the frictional force outside the cylinder wall is close to zero and the shaft is suddenly protruding or inclined.

(8) When the shaft sinks close to the design elevation, observation should be strengthened to prevent overheating. Brick piers or bolsters can be made at the intersection of the four corners or the wall and the bottom beam, so that the shaft is pressed against the brick pier or sleeper to stabilize it.

(9) When the shaft sinks, there is inclination. If the excavation cannot be corrected, the load can be adjusted. However, if one side has reached the design elevation, only the method of rotating high-pressure water is used to assist the sinking to correct the deviation.

(10) The earthwork excavated by the shaft shall be hoisted with a bucket and transported to the spoil ground, and shall not be piled near the shaft.

3.2 Reinforcement Stratum Tool-Changing Technology

5) Shaft back cover

(1) The shaft sinks to the design elevation, and then become stabilized by sinking for 2–3 days, or the accumulated subsidence is less than 10 mm within 8 h after observation, the back cover can be performed.
(2) Before the back cover, the new concrete contact surface at the edge of the blade should be rinsed or roughened, and the bottom of the well should be trimmed to make it a pot bottom shape. The radial drain should be dug by the blade foot to the center and filled with pebbles to make the blind water ditch. Two or three collecting wells in the middle are connected with the blind ditch, so that the bottom groundwater is collected in the collecting well and discharged by the submersible pump, keeping the water level below 0.5 m below the base surface.
(3) The back cover is usually covered with a layer of 150–500 mm thick pebbles or gravel, and then a layer of concrete cushion is poured on it, which is firmly filled under the blade and vibrated to ensure the final stability of the shaft. After reaching 50% strength, the waterproof layer of the coil is laid on the cushion layer, and the steel bars are tied, and both ends are extended into the cutting feet.
(4) Concrete pouring should be layered in the whole shaft area, uninterrupted, from all around to the central push, and with a vibrator tamp, when there are walls in the well, should be symmetrical before and after the left and right pouring hole.
(5) Water pumping shall continue during the concrete curing period. When the strength of concrete on bottom slab reaches 70%, water pumping shall be stopped one by one and blocked one by one for collection Wells. The plugging method is to drain the water in the catchment well, quickly fill and tamp with hard and dry concrete in the casing, and then tighten the flange plate with bolts or close by welding around, and the upper part is tamped with concrete pad. The construction process of shaft excavation is shown in Fig. 3.21.

6) Shaft construction case

The repair of the cutter head (with) in the waterless sand pebble stratum of the section of the Keyi Road Station-Fengtai South Road of Beijing Metro Line 9 is carried out by the ground in the predetermined position. The personnel are protected in the small shaft to repair and replace the cutter head (with) in front of the cutter head to ensure the continuous tunneling of the shield.

(1) Engineering geological conditions. The main section of Keyi Road Station-Fengtai South Road Station mainly passes through the pebble layer and the boulders layer. The maximum particle size is not less than 560 mm, the general particle size is 200–290 mm, and the percentage of the total mass of the pebble with the particle size larger than 20 mm is 85–98%. The bollard content is locally greater than 200 mm is 30–60%.

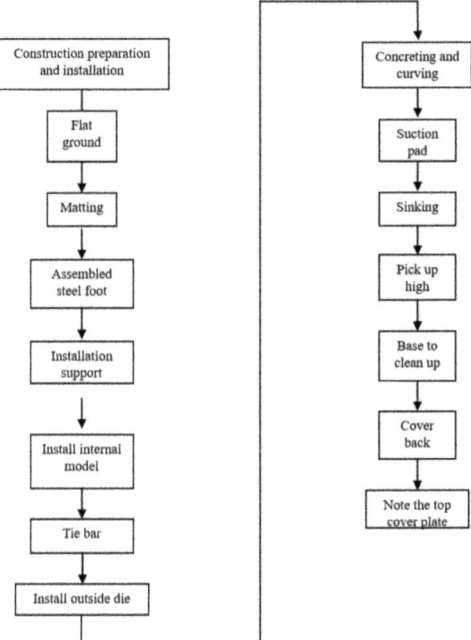

Fig. 3.21 Flow chart of shaft excavation

(2) Small shaft size selection. According to the working space required by a professional tool changer and the depth of the formation where the shield is located, the diameter of the working well is 1.6 m and the depth is 13.5 m. The external toothed retaining wall is used. After the thickness of the working wall is determined, the thickness of the retaining wall is 150 mm. The main tendon is 8 mm@200 mm, and the round stirrup is 8 mm@200 mm; the retaining wall is cast C25 self-mixing concrete.

(3) Release the well position. The small shaft is placed directly above the center line of the tunnel, the center coincides with the center of the cutter head, half is located above the shield (cutter), and the other half is in front of the cutterhead. The deviation of the well axis should be controlled within 20 mm. The relationship between the position of the small shaft and the cutter head is shown in Fig. 3.22.

(4) Earthwork construction and pouring retaining walls.

① Pour 200 mm thick concrete on the ground of the wellhead, and mark the axis and elevation.

② The excavation depth of each section of earthwork is 1000 mm, and the surface of the excavation face should be dug down from the inside of the well circle. The shape of the lower retaining wall is eight-shaped.

3.2 Reinforcement Stratum Tool-Changing Technology

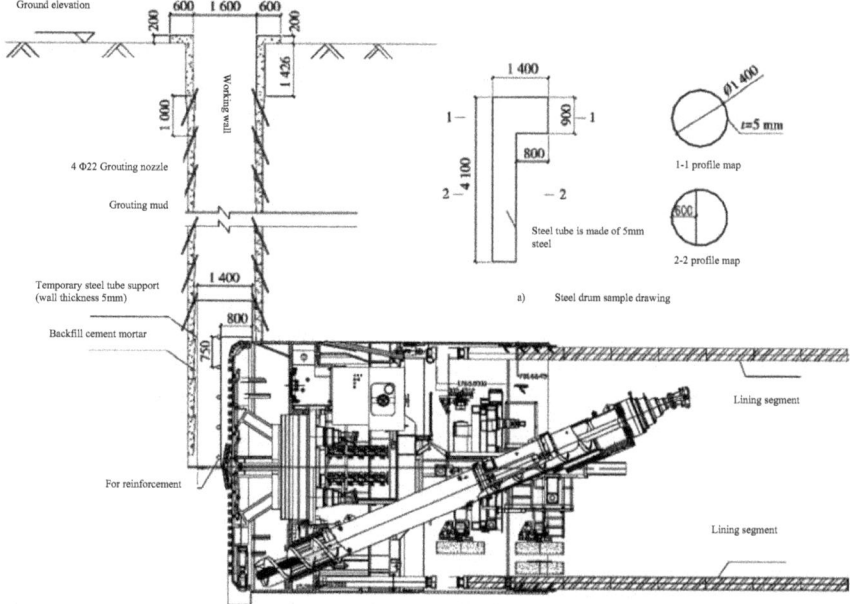

Fig. 3.22 Diagram of the relationship between small shaft and cutter head

③ Prepare two sets of wall template to remove the section and the lower section for repeated use. The template is fixed by fixtures and fasteners. If necessary, an arc-shaped inner steel ring made of channel steel or angle steel can be set at the upper and lower ends of each template as the inner support to prevent deformation caused by internal meme tension.

④ Immediately cast concrete at each section to ensure the stability of the hole wall. C25 concrete is artificially mixed in the field, artificially poured and compacted, and the slump is controlled at 80–100 mm. The demoulding time is controlled above 10 h, and in principle, the concrete demoulding strength is more than 70% of the concrete design strength.

⑤ The well axis and the elevation reference point are set at the wellhead position. After each wall is completed, the cross line is centered, the hanging line is projected to the bottom of the well, and the well deviation is checked centering on the hanging line. The well depth is based on the reference point and is measured by section.

(5) Grouting pipe installation. During the construction of each wall, four grouting pipes are reserved in the wall of the well. The grouting pipe is inserted into the surrounding soil by 500 mm, which is led upward from the inside of the formwork and temporarily sealed. After the construction of the small shaft is completed, the wall is grouted to enhance the stability of the small shaft.

(6) Steel casing installation. The steel casing adopts a full-circle casing with an inner diameter of 1400 mm, a wall thickness of 5 mm and a top height of 0.9 m.

The bottom portion is cut into a semicircular shape by 3.2 m, and the opening is welded with a scaffold tube as a strut. After the small shaft is excavated to the position of the cutter head, the excavation continues along the shaft in front of the cutter head to support the soil in front of the cutter head to form a semi-circular wellhead, and the overall protection of the casing is lowered to the excavation face by the crane. The outer side of the steel casing is fixed with a wedge-shaped square wood between the well wall, and the inner side is supported by a steel pipe. The longitudinal wire support and the steel pipe are temporarily supported in front of the cutter head, and keep a distance of 200 mm between the opening of the steel guard tube and the knife head (tool) according to the actual size of the site. 300 mm wide and 3200 mm long steel strips are welded radially outwards at the opening of the steel casing (the width can be adjusted according to the actual situation on site), and the gap between the steel casing and the well wall is closed. Semicircular. The distance between the semicircular guard tube and the cutter shall not be less than 200 mm. Finally, the outer part of the steel guard tube shall be voids and poured concrete.

(7) Tool changing in the small shaft.

① Tool cleaning. The staff clean the cutter spokes and the soil around the cutters in the small shaft to expose the cutters, welds and bolts.
② Tool inspection. The wear of the tool is checked in detail from the outside of the cutter head inwardly, and when it is determined that replacement is required, the tool with the corresponding number is replaced.
③ Tool replacement. Position the new tool in its original position and tighten the mounting bolts. The cutting knife welded on the cutter head panel is cut by gas cutting, the cutting surface is polished with a sander, and the corresponding numbered leading knife is welded to the cutter panel. After each tool is removed, new tools need to be replaced in time. After all the tools that need to be replaced in the small shaft space are replaced, the personnel in the well exits to the ground, and the duty engineer rotates the cutter head with the manual control button. After the cutter head is turned to the next tool change position, the cutterhead is stopped.
④ Quality inspection. After each batch of tools is replaced, the mechanical engineer on duty checks the installation quality and checks for missing or not fixed. After the mechanical engineer confirms the correctness, the follow-up operation can be continued, and all the tools can be inspected, repaired and replaced in the same way.
⑤ Resume the excavation. After the repair of the cutterhead, the inner shaft of the small shaft in front of the tunnel excavation section (within the height of the cutterhead) is cut off from the bottom to the top, and the backfill is layered. During the cutting process, the stability of the face is observed at all times to ensure the safety of the construction personnel. Finally, the clay is backfilled into the small shaft until the ground, and the earth pressure is established in the earthen tank to resume normal tunneling.

3.2 Reinforcement Stratum Tool-Changing Technology

(8) Main points of shaft changing and overhaul control:

① Do well in monitoring the settlement of small shafts and their surrounding areas. During the small shaft change, the ground subsidence monitoring is carried out once every 4 h, and the warning line is set on the ground within 20 m around the small shaft. The special person is on duty 24 h, all vehicles and personnel are prohibited from passing through, and the inspection records are made. The maximum settlement of the surface during the tool change is controlled within 5 mm, ensuring safe and smooth operation of the opening work.

② Do well in strengthening the wall of small shaft to ensure the well wall is stable and stable. The stability of the front face of the cutter head and the stability of the small shaft wall are the prerequisites for ensuring the smooth operation of the small shaft change and the safety of the changer. When the small shaft is applied, the shaft must be grouted and reinforced outside the shaft so that the small shaft wall and the surrounding soil become a solid whole. At the same time, the whole steel casing in front of the cutter head is installed and fixed, and the steel guard is ensured. The cylinder and the cutter head are kept at a distance of not less than 200 mm to prevent the large pebbles from damaging the steel casing and the small shaft support system during the rotation of the cutter head; according to the actual situation in the field, temporary baffle can be welded in the gap between steel tube and shield above the cutter head to block the soil above the cutterhead and prevent the soil layer from falling off the side wall of small shaft when the cutter head rotates.

③ Do well in the whole process of small shaft maintenance and tool change. In the whole process of small shaft repair and tool change, an experienced engineer must be assigned to observe and monitor the stability of the face in front of the cutter head and the stability of the small shaft. Every time before entering the work, the duty engineer needs to observe the stability of the soil in front of the small shaft and its cutterhead. After the safety is determined, the operator can enter the well to carry out the inspection of the cutter head. Each inspection can only accommodate 2 people at the same time working in the well. Before the cutter head is rotated, it is necessary to ensure that there are no personnel, tools, etc. in front of the cutterhead in the small shaft.

④ Do well in ventilation and ventilation in the well. In the small shaft, the repairing cutter head has a small working space, and the electric welding operation has poor air circulation. Therefore, every time the personnel enters the shaft, the gas detection work must be done, and the well ventilation is performed 1 h in advance. And continue to ventilate throughout the overhaul.

⑤ Establish a registration system for personnel entering and leaving small shafts. During the entire inspection and replacement of the knife, the registration system for entering the wells is implemented. Only two persons can be allowed to perform the inspection of the cutterhead at the same time. Due to the small working space in the small shaft, the staff should not work more than 2 h in front of the cutterhead in the small shaft, and the stability of the face must be

monitored by the special person. Ground duty personnel must always maintain communication with the bottom hole staff to ensure that the underground personnel are in a safe state.
⑥ Do well in quality inspection of tool replacement and maintenance. The quality of the tool welding and bolting is directly related to whether the subsequent shield can be properly boring. For each repair, after completing the cutter head on a spoke, the mechanical engineer on duty checks the installation quality and checks whether there is any missing installation or the installation is not firm. All the tools, materials, etc. in the well are checked for shipment. After the mechanical engineer confirms there is no error, the subsequent cutter can be repaired.
⑦ Make a good reserve for emergency supplies. Prepare anti-virus masks, oxygen bags, stretchers, medicine cabinets and other relief materials. During the repair of small shafts, telephones, walkie-talkies, signal lights, etc. can be used to ensure effective communication in the well, in the tunnel and on the ground, so as to ensure smooth communication between inside and outside in emergency situations. And immediately start an emergency rescue program to ensure human safety.

3.2.5 Underground Continuous Wall Reinforcement Technology

The underground continuous wall is a foundation project. A trenching machine is used on the ground. Along the peripheral axis of the deep excavation project, a long and narrow deep trough is excavated under the mud retaining wall condition. After clearing the trough, the steel bar is suspended in the trough. The cage is then constructed by using pipe method to construct underwater concrete sections into a unit trough section. This is done step by step, and a continuous reinforced concrete wall is built underground to serve as water interception, seepage prevention, load bearing and water retaining structure.

According to the wall form, it can be divided into: pile type; slot type; combined type.

According to the use of the wall, it can be divided into: impervious wall; temporary retaining wall; permanent retaining soil (bearing weight); as the basis.

According to the wall material can be divided into: reinforced concrete wall; plastic concrete wall; solidified mortar wall; self-hardening mud wall; prefabricated wall; mud groove wall; post-tensioned prestressed wall; steel wall.

According to the excavation situation, it can be divided into: underground retaining wall (excavation) and underground anti-seepage wall (without excavation).

Due to the limitation of construction machinery, the thickness of the underground continuous wall has a fixed modulus and cannot be flexibly adjusted according to the pile diameter and rigidity like the pile. Therefore, underground continuous walls can only show economic and unique advantages under certain depth of foundation

3.2 Reinforcement Stratum Tool-Changing Technology

pit engineering or other special conditions. Generally applicable to the following conditions:

(1) Deep foundation pit excavation depth exceeding 10 m.
(2) The retaining structure is also part of the main structure and has strict requirements for waterproofing and impermeability.
(3) Construction by reverse method is adopted. When the ground and underground are constructed synchronously, the underground continuous wall is generally used as the retaining wall.
(4) Projects with high requirements for deformation and waterproofing of the foundation pit itself, which are adjacent to the building (structure) with high protection requirements.
(5) The space inside the foundation pit is limited, and the distance between the underground outdoor wall and the red line is very close, and other construction forms that cannot meet the construction operation requirements.
(6) In the ultra-deep foundation pit, such as the deep foundation pit of 30–50 m, when the other enclosures can not meet the requirements, the underground continuous wall is often used as the envelope structure.

1) The role of underground continuous walls

(1) Earth retaining effect. When digging the trenches of the underground continuous wall, the soil close to the surface is extremely unstable and easy to collapse, and the mud can not function as a retaining wall. Therefore, the guide wall acts as a retaining wall before the unit slot section is dug.
(2) As a benchmark for measurement. It specifies the position of the groove, indicating the division of the groove section of the unit, and also serves as a reference for measuring the elevation, verticality and accuracy of the groove.
(3) As a support for a heavy object. It is not only the support of trenching mechanical track, but also the support of steel cage, joint pipe and so on, and sometimes also bear the load of other construction equipment.
(4) Depositing mud. The guide wall can store mud and stabilize the mud level in the tank. The mud level should always be kept 20 cm below the guide wall and 1.0 m above the water table to stabilize the tank wall.
(5) Prevent mud from leaking; prevent surface water such as rainwater from flowing into the tank.

2) Characteristics of underground continuous wall

(1) The advantages are as follows:

① High efficiency, short construction period, reliable quality and high economic efficiency.
② The vibration is small during construction and the noise is low, which is very suitable for urban construction.
③ It occupies less space and can make full use of the limited ground and space within the building red line to give full play to the investment benefits.

④ The anti-seepage performance is good. Due to the improvement of the wall joint form and construction method, the underground continuous wall is almost impervious to water.
⑤ Can be used for reverse construction. The underground continuous wall has high rigidity and is easy to install embedded parts, which is very suitable for reverse construction.
⑥ Can be close to the construction. Thanks to the above several advantages, it is possible to construct an underground continuous wall close to the original building.
⑦ The use of underground continuous walls as a vertical anti-seepage structure for hydraulic structures such as earth dams, tailings dams and sluices is very safe and economical.
⑧ The wall has high rigidity. It can withstand great earth pressure when foundation pit excavation, and rarely occurs foundation settlement or landslide accidents. It has become an indispensable retaining structure in deep foundation pit support engineering.
⑨ It is suitable for a variety of foundation conditions. Underground continuous walls have a wide range of applications for foundations, from weak alluvial strata to medium-hard strata, dense gravel layers, and all foundations such as soft rock and hard rock.
⑩ Can be used as a rigid foundation. The underground continuous wall is no longer simply used as an anti-seepage waterproofing and deep foundation pit retaining wall, and more and more underground continuous walls are used instead of pile foundations, shafts or caissons to withstand larger loads.

(1) The disadvantages are as follows:

① When the city is constructed, the treatment of the waste mud is troublesome.
② If the underground continuous wall is used as a temporary retaining structure, it will cost more than other methods.
③ If the construction method is improper or the construction geological conditions are special, the problem that the adjacent wall segments cannot be aligned and leaks may occur.
④ Under some special geological conditions (such as very soft silty soil, alluvial layers containing boulders and super-hard rocks, etc.), construction is very difficult.

3) Construction steps of underground continuous wall

(1) Guide wall construction (Fig. 3.23). Before the trench section is excavated, a guide wall is constructed along the longitudinal axis of the continuous wall, and cast-in-place concrete or reinforced concrete is used. The depth of the guide wall is generally 1.2–1.5 m, and the top surface is slightly 10–15 cm above the ground to prevent surface water from flowing into the guide groove. The thickness of the guide wall is generally 100–200 mm, the inner wall surface should be vertical, and the inner wall clearance should be the continuous wall design

Fig. 3.23 Construction of underground diaphragm

thickness plus construction allowance (generally 40–60 mm). The allowable deviation of the distance between the wall and the longitudinal axis is ±10 mm, the tolerance of the inner and outer guide walls is ±5 mm, and the top surface of the guide wall should be kept horizontal. The guide wall should be built on a dense sticky ground. The wall back should be replaced by a soil wall to prevent the surface water outside the tank from seeping into the tank. If the back side of the wall needs to be backfilled, apply viscous soil layering to avoid leakage. The guide wall in each slot section shall be provided with an overflow hole. Before the foundation groove is cut, the guide wall for protecting the upper part of the base groove should be built, and the wall is protected by mud.

According to the designed wall width and deep section, the steel frame is placed, and continuously to form a continuous wall. The main construction techniques are guide wall, mud retaining wall, slotting construction, underwater concrete pouring, wall joint processing and so on. The guide wall is usually a reinforced concrete structure that is infused in place. The main functions are: to ensure the geometric size and shape of the underground continuous wall design; to store part of the mud, to ensure the stability of the liquid level during the construction of the groove; to bear the load of the grooved machine, to protect the notch from being damaged, and to serve as a steel frame for installation. The depth of the guide wall is generally 1.2–1.5 m. The top of the wall is 10–15 cm above the ground to prevent the inflow of surface water and affect the quality of the mud. The bottom of the guide wall cannot be located in a loose soil layer or a place where the groundwater level fluctuates.

(2) Mud protection wall. The mud wall is pressed against the wall to protect the shape of the deep groove, and the concrete is replaced by the slurry. Mud materials usually consist of bentonite, water, chemical treatments and some inert materials. The role of the mud is to form a water-tight mud on the wall of the tank, so that the hydrostatic pressure of the mud effectively acts on the tank wall to prevent water seepage and spalling of the tank wall, and to

maintain the stability of the wall surface. At the same time is can carry the mud and suspended soil out of the ground. Slotting in the gravel layer may be carried out by using an anti-blocking agent such as wood chips or vermiculite to prevent slurry leakage. The mud is divided into two types: static type and cyclic type. When the slurry is used in a circulating way, a purifying device such as a vibrating screen or a cyclone is applied. After the indicator has deteriorated, chemical treatment should be considered, or the old slurry should be discarded and replaced with a new slurry.

(3) Groove construction. The use of slot into the special machinery: rotary cutting multi-head drill, guide plate grab, impact drill, and so on. Construction should be selected according to geological conditions and wall depth. Generally, when the soil is soft and the depth is about 15 m, the common guide grab can be used; for the dense sand layer or gravel layer, the multi-head drill or the heavy-duty hydraulic guide grab can be used; in the case of large gravel or the groove in the rock foundation, it is better to use impact drill. The unit length of the trough section is generally 6–8 m, which is usually determined by combining the soil condition, the weight of the steel skeleton, the structural size, and the division paragraph. After standing in the tank, it should be allowed to stand for 4 h, and the relative density of the mud in the tank should be less than 1.3.

(4) Underwater concrete. The pipe method is used according to the underwater concrete filling method. However, before the concrete is started to be poured into the concrete, a pipe plug can be suspended in the pipe and the mud in the pipe can be extruded by the pressure of the poured concrete. The concrete should be continuously poured and the concrete filling amount and rising height should be measured. The spilled mud is sent back to the mud sedimentation tank.

(5) Wall joint processing. The underground continuous wall is composed of a plurality of wall sections. In order to maintain continuous construction between the wall sections, the joint adopts a lock pipe process. That is, steel pips of equal diameter to the groove, that is, the lock pipes are pre-inserted at the end of the groove section before the concrete of the groove section is poured. After the initial setting of concrete, they should be slowly pulled out, so that the end portion is formed into a semi-concave shape. There are also rigid joints according to the force requirements of the wall structure, so that the two wall sections are joined together.

The construction process of the underground continuous wall is shown in Fig. 3.24.

4) Underground continuous wall construction case

The first phase of Lanzhou Metro Line 1 Yingmentan-Matan section is the first underwater tunnel project in China to penetrate the complex of the Yellow River. The section passes through the bottom of the Yellow River at about 45 m upstream of the Yintan Bridge. The length of the section is 1907.3 m, and the length of the river channel below the Yellow River is 404.5 m. The water surface of the Yellow River is about 200.5 m wide, the water depth is generally 3–7 m, and the deepest place is

3.2 Reinforcement Stratum Tool-Changing Technology

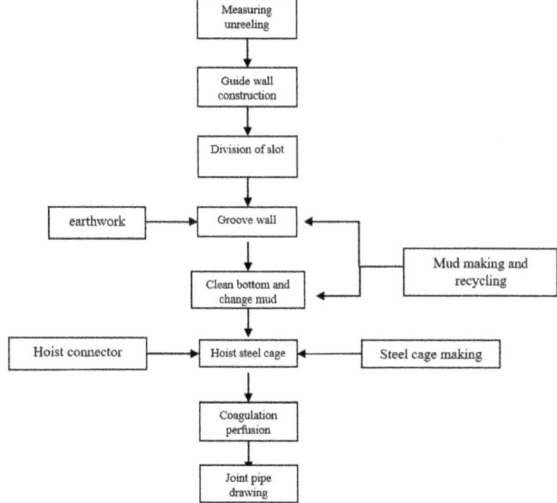

Fig. 3.24 Construction flow chart of underground diaphragm wall

9.0 m. The main body of the interval tunnel is located in the 311 strong permeable sand and gravel stratum with a permeability coefficient of about 55 m/d. The layer of boulders and pebbles account for 55–70%, the general particle size is 20–50 mm, the boulders are less, the maximum particle size can reach 450 mm, the gravel content is 10–25%, and the medium coarse sand is filled; it has poor grading, netter rounded, poor sorting, compact structure; poor tunnel surrounding rock self-stable ability, easy collapsing. The comprehensive classification is VI. The distribution law of each layer of soil is shown in Fig. 3.25.

The choice of atmospheric pressure opening is carried out in the air well (mileage YDK13 + 755.7) at normal pressure into the cabin for the repair and replacement of tools and equipment, as well as the modification of the cutterhead, so that the

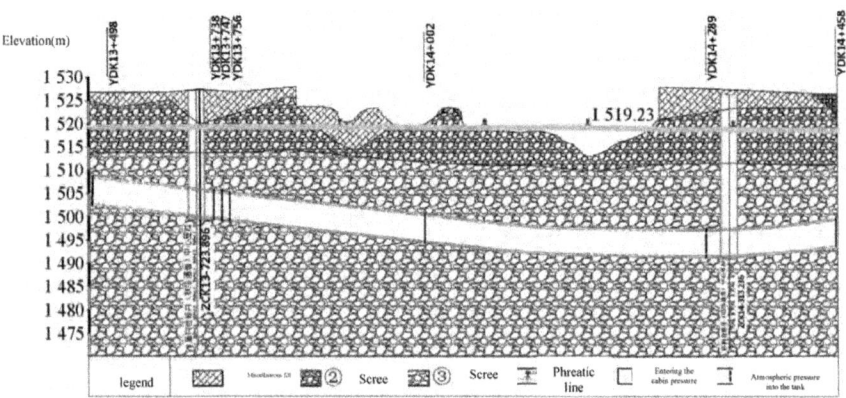

Fig. 3.25 Geological profile between yingmen beach and yima

shield can be safely crossed at one time and the maintenance is not stopped as much as possible. The wind well retaining structure adopts the reinforced concrete underground continuous wall, and the shield tunnel area adopts the transition of glass fiber ribs. The main body of the wind shaft in the pit is a cast-in-place reinforced concrete frame structure, which is constructed in reverse. The sleeve is used in the well, and the single-layer slurry is used for foundation reinforcement from bottom to top. The reinforcement depth is 8 m below the basement of the second floor to the bottom of the floor to form an effective shaft reinforcement area and a bottom reinforcement area. The vertical section of the wind well grouting reinforcement is shown in Fig. 3.26.

According to the actual situation, the aquifer is considered to be a homogeneous aquifer. Since the aquifer is thick and the foundation pit is close to the river bank, the construction manual is used to calculate the water inflow. A total of 36 precipitation wells are designed in the wind well area, including 30 wells outside the well whose depth is 56 m. There are 6 dry wells in the well, whose depth is 42 m. The layout of the precipitation well is shown in Fig. 3.27.

When the shield tunneling reaches the 15th ring in front of the underground continuous wall, increase the frequency of contact measurement, ensuring that the shield enters the wind well area with the correct posture and elevation position. When the right-line shield tunnels into the mileage YDK13 + 738.8 (about 486 ring), the cutterhead starts to cut the underground continuous wall. At this time, in order to

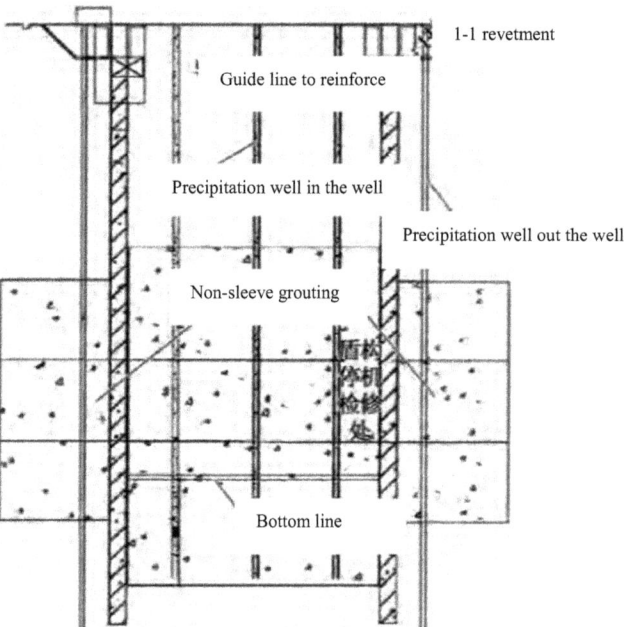

Fig. 3.26 Longitudinal section of reinforcement of air wall

3.2 Reinforcement Stratum Tool-Changing Technology

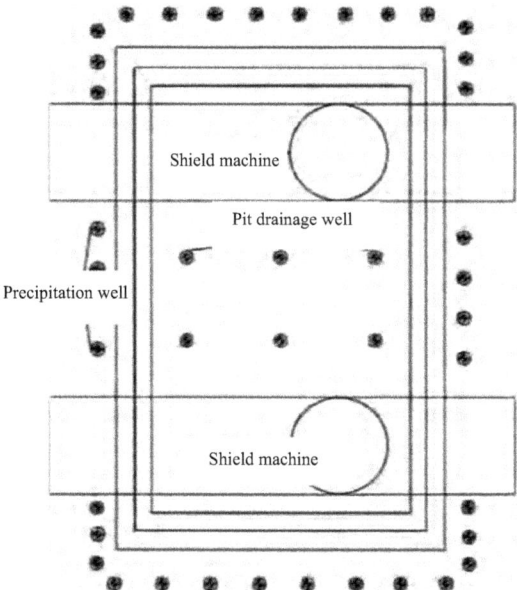

Fig. 3.27 Precipitation well arrangement

ensure the safety of the tunneling construction, the excavation parameters need to be adjusted during construction: reduce the tunneling speed to 10 mm/min; Increase the cutter speed to 1.2 r/min; increase the synchronous grouting volume to 8 m^3/ring. At the same time, you should still pay attention to the following issues:

(1) Since the underground continuous wall is 1.2 m thick, the glass door rib + C30 concrete is used in the hole door area. During the excavation process, the glass fiber reinforced slag will block the mud water treatment equipment, and the mud will also be doped with concrete. As a result, the quality has dropped sharply, so it is necessary to take emergency measures in advance.

(2) After the shield enters the wind well area through the small continuous underground wall, the precipitation well in the wind well stops the precipitation until the shield reaches the stop position.

(3) After the shield passes through the small mile underground underground wall, it resumes normal tunneling until it reaches the stop position.

(4) When the shield begins to cut the underground continuous wall, the precipitation in the wind well stops, until the shield reaches the stop position, and then the precipitation is restored. The shield tunneling stops when the mileage position is YDK13 + 756.1 (the 499 ring is pushed forward). Subsequent advancement measures the distance of the jack cylinder by measuring the stroke of the jack cylinder one by one, and slowly approaches the final position step by step. The last loop piece after reaching the stop position is not assembled, so as to follow the replacement of the shield tail brush and the replacement of the jack cylinder. Due to the diameter of the shield cutterhead excavation is 6.48 m, the outer diameter of the lining segment is 6.2 m, and the construction gap is 14

cm, this construction requires two additional grouting, namely the underground continuous wall and the lining pipe ring at the end of the shield.

Since the shield passes through the underground continuous wall, the groundwater can enter the wind well through the gap between the underground continuous wall and the lining pipe ring. Therefore, it is intended to after passing through the small mile end underground continuous wall (486 ring) at the end of the shield, that is the range of 6 m (5 rings) on the north side of the wall, the gap between the underground continuous wall and the segment lining ring is filled and sealed by reinforcing grouting.

According to the construction plan, it is planned to check and replace the first two sealing brushes at the tail of the shield at the wind well. In order to ensure the safety during replacement, it is necessary to reinforce the grouting at the tail of the shield under the premise of precipitation. Reinforced grouting uses double slurry and polyurethane. The refill position is set to the second loop piece after the shield is stopped.

At the time of grouting, it should be noted that since the tunnel in this section is 2.8% downhill, the slurry will spread to the excavation surface under the action of gravity along the water flow. Therefore, the cutter head should be rotated once every 2 h, and the mud water circulation is started to replace the slurry of the excavation chamber to prevent the cutterhead from being consolidated due to the slurry being smashed to the excavation surface.

According to the plan, the cutterhead should be partially strengthened and rebuilt in the Yingma section of the wind well to ensure that the shield can cross the Yellow River at one time. It is planned to excavate a working space from the front of the cutter head through the manual digging piles to replace the cutter and partially modify the cutterhead. The manual digging pile is shown in Fig. 3.28. The pile diameter is 1200 mm, the pile depth is 27.5 m, and the center distance between the piles is 2.7 m. The center position of the manual digging pile is located at the front left and front right sides in the center of the shield cutter, which is basically at the maximum excavation diameter of the shield cutterhead and convenient for the construction work. The artificial digging pile is made of concrete support with a thickness of 10

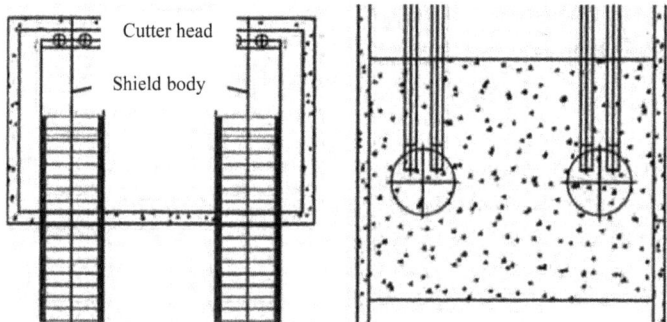

Fig. 3.28 Schematic diagram of manual digging pile

cm. According to the experience of the previous shield construction, it is planned to replace all the tools in the normal pressure condition. The type and quantity of the replacement tools are determined according to the actual situation.

3.2.6 Comparison of Several Techniques

For a list comparison of each reinforced formation technology, which in Table 3.3.

Table 3.3 Comparison of reinforcement schemes

Scheme	Advantages	Disadvantages	Remark
Spinner reinforcement	Suitable for artificial fill, sand, clay, loess, silt and other soft soil Environmental and groundwater pollution is small Better durability Low construction noise	For areas with large diameter of one gravel or large groundwater flow rate, the injection slurry cannot condense around the filling pipe, so it is not suitable to use the jet grouting method	
Deep mixed pile	Three piles in one construction, the speed of pile formation is fast With an automatic slant measuring system, its verticality can be adjusted constantly High pile strength	For common triaxial mixing pile, reinforcement depth is within 30 m After more than 30 m, the stability of pile frame is weakened, and the connection of drill pipe is more difficult. Moreover, stirring power and other equipment also limit the application of this technology in the field of deep foundation reinforcement	
Freezing method	The freezing effect of freezing body can be guaranteed, There is no pollution to foundation and ground water	Due to the strange shape of the reinforcement, some areas of frozen pipe can not reach, and are easy to form the phenomenon of incomplete cementation The construction period is long, and it needs to wait until it is completely frozen and reaches the strength, then tools can be changed Through long-term freeze, the cost is considerable	

(continued)

Table 3.3 (continued)

Scheme	Advantages	Disadvantages	Remark
Shaft	The embedding depth can be large, the integrity is strong, the stability is good When shaft is constructed, the impact on the adjacent buildings is small	Construction time is long; quicksand is easy to occur during construction, which makes it difficult for the shaft to tilt or sink	
Underground diaphragm wall method	Small vibration and low noise during construction Being practical for all kinds of geological conditions Good anti-seepage effect	The underground diaphragm wall is used as the stop water curtain, because it is too close to the shield machine and the forming tunnel ring, the construction risk is high If the insertion depth is not enough, it is easy to form piping at the bottom of diaphragm wall, causing engineering accidents The cost is high, and it needs many supporting equipment	

3.3 Normal Pressure Tool Change Technology

3.3.1 Tool Change Process

(1) Determine the location of the reinforcement. Carry out a detailed survey of the stratum where the shield is stopped, and select a stratum with high soil strength and good self-stability.
(2) Determine the formation reinforcement technology. Carefully study the properties of the stratum rock and soil, compare and select various reinforcement methods, and choose the optimal solution.
(3) Precipitation in the well. Drain the reinforced formation until the construction personnel can go down the well.
(4) Check the cutter head and the cutter. After the construction personnel go down the well, carry out a comprehensive inspection of the shield cutterhead and the tools. Mark and record the position of the tools and the cutterhead with excessive wear value for easy replacement.
(5) Pull out the tool. Pull out the tools with too much wear value.
(6) Welding cutter head. Repair the excessively worn parts of the cutter head.
(7) Install the tool, for the tools that has been pulled out.
(8) Overall inspection. Perform a comprehensive inspection of the shield cutter and tools to check the integrity of the cutter repair, the tightness of the tool bolts, and the quality of the tool weld.

(9) Resume the excavation. The shield first conducts trial operation, checks the performance of the shield equipment, and then restores the shield. Among them, stratum reinforcement is the most critical part of the atmospheric pressure changing process. Determining the appropriate reinforcement location and stratum reinforcement technology is the basic guarantee for the subsequent tool-changing process, which plays a vital role in the tool change process.

3.3.2 Shield Cutter Tool Welding Process

The damaged part of the cutter head is mainly concentrated on the outer edge of each arm and the outer circular arc plate of the cutter head. The repair process is mainly carried out by planning, grinding and repair welding. In particular, the circular arc plate on the outer edge of the cutter head cannot be replaced due to the integrity of its own structure. Therefore, in the repair, only the welding repair method of the steel plate can be used, that is, the steel plate of the appropriate form is used to fill the damaged portion. According to the actual situation, select the appropriate break form for welding.

3.3.2.1 Types of Welding Processes

The commonly used welding methods can be broadly classified into three categories: wet underwater welding, dry underwater welding, and partial dry underwater welding.

1) Wet underwater welding

Wet underwater welding is a method in which welds are welded in water without special drainage measures. The problem of wet underwater welding is the most prominent. It is difficult to obtain welded joints with good quality by this method. Especially in the case of poor construction conditions, the quality of wet underwater welding is more difficult. However, wet underwater welding has the advantages of simple equipment, low cost, flexible operation and strong adaptability, so it will be widely used in the future.

The most common methods used in wet underwater welding are electrode arc welding and flux cored arc welding. When welding, the submersible welder should use an electrode with waterproof coating and a welding tong specially designed or modified for underwater welding. In the case of high quality requirements, the electrode can be placed in the inflatable container before use to prevent the electrode from absorbing moisture.

2) Dry underwater welding

Dry underwater welding refers to the method in which a submersible welder and a workpiece are welded under completely dry or semi-dry conditions. When performing dry underwater welding, it is necessary to design and manufacture complex pressure cabins or working cabins. According to the pressure in the pressure cabin or the working cabin, the dry underwater welding is divided into high pressure dry underwater welding and normal pressure dry underwater welding.

(1) High pressure dry underwater welding. The choice of high pressure gas is very important in the implementation of high pressure dry underwater welding. Air has many advantages as an ambient gas, low cost and available anywhere, anytime, and only needs to be compressed to the proper pressure. However, the oxygen content of the air is high (about 30%), which means that when the pressure is under a few atmospheres, the flammability of the objects in the cabin will increase significantly, and there is a big safety hazard. In addition, in order to ensure the quality of the weld, it is necessary to protect the molten pool from contact with nitrogen and oxygen in the air, which is difficult to ensure in a high pressure environment of compressed air. Therefore, air can be used as an ambient gas when the pressure is low, and it should not be used when the pressure is high.

The argon supply is convenient, the density is close to that of air, and the thermal conductivity is lower than that of helium, which can reduce the cooling rate of the weld metal. However, in actual underwater welding, argon gas has an anesthetic effect on the diver under high pressure environment, so the use of argon as the ambient gas is more dangerous to the construction workers. Because the anesthetic effect of argon is only a physiological reaction to human, it is a good choice to use argon as the cabin gas for underwater welding without divers. Helium is much more expensive than argon and has a density of one tenth of argon. A typical ambient gas consists of helium and oxygen, independent of pressure. Such a gas has no effect on flammability due to low oxygen content, and does not cause damage to the diver during actual underwater welding. However, the high thermal conductivity of niobium increases the cooling rate of the weld metal.

(2) Dry normal pressure underwater welding. The welding takes place in a sealed pressure cabin, and its pressure in the pressure cabin is equal to the atmospheric pressure on the ground, independent of the ambient water pressure outside the pressure cabin. This type of welding is neither affected by water depth nor by water, and the welding process and welding quality are the same as when welding on the ground. The biggest advantage of this method is that it can effectively eliminate the influence of water on the welding process. The welding conditions are exactly the same as those of the ground welding, so the welding quality is also the most secure. However, the cost of dry pressure underwater welding equipment is more expensive than that of high-pressure dry-process underwater welding, and there are more welding assistants, so it is generally only used for important structures of deep-water welding. At present, atmospheric pressure dry underwater welding technology is rarely used.

3) Partial dry underwater welding

Partial dry underwater welding means that the submersible welder and the workpiece are directly in the water. A specially constructed drainage cover is placed on the part to be welded, and the water in the cover is drained by air or shielding gas to form a partial gas phase space for welding. Since local dry underwater welding reduces the harmful effects of water, the quality of welded joints is significantly improved over wet welding. There are many types of partial dry underwater welding, including dry box welding, dry point welding, water curtain dry welding, and steel brush underwater welding. Compared with dry welding, the partial dry method does not require a large and expensive drainage cabin, and the adaptability is significantly increased. It combines the advantages of both wet and dry methods. It is a more advanced underwater welding method. Partial dry underwater welding can directly obtain the joint quality close to dry welding. At the same time, due to the simple equipment and low cost, it has the flexibility of wet welding and is therefore a promising underwater.

3.3.2.2 Welding Technology and Process

1) Process

Job preparation → cutting groove → positioning welding → bottoming welding → CO_2 → protective welding → low temperature post heat treatment (when required) → weld inspection.

2) Welding control

(1) The welding procedure specification shall be prepared in accordance with the qualified welding procedure qualification report.
(2) The welding rod or wire diameter, welding current, welding speed, welding arc length, etc. shall be selected in accordance with the welding procedure specification.
(3) The baking electrode should meet the specified temperature and time. The electrode removed from the oven is placed in the electrode holding barrel and used as needed. The insulated barrel shall be insulated according to the required load. The electrode after 10 min from the holding barrel must be re-baked, and the number of continuous baking shall not exceed two.
(4) Leading arc. The arcing point of the fillet weld should be at the end of the weld, preferably greater than 10 mm. It should not be arced casually. After the fire is triggered, the electrode should be pulled away from the weld zone immediately, so that the gap between the electrode and the component is maintained at 2–4 mm to produce arc. Butt welds and butt joints and angle joint welds shall be provided with arc runner plates and lead plates at both ends of the weld. They shall be welded to the weld zone after arcing on the arc

runner plate, and the joints in the middle shall be in front of the weld joints. Fire at 15–20 mm, and preheat the weldment before returning the electrode to the beginning of the weld. Fill the weld pool to the required thickness before welding.

(5) Close the arc. After each weld is welded to the end, the arc should be filled and arced in the opposite direction of the welding direction so that the arc pit is inside the weld bead to prevent the arc pit from biting. After the welding is completed, the arc cutting plate should be cut by gas cutting, and the grinding should be smooth, and it is not allowed to hit with a hammer.

(6) Clearing slag. After the entire weld is completed, the slag is removed. After the welder self-test (including appearance and weld size) confirms there is no problem, the weld can be transferred.

(7) The welded steel structure of 16Mn/16Mn and 16Mn/Hardox400 materials is used to prevent welding cracks. Local preheating is carried out by oxygen and acetylene before welding. The local preheating heating width is not less than 100 mm on each side, and must be kept warm after welding to slow down the cooling. Speed (16Mn/15CrWMo <Abrasion-resistant block> material welds should be pre-weld preheating, post-weld insulation measures, and if necessary, eliminate welding stress by equipment that eliminates welding stress).

(8) When welding multi-layer welding, the surface of the previous layer of weld must be cleaned, the oxide layer should be polished away with a grinding wheel, and the defect should be processed before filling the next layer of weld.

(9) When defects are found by non-destructive testing, the position of the defects must be accurately identified and repaired after cleaning.

(10) When the wind speed exceeds 2 m/s, the welding should be stopped or windproof measures should be taken.

(11) The relative humidity of the work area should be less than 80%.

(12) The welding material used for tack welding should be the same as the welding material for formal welding. The tack weld should have the same quality requirements as the final weld. If the butt weld is completed by steel liner, the tack welding should be welded in the joint groove. The thickness of the tack weld should not exceed 6 mm of the design weld thickness, the length of the tack weld should not be greater than 40 mm, and the spacing of the location weld should be 200–500 mm. The heat is higher than the formal preheating temperature. When there are pores and cracks in the positioning weld, rewelding must be removed.

(13) In addition to the process or inspection requirements, it is necessary to weld in series, and each weld should be continuously welded once. When the welding is interrupted for any reason, measures to prevent cracking should be adopted according to the process requirements. The surface of the weld layer should be inspected before re-welding. After confirming that there is no crack, the welding can be continued according to the original process requirements.

3.3 Normal Pressure Tool Change Technology

When welding is applied, the pre-heat treatment of the welded seam should be partially performed.

(14) The top layers must be welded from the edge of the underlying material so that the topmost layer of weld can cover the entire area of the weld. The shielding gas should have sufficient flow and maintain laminar flow. The spatter attached to the contact tip and nozzle should be removed in time to ensure good protection.

(15) The dirt in the wire feeding hose should be cleaned frequently. For semi-automatic welding, the wire feeding hose shall have a radius of curvature of not less than 150 mm.

(16) The length of the wire extending from the tip of the wire should be 10–12 times the diameter of the wire (generally in the range of 10–20 mm). If the dry elongation of the wire is too short, it will increase the splash metal blockage due to the proximity of the nozzle of the torch to the workpiece. If the nozzle has long the dry elongation of the wire, it will increase the splash, resulting in unstable welding, poor gas protection effect. In actual work, generally select the wire diameter according to the thickness of the workpiece, the shape of the groove, the welding position, etc., then determine the welding current, adjust the loop inductance, and minimize the splash.

(17) Welding procedure specification. When welding, the parameters of welding should be controlled strictly according to the welding procedure, including weld size, weld penetration, form of welding, current and voltage. The reference standards are as follows:

① "Code for acceptance of construction quality of steel structure engineering" (GB50205-2001).
② "Code for Welding of Steel Structures" (GB50661-2011).
③ "Carbon steel for low-alloy steel for gas-shielded arc welding" (GB/T8110-2008).
④ "heat-strength steel electrode" (GB/T5118-2012).
⑤ "surfacing welding rod" (GB/T984-2001).
⑥ "stainless steel electrode" (GB/T983-2012).
⑦ "Non-destructive testing of welds, ultrasonic testing technology, inspection grade and assessment" (GB/T11345-2013).
⑧ "Nondestructive Testing of Weld Penetration Testing" (JB/T6062-2007).

3.3.2.3 Quality Assurance Measures for Welding

(1) Welding materials should meet the design requirements and relevant standards, and the quality certificate and baking records should be checked.

(2) The welder must pass the examination and have the certificate of conformity and the date of assessment.

(3) Class I and Class II welds must be inspected by flaw detection and shall comply with the design requirements and construction and acceptance specifications.

The intrinsic quality is detected by ultrasonic flaw detection, and the external quality is added by penetrant inspection.

(4) There must be no defects such as cracks and welds on the surface of the weld. Class I and Class II welds shall not have defects such as surface pores, slag inclusions, crater cracks, arc scratches, etc., and Class I welds shall not have defects such as undercut, insufficient weld, and root shrinkage.

(5) When it is required to weld through the weld, if the back cannot be cleaned and other measures cannot be taken, the bottom layer must be completed by single-sided welding double-sided forming welding technology.

(6) The width of a single weld is not more than 15 mm and the thickness is not more than 5 mm, and it is welded with a small current.

(7) The detection method of weld quality is generally divided into two types: non-destructive testing and lossy testing. In order to ensure that all hoop weld defects and longitudinal weld defects are detected during the test, all welds should be probed if the test conditions permit. After the welding is completed, the weld is subjected to ultrasonic flaw detection.

(8) Temperature and humidity control measures in the welding area and wind speed control measures.

① According to the welding procedure, each weld is heated by a torch heater before welding, and the temperature within 100 mm around the weld should be controlled at 50–80 °C; the weld temperature should be checked at any time if the temperature. If the requirements are not met, the welding shall be stopped and the welding shall be continued after heating; if the welding is stopped, the welding must be started before the next welding. Each weld has a welding temperature control record sheet.

② Because the front excavation space is underground and the environment is humid, the weld area should be heated frequently to ensure that the humidity in the weld area is not more than 80%. At the same time, waterproof measures should be set around the weld to prevent water from splashing onto the weld causing cracks.

③ The wind speed around the weld should be controlled below 2 m/s. Consider the need of front ventilation, the air outlet cannot be facing the weld working area during ventilation design, and windproof enclosure is provided in the weld area to ensure in the welding area the wind speed meets the requirements.

The allowable deviation of the weld dimensions is shown in Table 3.4. Appearance quality standards are shown in Table 3.5.

3.3.2.4 Welding Safety Measures Guarantee

(1) Construction electricity meets the "Safety Technical Specifications for Temporary Electricity Use at Construction Sites" (JGJ46-2012).

3.3 Normal Pressure Tool Change Technology

Table 3.4 Allowable deviation of weld dimension

Number	Project		Allowable deviation (mm)			Test method
			Class I	Class II	Class III	
1	Butt weld residual height (mm)	$b < 20$	0–3	0–3	0–4	Check with weld gauge
		$b \geq 20$	0–4	0–4	0–5	
	Butt weld wrong side (mm)		$<0.15t$ and ≤ 2.0	$<0.15t$ and ≤ 2.0	$<0.15t$ and ≤ 3.0	
2	Fillet weld size (mm)	$h_f \leq 6$	0–1.5			
		$h_f > 6$	0–3			
	Fillet weld residual height (mm)	$h_f \leq 6$	0–1.5			
		$h_f > 6$	0–3			
3	Generally, the welding Angle size of full penetration butt and corner weld required	T joint, cross joint, corner joint	$h_f \geq (t/4 + 0–4)$ and ≤ 10			
	Fillet dimensions for full penetration butt and corner welds to be fatigue checked		$h_f \geq (t/2 + 0–4)$ and ≤ 10			

Note b is weld width, t is a thinnest thickness at the joint and h_f is size of a fillet weld

(2) The welding machine should be placed in a place that is waterproof, dry and well ventilated, and should be reliably grounded. There shall be no flammable or explosive materials at the welding site.

(3) The leakage protector in the electric welding mechanical switch box must meet the requirements of relevant regulations.

(4) The secondary wire of the electric welding machine shall be made of waterproof rubber sheathed copper core flexible cable. The length of the cable shall not exceed 30 m. Metal parts or structural steel bars shall not be used instead of the ground wire of the secondary line.

(5) Protective equipment must be worn when welding with electric welding machinery to prevent electric sparks and arcs from injuring people.

Table 3.5 Allowable deviation of appearance of grade ii and iii welding seams

Defect type	Type two	Type three
Not fully welded (not up to design requirements)	$\leq 0.2 + 0.02t$ and ≤ 1.0 mm	$\leq 0.2 + 0.04t$ and ≤ 2.0 mm
	Total unwelded length per 100.0 mm weld ≤ 25.0 mm	
The roots of contraction	$\leq 0.2 + 0.02t$ and ≤ 1.0 mm	$\leq 0.2 + 0.04t$ and ≤ 2.0 mm
Undercut	$\leq 0.05t$ and ≤ 0.5 mm Continuous length ≤ 100.0 mm and total length of edge bite on both sides of weld $\leq 10\%$ * weld length	$\leq 0.1t$ and ≤ 1.0 mm
Crater crack	Not allowed	Few arc pit cracks with length ≤ 5.0 mm are allowed
Arc scratch	Not allowed	Few arc scratches are allowed
Bad joint	The depth of the hole 0.05t and ≤ 0.5 mm	The depth of the hole 0.1t and ≤ 1.0 mm
	No more than one per 1000 m	
Slag on the surface	Not allowed	Width $\leq 0.2t$, length $\leq 0.5t$ and ≤ 20 mm
Surface porosity	Not allowed	Allowable diameter per 50 mm weld length $\leq 0.4t$ Two holes, hole spacing ≤ 6 * hole diameter

Note t is a thinnest thickness at the joint

(6) The ventilation in the construction site should be absolutely good, and the welder should have the protection person in the welding process.
(7) When welding, the connecting line of the welding machine should be connected as close as possible to the welding point. The welder cable must never be connected to the shield because this type of connection allows the welding current to pass directly through the main drive of the cutterhead, which will result in damage to the main drive of the cutterhead.
(8) The ultimate goal of the shield machine to open the cabin is to repair the cutterhead in time, and to modify the excavation performance of the cutter to adapt it to the current excavation stratum and ensure the smooth progress of the next stage of excavation.

3.3.3 The Contents of the Cutter Repair

For the repair of the cutterhead, in principle, the cutterhead is repaired to the original appearance, but due to the limitations of underground space, working time, and stability of the front soil, it is very difficult to restore the cutterhead to its original appearance. The principle of the cutterhead repair is to repair its strength, rigidity and function. The key point is to repair the cutter holder and the cutter box of the cutterhead so that it can meet the strength requirements required for the cutter to cut the soil. For the damaged steel structure, mainly reinforce and increase wear protection.

3.3.3.1 Repair of Cutter Steel Structure

In order to ensure the repair of each tool holder and knife box of the cutter head, it is necessary to repair the damaged steel structures. The repair of the cutter steel structure is aimed at ensuring the strength and rigidity of the cutter head and the wear resistance of the edge of the cutter head, and is mainly repaired by means of repair welding steel plates and wear plates.

The damaged part of the cutter box type main structure is generally concentrated on the outer edge of each arm and the outer ring of the cutter head. Figure 3.29 shows the wear and repair. In the repair, the main work is planing, grinding and repair welding. In particular, the circular arc plate on the outer edge of the cutter head has its own structure and cannot be replaced. Therefore, the repair method is to use a steel plate of proper form to repair. The damaged part is filled with a suitable form of steel plate, and the appropriate breaking form is selected according to the actual situation for welding, and then the knife seat, the outer edge wear strip, the wear plate and the like are welded thereon.

3.3.3.2 Repair of Edge Blade Holders

The blade seat of the blade is divided into a main tool holder and a support block, and is mainly welded to the front panel and the side panel of the cutter arm frame. When doing the repair work, fix it in the following order:

(1) Cutter positioning. The repair of the blade holder will be carried out step by step, and only one blade in one arm will be repaired in one operation. Therefore, before entering the excavation cabin, the cutter head must be fixed to the arm to be repaired, located directly above.
(2) Dig out the damaged seat. The old blade must be removed and transported out of the excavation chamber before repair. The removed part should be marked according to the installation location.
(3) Repair the steel structure of the knife seat and polish the groove that meets the welding requirements. When positioning a new blade holder, secure the template to the remaining screw holes in the original blade holder and secure the holder to the template. At this point, the tool holder should be fine-tuned in place until the seat is fully aligned with the worn steel structure. After the adjustment is completed, the tool holder is spot welded to the cutter head. Then remove the template to make up the welding position.
(4) Weld the main tool holder.

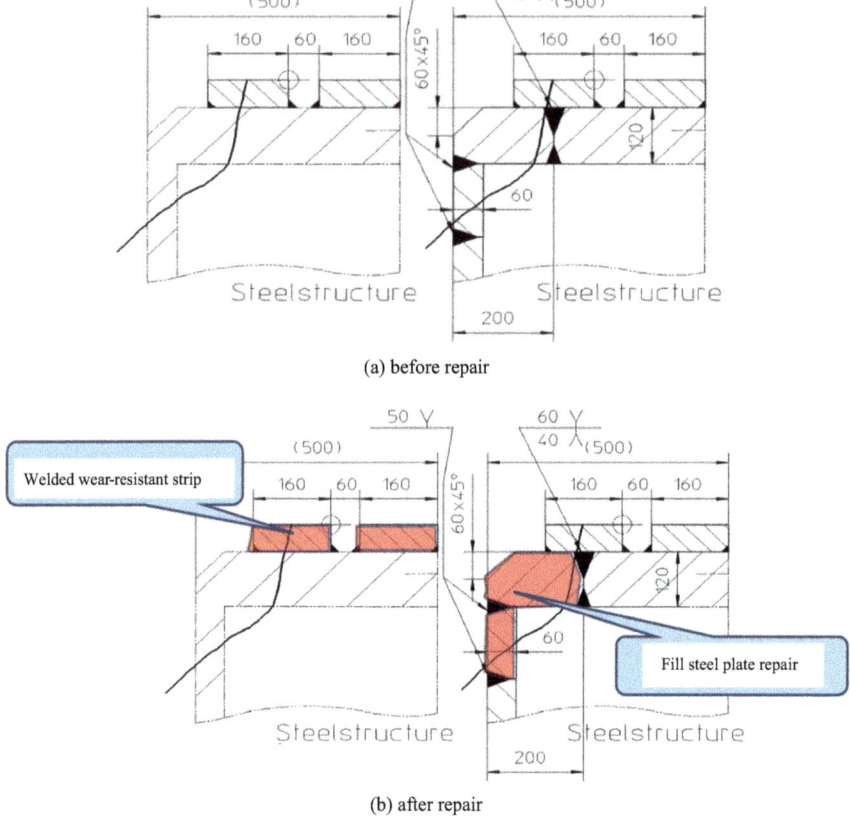

Fig. 3.29 Schematic diagram of repairing the steel of the outer edge of the blade width arm (unit mm)

(5) Weld support block.
(6) Install a new blade.

3.3.3.3 Repair of the Hob

The hob cutter repair process is as follows: (1) Plan the damaged hob and the steel structure of the cutter that needs to be repaired. (2) After grinding the groove form that meets the requirements, the steel structure of the knife box is repaired. (3) Welding a new hob box. The specific repairing measures and weld structure of the three different hobs are shown in Figs 3.30 and 3.31.

3.3.3.4 Repair of the Wear-Resistant Strips and Wear Blocks of the Cutterhead

3.3 Normal Pressure Tool Change Technology

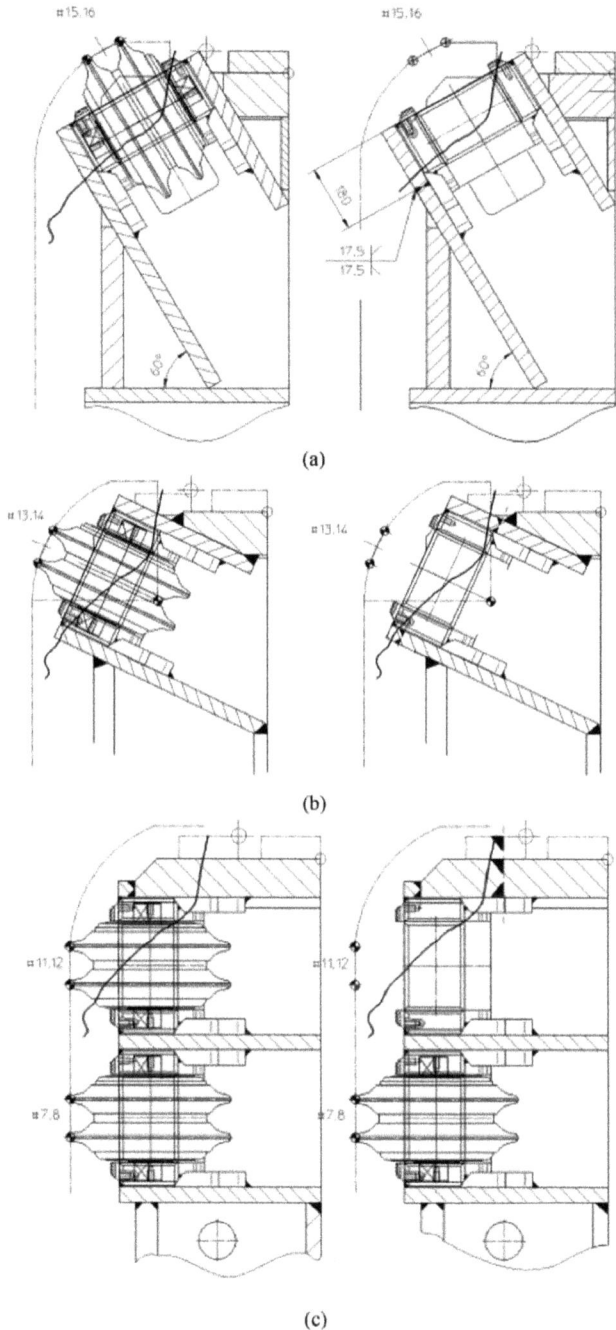

Fig. 3.30 Steel structure diagram of three hob boxes

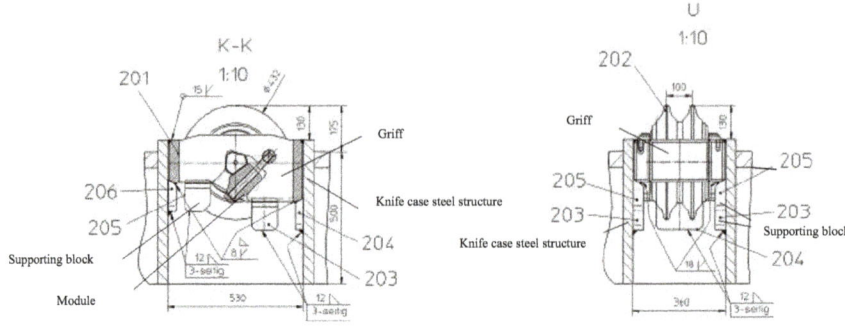

Fig. 3.31 Weld new hob box

There is no tool arrangement on the outer edge of the auxiliary arm of the cutter head. In order to protect the outer edge of each auxiliary arm from damage, the original cutter disc is designed with a wear-resistant plate at the corners of the outer edge to prevent damage, and the thickness is 12 mm or 20 mm. The size and distribution are shown in Fig. 3.32.

Repair the wear strip on the curved steel structure on the outer edge of the cutter head that has been repaired and polished. However, in order to better protect the cutter

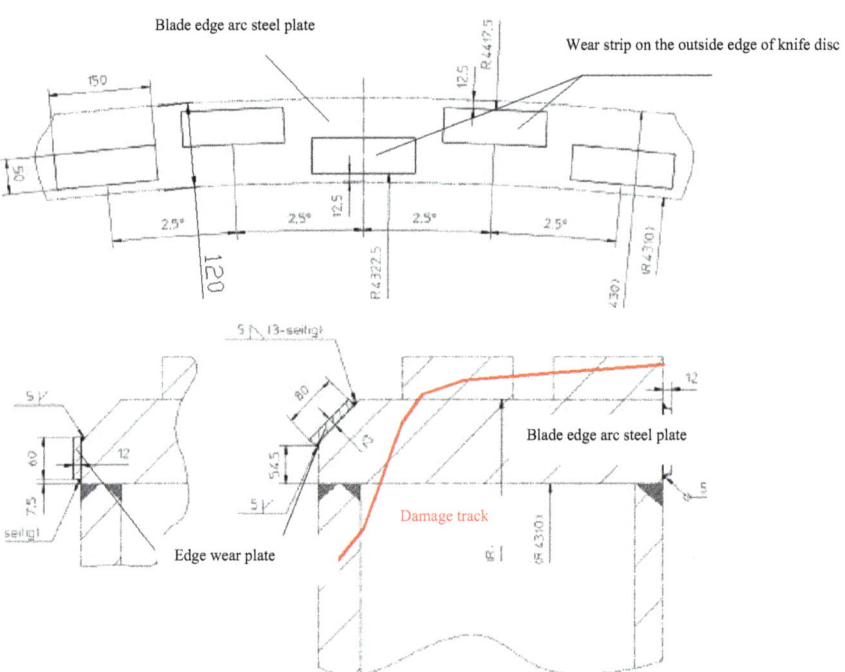

Fig. 3.32 Schematic diagram of wear-resisting block repair

steel structure. When repairing the steel structure, 12 mm or 20 mm wear-resistant steel plate is used for protection on the steel structure. This step can be selected according to the actual situation.

3.3.3.5 Repair of Super-Cutting Knife

According to the wear of the over-excavation knife, if only the nearby steel structure is worn, it is only necessary to repair the steel structure part to return to the state before the wear. If the over-excavation knife is found to be worn when repairing the cutter head, whether it is a tool or a tool holder, it needs to be repaired in time.

3.4 Example of the Yellow River Tunnel in the Middle Route of the South-To-North Water Transfer Project

3.4.1 Project Overview

The South-to-North Water Diversion Middle Line Crossing Yellow Tunnel Project (Fig. 3.33) is a key project for the main channel of the South-to-North Water Transfer Project to cross the Yellow River. The total length of the tunnel is 4,250 m, including 3,450 m for the cross-river tunnel and 800 m for the Lushan tunnel. The double-hole arrangement is used, and the tunnel axis spacing is 28 m. Two mud-water pressurized shields are used to advance from north to south. The maximum depth of the tunnel is 35 m, the minimum depth is 23 m, the maximum water pressure is 0.45 MPa, the minimum curve radius is 800 m, the slope of the tunnel is 0.1 and 0.2%, and the slope of the Lushan tunnel is 4.9107%.

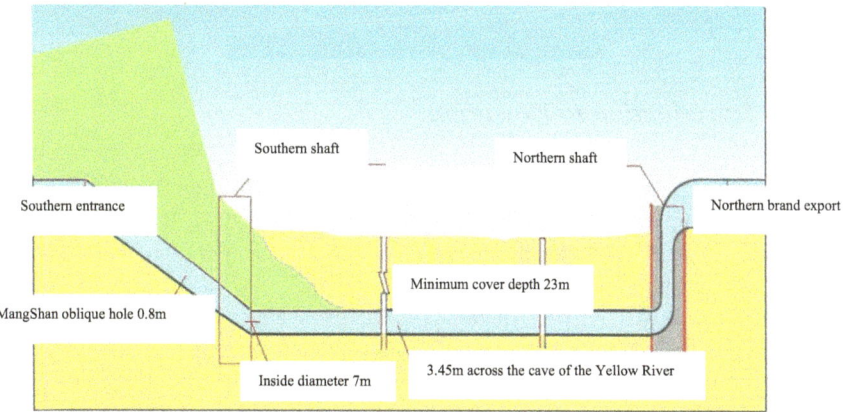

Fig. 3.33 Location diagram of yellow crossing project

Table 3.6 Quartz content of fine sand and medium sand

Sample depth (m)	Sand name	Quartz content (%)
35.8–35.9	Fine sand	40.0
34.0–34.3	Medium sand	45.0
39.4–39.8	Medium sand	69.5

The Quaternary is widely distributed in the stratum of the project, and the underlying bedrock is the upper Tertiary. The two are unconformity contacts, and the bedrock is buried at 37.60–89.70 m. The Quaternary strata specifically include: the Holocene alluvium, mainly fine sand and medium sand; the Upper Pleistocene alluvium, mainly sand gravel layer; the Middle Pleistocene and floodplain, mainly brownish yellow, brownish yellow or light brown red silty loam. The underlying bedrocks of the Quaternary strata are mainly claystones, sandstones, argillaceous siltstones, glutenite layers and partially interbedded marlstones. In addition, detailed geological reports from various layers show that the quartz content of the sandstone in the various layers that the shield passes through is generally higher, and the quartz particles in the medium sand formation are as high as 70%, which is very resistant to the wear resistance of the cutterhead and the tools. The quartz content of each layer is shown in Table 3.6.

For the geological conditions and engineering requirements of the project, two mud-water pressurized shield machines were specially designed by Herrenknecht, Germany. The cutter plate form and tools arrangement of the shield are shown in Fig. 3.34.

The cutterhead is in the form of a spoke cutter with an excavation diameter of 9000 mm. It has a rotary joint for bentonite, hydraulic pipe, electrical circuit. It also has a central flushing device, an opening ratio of 36%, and 24 advanced knives (difference from the blade height) 30 mm), 6 tooth cutters (with a blade height difference of 30 mm), 90 blades (multiple tungsten inserts), 16 blades, one profile knife with soft soil cutter, 2 double edge scrapers and 3 scraper wear detection devices; the tool change mode is back mount.

3.4.2 Introduction to Downtime

The South-North Water Transfer Middle Line crossed the Yellow IIB shield to begin the assembly and commissioning of the shield on May 1, 2007. It was successfully launched on July 8, 2007. Until September 2, 2008, the shield tunneling was about 1360 m (849 rings).

The 875 m (the first 546 ring) of the shield was excavated in a single sand layer. The tunneling parameters of the shield were relatively stable, with no major changes, the average torque was 0.78 MN m, and the average tunneling speed was 37.1 mm/min. The average thrust is 28204 kN. After 875 m (after 546th ring), it entered the soil layer, and the tunneling parameters such as thrust and torque increased, and the

3.4 Example of the Yellow River Tunnel in the Middle ...   143

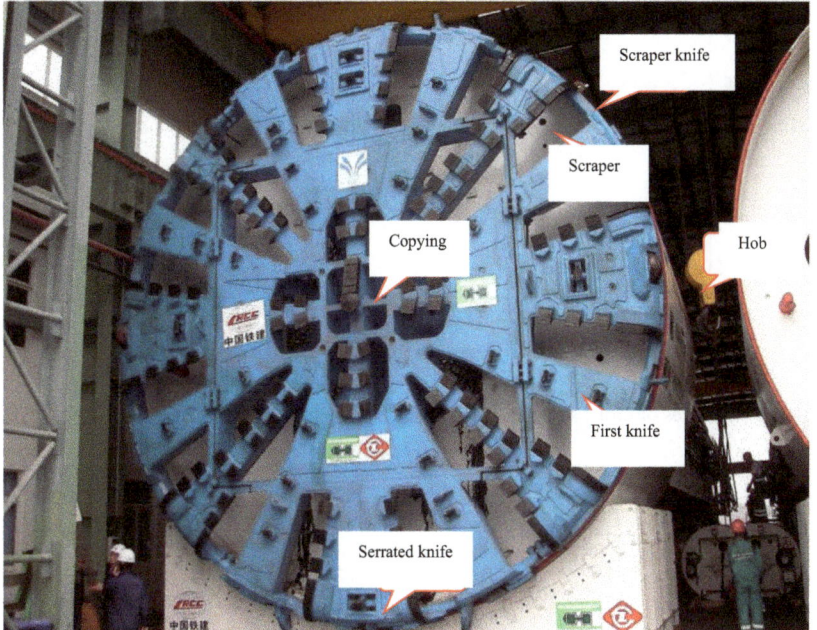

Fig. 3.34 Shield machine through yellow tunnel

tunneling speed decreased significantly. In particular, when boring between the 810 ring and the 845 rings, the cutter torque is increased to 5 MN m, and the boring speed is reduced to less than 10 mm/min.

When the 848 rings were drilled, the torque suddenly increased to 6.5 MN m, and the tunneling speed dropped to 3 mm/min, and normal tunneling was impossible. In order to ensure the safety of machinery and equipment and long-distance construction in the future, it is decided to stop the inspection at the 849 rings. Figure 3.35 shows the change of the tunneling parameters of the shield after the 810 rings.

The shield is parked at the Yellow River beach, with flat terrain and open areas. After investigation, the axis of the shield tunnel is 33.5 m away from the ground, and the groundwater level of the construction site is 4.00 m below the natural ground, as shown in Fig. 3.36.

According to the detailed lithology data of the reclaimed borehole, and referring to the relevant geological data in the Yellow River Middle Route of the South-to-North Water Diversion Project, the stratigraphic structure is layered. The top-down distribution of the lithology of the stratum is shown in Table 3.7. The loam layer of the shield is tightly structured. The plastic silty loam has a small permeability coefficient and high strength. The hard plastic silty loam has a large permeability coefficient and contains pebble and calcareous nodules.

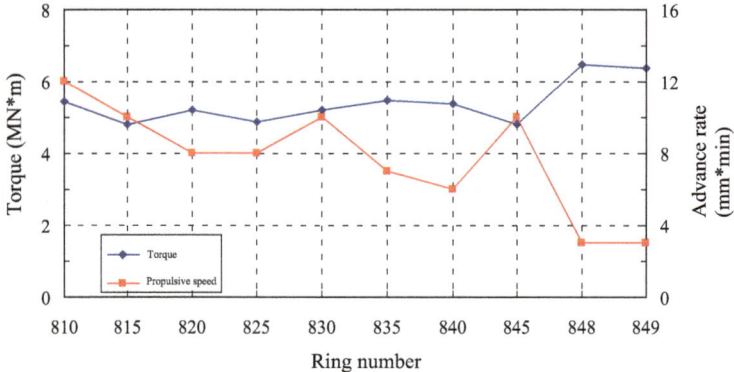

Fig. 3.35 Changes of thrust speed of shield and torque of cutter before shutdown

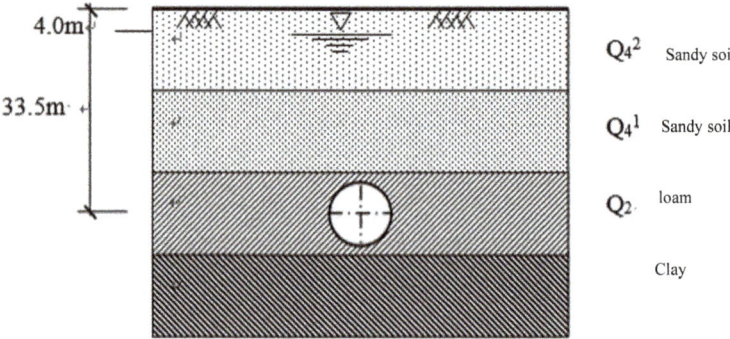

Fig. 3.36 Stop position profile

3.4.3 Guarantee the Stability Measures of the Excavation Face

The methods for ensuring the stability of the excavation face can be roughly divided into two types: a method of pre-reinforcing surrounding rock under normal pressure and a method of pressing the air pressure support into the cabin. Because the repair and modification of the cutterhead requires hot work, the current hot work under high pressure is very dangerous and difficult to operate, so the pressure-injection operation has greater risk. At the same time, the depth of the soil above the shield is within the allowable range of surface reinforcement; the surface is open, and there is no surface water, and the conditions for pre-reinforcement are available. Considering the above situation, it is decided to use the method of pre-stressing the surrounding rock under normal pressure to ensure the stability of the excavation face. There are many common methods for strengthening surrounding rock, and it is decided to use three-axis mixing pile method to strengthen the surrounding rock. In order

Table 3.7 Geologic stratification

Level number	Soil mass name	Thickness	Soil mass description
1	Planting soil	0.6	Gray, mainly silty, loose, locally unsandy loam
2	Silt	1.8	Pale yellow, relatively pure, loose—slightly dense zhua
3	Medium-coarse sand	11.6	Pale yellow, relatively pure, saturated, slightly dense—compact
4	Coarse sand	11.7–13.9	Light brown gray, saturated, compact, containing mica, quartz, feldspar and other minerals
5	Silty loam	2–5.1	Brown yellow, compact structure, small permeability coefficient of plastic-shaped silty loam, high strength, large permeability coefficient of hard plastic-shaped silty loam, locally containing many calcium nodules
6	Silty clay	1.1–5.1	Dark brown, plastic—hard plastic
7	Silty loam	1.8	Brown yellow, relatively compact, plastic, containing a small amount of mud calcareous mass, up to 9 m
8	Silty clay	1.1–1.3	Light brown red, plastic—hard plastic shape, containing a large number of calcium nodules
9	Calcareous tuberculous cement layer	1.5–1.5	Compact structure, low permeability coefficient, high strength
10	Silty clay	1.1–2.4	Brown yellow, more compact structure, plastic shape, high sand and gravel content
11	Sandstorm	0.75–2.5	Light white, high strength
12	Fine sand	0.3–2.95	Dense

to prevent groundwater from penetrating into the working area of the shield cutter during repairing the cutter head, accidents such as water inrush around the cutter head and collapse of the excavation surface occur, and well point precipitation is used at the same time. After the construction of the triaxial mixing pile is completed, the weakened part of the triaxial mixing pile (such as the joint of the mixing pile and the forming tunnel) is added with a precipitation well, and then the water is continuously discharged for 24 h to reduce the groundwater level in the tool changing area. The overall scheme is shown in Fig. 3.37.

Firstly, a circle of water curtains is arranged around the shield. The front and sides of the shield are constructed with a mixing pile anti-seepage wall. The reinforcement

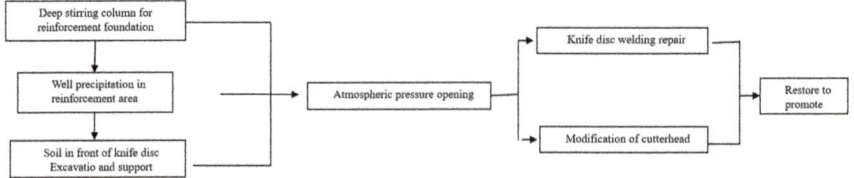

Fig. 3.37 Overall scheme

depth is 42 m. The single-row three-axis mixing pile is used for construction. At the back of the rare-proof shield is a seepage wall using a row of three-axis mixing piles. At the same time, in order to ensure the normal pressure to repair the cutter, the soil above and behind the cutterhead is stable, and the soil in front of the shield is reinforced with 3 rows of mixing piles. The reinforcement depth is 42 m, and 5 rows of mixing piles are arranged above the shield. The detailed dimensions of reinforcement and reinforcement are shown in Figs. 3.38, 3.39 and 3.40.

After the construction of the mixing pile is completed for 15 days, the soil in the anti-seepage wall began to be precipitated. According to the working conditions and hydrogeological conditions of the project, it is proposed to construct 5 precipitation wells and 3 observation wells, in which 30 m precipitation wells, one 30 m observation well and one 40 m precipitation well, one 40 m are arranged in the

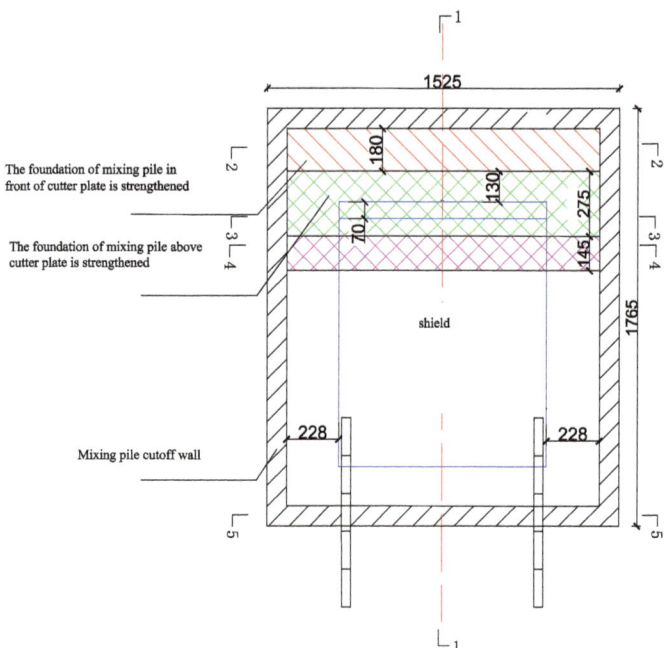

Fig. 3.38 Plane of reinforcement of surrounding rock and cutoff wall of mixing pile (unit: cm)

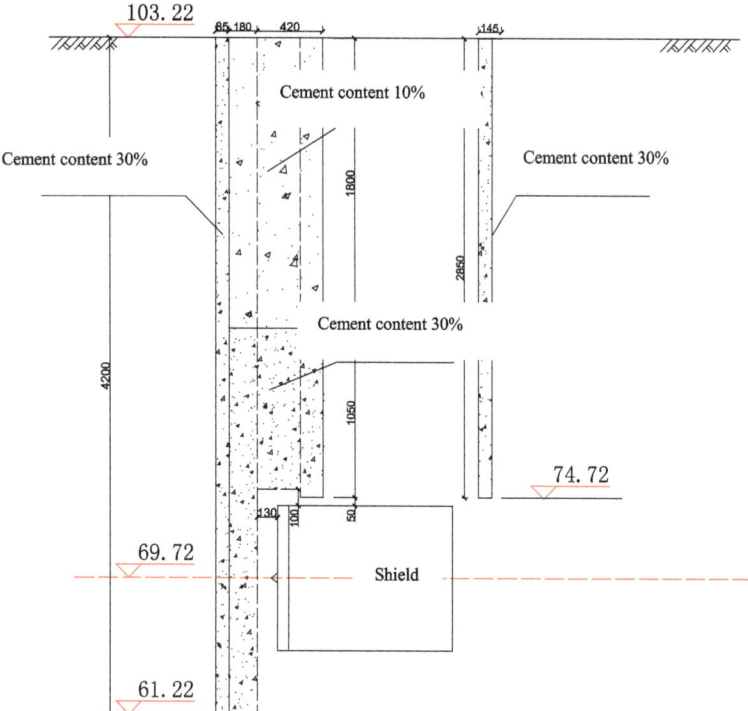

Fig. 3.39 Reinforcement of mixed pile surrounding rock and longitudinal section of cutoff wall (unit: cm)

reinforcement area. two precipitation wells and one observation well were arranged outside the reinforcement area, and the depth was 30 m. Figure 3.41 is a layout plan of the precipitation. The precipitation well point system consists of a submersible pump and a well pipe filter. The diameter of the wellbore of the precipitation well is 700 mm, the diameter of the well pipe is 315 mm and the wall thickness is 5 mm. The diameter of the observation well hole is 300 mm, and the diameter of the well pipe is 150 mm and the wall thickness is 5 mm.

3.4.4 Excavation Support

When the surrounding rock reinforcement is completed and the precipitation work is ready, the cutter can be repaired under normal pressure. However, because the cutter repair and tool change work lasts for a long time, a stable and long-lasting excavation space is required, and the soil on the excavation surface is unstable, and there may be rich groundwater, so it is necessary to excavate the front of the cutterhead and make a certain amount of space and carry out the necessary support. The excavation support

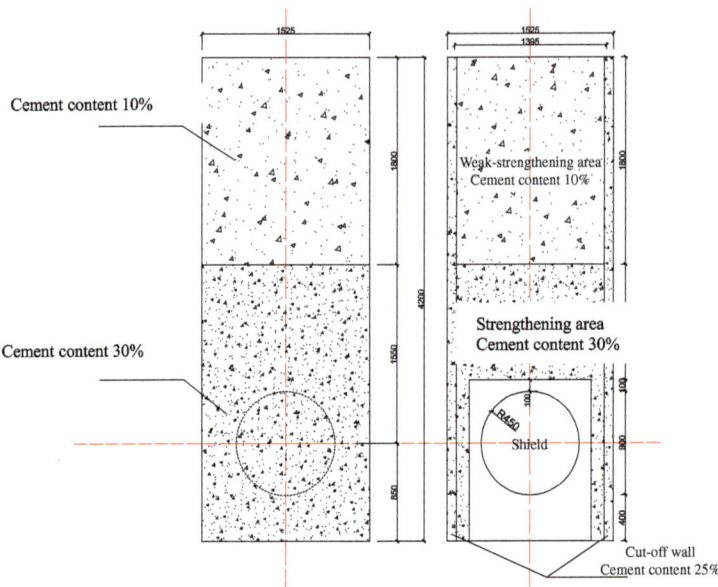

Fig. 3.40 Mixing pile perimeter reinforcement and impermeable wall cross section

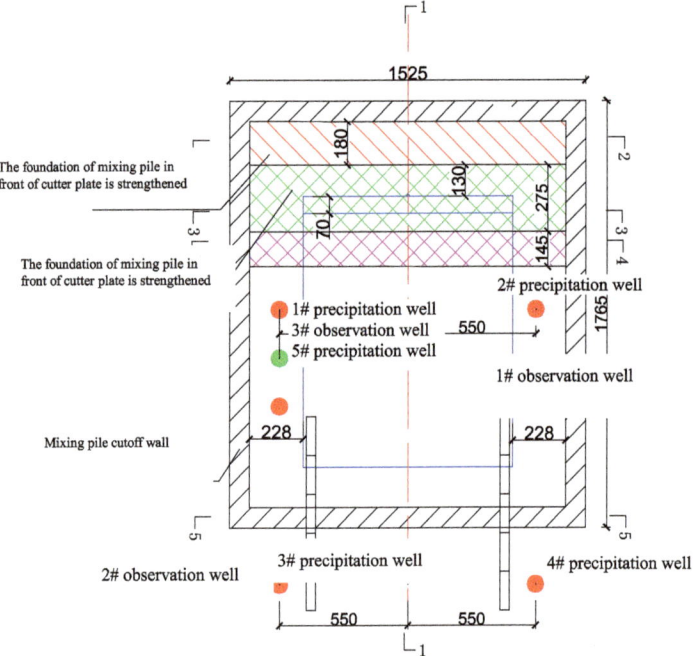

Fig. 3.41 Layout of precipitation well (unit: cm)

3.4 Example of the Yellow River Tunnel in the Middle ...

adopts the scheme of spray anchor support. The spray anchor support works together with the anchor rod, the steel mesh and the concrete to improve the structural strength and deformation rigidity of the excavation surface soil, reduce the soil deformation and strengthen the soil overall stability.

1) Excavation of earthwork.

To reduce the disturbance of the excavation surface soil during excavation, uses step-by-step excavation (Fig. 3.42). First excavate the upper soil, and partially loosen the uneven soil after excavation. Clean the corners of the wall trimmed and smooth, and then spray anchor support is carried out. After the upper shotcrete is finally set, the lower soil is excavated and supported.

2) Shotcrete

(1) Before the spraying operation, it is necessary to carry out comprehensive inspection and trial operation on mechanical equipment, as well as wind, water pipelines and electric wires.
(2) Before spraying the concrete, the cement slurry is sprayed on the surface to ensure good adhesion between the shotcrete and the soil on the excavation surface.
(3) A flag that controls the thickness of the shotcrete is buried to ensure the thickness of the concrete jet.
(4) The spraying operation should be carried out in stages, and the spraying sequence is from bottom to top.
(5) After spraying the first layer of concrete, construct the anchor rod and hang the steel mesh. The thickness of the first layer of concrete is 4–5 cm. Nozzle and spray surface should be perpendicular, and should be kept at a distance of 0.6–1.0 m. The second layer of shotcrete should be carried out after the first layer of concrete is finally set. The final shotcrete thickness shall not be less than 20 cm. If the first layer of concrete is finally condensed for 1 h and then the second layer of concrete is sprayed, the surface of the sprayed layer should be cleaned with water. The thickness and the light perception of the surface

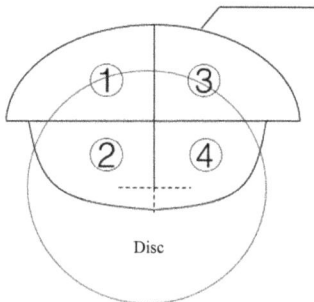

Fig. 3.42 Earth excavation space

must be ensured during the second spray. The concrete is sprayed for 24 h and then watered to ensure the quality of the concrete.
(6) The water-cement ratio should be controlled during spraying to keep the concrete surface flat, with a wet luster, no dry spots or slippery flow.
(7) The jet machine is provided with a shield to connect the bridge aisle plate.

3) The anchor bolt of the anchor construction is driven into the soil layer by 3 m 20 mm reinforced steel.
4) Hanging net

(1) First straighten the disc steel bar (6.5 mm), take the material according to the shape of the excavation surface, and weave the steel mesh according to the specifications of the mesh 15 cm × 15 cm. The distribution should be even and the binding should be firm.
(2) After the steel mesh is compiled, the joint with the anchor rod is welded firmly with the 16 mm steel bar and the anchor rod to ensure that the steel bar does not shake when the concrete is sprayed.
(3) The steel mesh must be close to the concrete surface to ensure the thickness of the protective layer of the steel mesh.

5) Maintenance

(1) After the last sprayed concrete is finally condensed for 2 h, spray water for maintenance immediately, and spray at least 4 times a day. The curing time must not be less than 7d.
(2) When the first water spray curing after final condensation, the pressure should not be too large to prevent the surface of the sprayed concrete protective layer from being damaged.
(3) When the temperature is lower than 5 °C, water spray maintenance is not allowed.
(4) If abnormal phenomena such as peeling, external drums, cracks, local moisture, uneven color, etc. are found during the curing process, the reasons should be analyzed and measures should be taken to repair them to prevent future problems.

3.4.5 Cutterhead and Tool Repair

The focus of the cutterhead repair is to repair the cutter holder and the cutter box of the cutterhead, so that it can meet the strength requirements for the cutter to cut the soil. For the damaged steel structure, it mainly strengthens and increases the wear protection and repair work.

In order to meet the needs of future excavation, and to adapt to the complicated geological conditions of wearing yellow, after repairing this cutterhead, some tools need to be repaired and modified to increase the rigidity, strength and wear resistance of the cutter.

3.4 Example of the Yellow River Tunnel in the Middle ...

The damaged part of the cutter head is mainly the outer edge of each arm and the circular arc plate of the outer ring of the cutter head. When repairing, the main work is to plan, grind and repair the weld. In particular, the circular arc plate on the outer edge of the cutter head is not integral due to its own structure, so it is repaired by welding steel plates during repair. Use the appropriate form of steel plate to fill the damaged part, select the appropriate groove form for welding according to the actual situation, and then weld the knife seat, the outer edge wear strip and the wear plate on it.

Due to the large amount of welding work and limited working surface, in order to ensure the quality and progress of the repair of the cutterhead, this cutterhead repair uses a combination of ordinary manual arc welding and gas shielded welding (carbon dioxide gas or mixed gas).

In view of the problems existing in the current tool size, material and welding process, in order to meet the needs of high content of calcium tuberculosis layer and sand pebble layer and prevent the tool from being damaged by impact, it is decided to transform some tools of the shield. Combined with the current situation of tool wear abnormality and the analysis of wear causes, the tool is improved from the following aspects:

1) Replace the cutting tool with a new type of tool, adjust the shape, setting and size of the alloy block, but retain its height and installation size. This tool change uses the following two types of tools:

(1) Shell knife. The alloy is distributed in the form of a triangular alloy with a cutting edge, which can better function as a cutting knife to divide the soil. Due to the arrangement of multiple alloys, the base and the alloy are arranged in a curved step, and the layers are mutually protected, which can increase the wear resistance of the tool. The shell knife is shown in Fig. 3.43.

(2) Improved type of advance knife. The original blade and cutting size are retained, and the original cutting edge is passivated to increase its impact

Fig. 3.43 Shell knife

resistance; however, a stepped alloy is added in the middle of the original tool to increase the wear resistance of the body. The improved front cutter is shown in Fig. 3.44.

2) Reconstruction of the hob

This time the tool change retains the No. 15 and No. 16 hobs on the outermost edge of the cutter head, and the remaining six hobs are all replaced with the cutters. The hob cutter ring is inlaid with alloy teeth, which increases the rolling ability of the hob in the clay layer, reduces the occurrence of eccentric wear of the hob cutter ring, and improves the anti-wear ability of the hob cutter ring. This replacement tooth cutter retains the original form of the Herrenknife tooth cutter. The replacement of the hob and the tooth cutter is shown in Figs. 3.45 and 3.46.

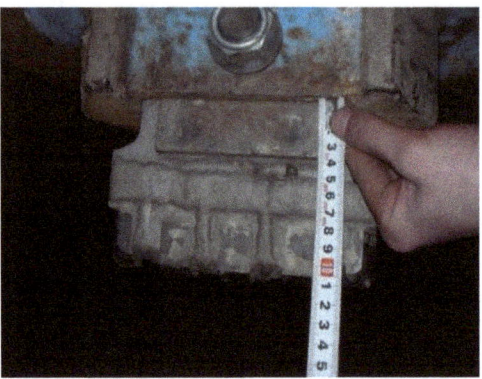

Fig. 3.44 Improved antecedent

Fig. 3.45 Schematic diagram of hob replacement

3.4 Example of the Yellow River Tunnel in the Middle ...

Fig. 3.46 Schematic diagram of tooth cutter replacement

According to the next stage of the tool change plan, before entering the sand pebble formation, it is planned to replace all the cutters and the alloy cutter hobs into a double-edged hob with a thickened cutter ring to increase the rigidity and impact resistance of the cutter ring to meet the driving needs of sand and gravel strata.

3) Reconstruction of edge blade

The main damage part of this cutterhead and tool is the cutter in the edge of cutterhead. The wear of various cutters is compared. At the same time, the form of the original cutter cutter is analyzed. According to the analysis results, the form of the existing cutter is changed to the degree of improvement. But it still retains the overall form and installation height of the blade, and only strengthens the wear resistance and strength of the blade itself. Take the following two forms of tools:

(1) The cutting edge is changed to the slotted brazing alloy form, the back of the blade is heat-fitted to the gold column, and the turbulent wear-resistant layer is covered in other parts of the cutter body to increase the wear resistance of the cutter body.
(2) The cutting edge adopts the form of "7" shaped welding alloy block. The alloy adopts passivated sheet alloy, and the size of the alloy is controlled at the same time, so that it can be tightly combined in the curved part to improve its overall wear and impact resistance.

4) Reconstruction of the front scraper

This tool change only replaces the tool damaged by the alloy head, retaining its original structure and form.

5) Transformation of the center knife

According to the experience of the subway construction site in the past, the central knives currently used on the shield are more susceptible to damage when digging into the gravel stratum, and the replacement of the central knives is very difficult. Therefore, the following measures were taken to strengthen the center knife during this tool change:

(1) Strengthen the strength and wear resistance of the center knife to prevent damage to the tool.
(2) Add a protective knife on the center knife path to reduce the wear on the center knife.
(3) Add 4 first-hand knives in the center to achieve the effect of tearing the soil first.

Increasing the outer edge protection tool

Because the tool working on the outer edge of the cutter head is faster than the tool line working in the inner area of the cutter head, the crushing load is heavier, the load condition is more severe, and the tool wears faster. In order to protect the outer edge of the cutter head, a protective cutter is installed on the outer edge of the cutter head. The height of the first cutter is 160 mm, which is lower than the height of the edge hob to reduce the force on the edge. The first cutter path is the same as that of the outer edge hob are first about the four trajectories. And the number of tools added to each trajectory is: #16 trajectory increases by 5, #15 increases by 4, #14 increases by 4, and #13 increases by 3. The outer edge of the cutter head is equipped with a protective cutter, as shown in Fig. 3.47.

6) In addition to adding tools to the outer edge of the reinforced cutterhead, it is planned to add protective measures such as wear blocks to ensure that the steel structure of the cutter is not subject to wear in the event of damage to the cutter.

Fig. 3.47 Edge guard knife

(1) When the steel structure of each arm of the outer edge of the cutter head is repaired, the wear plate or the wear block (the plate thickness is determined according to the actual) is welded on the steel structure to increase the wear resistance of the steel structure.
(2) Add the hob protection block to ensure that the tool box will not be damaged after the tool is damaged. After the hob is repaired, install a protective block around the hob.

Increase the outer edge protection block of the cutter to protect the outer ring of the cutter from damage. Install eight protective knives on the outer edge of the cutter head with eight auxiliary arms. The height of the protection knives is the same as the height of the wear-resistant strips on the outer edge of the cutter head. The edge gauge is shown in Fig. 3.48. Through the analysis and discussion of the whole process of repairing the tunnel of the Yellow River Tunnel, combined with the previous experience of shield construction, it can be explained that the stability of the excavation surface plays a key role in the success of the normal pressure opening. And the mixing pile water curtain's permeability, reinforcement strength of the surrounding rock above the shield, precipitation in the reinforcement zone, and support effect after the excavation of the soil in front of the cutter head determine the stability of the excavation face. At the same time, the modification of the tool is also made for the strata of different stages. Only based on the use of the tool and the actual damage, the performance of the tool is fully evaluated, and the relevant pressure checker tool inspection and tool change plan are formulated to ensure the long The tunnel construction of the distance and composite stratum is smoothly carried out.

Fig. 3.48 Edge caliper

Chapter 4
Normal Pressure Tool Change Technology Based on Basic Pressure and Changeable Knife Design

The realization of the conventional cutterhead atmospheric pressure tool requires reinforcement of the excavation face and corresponding precipitation measures. Although this technology is safe and can help to protect the safety of the workers, the construction period required for reinforcement and precipitation of the formation is usually required. Longer, and the cost is relatively high. For projects that require only short-term inspection and repair and serious tool wear and need to be replaced urgently to avoid delays, the conventional cutter-to-cylinder tool change technique is a bit awkward. To this end, the designer adjusts the structure of the cutterhead, a part of the retractable tool is installed on it to form a normal pressure tool change technology based on the atmospheric pressure changeable knife design, which solves such problems well and has been successfully applied in many large shield construction processes. This chapter will focus on the principle and construction process of the tool change technology to better demonstrate its specific application.

4.1 Introduction of Normal Pressure Changer Technology Based on Normal Pressure Replaceable Knife Design

4.1.1 Principle of the Process of Changing the Knife Under Constant Pressure

The atmospheric pressure changeable shield cutter disc is in the form of a hollow body. According to the different position of the cutter, some of the cutter seats are used the back-mounted type, the cutting path of the tool on this part of the holder covers the entire cutter surface. The ram is set in the back-mounted tool knives. The personnel can directly enter the spoke arm, extract the tool from the knives, and then close the broach gate to open the high pressure chamber and the arm in front

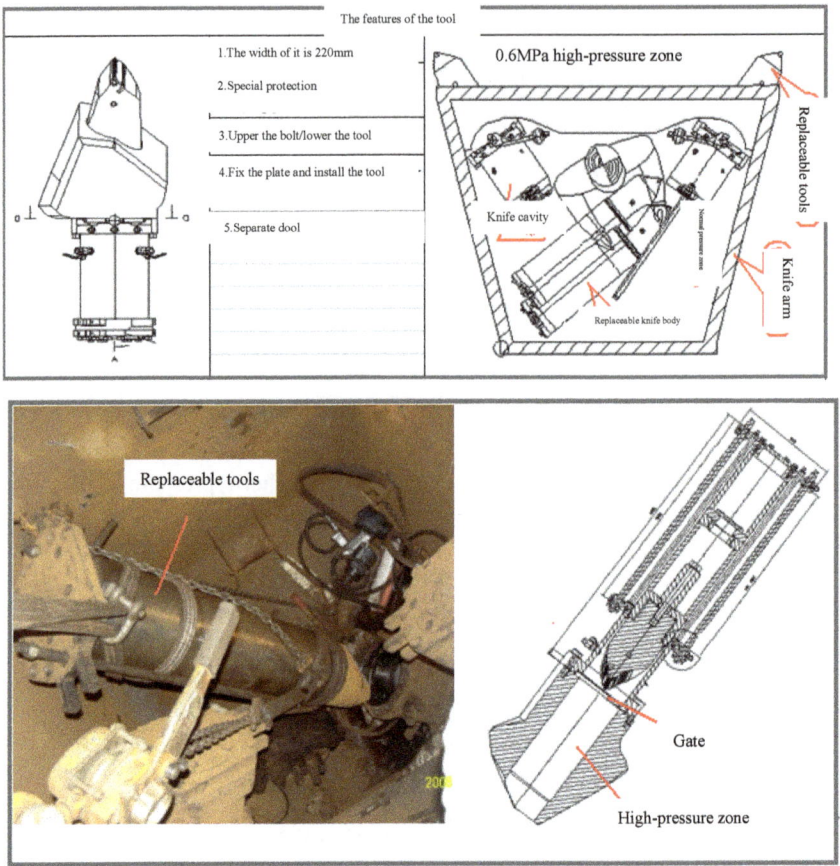

Fig. 4.1 Schematic diagram of the replaceable tool

of the cutter head. The normal pressure chamber is separated. After the new tool is replaced, the shutter is opened and the tool is replaced to realize the tool replacement at normal pressure. This is the principle of shield normal pressure tool change, as shown in Fig. 4.1.

The characteristics of the replaceable tool are: Tool width 20 mm; special seal protects the arm from water attack; Guide bolt lift /Lower the tool; fixed plate for final tool installation; pressure barrier.

4.1.2 Features of the Normal Pressure-for-Knife Method

Normal pressure tool change is a new type of tool change process in shield construction. Its main features are as follows:

4.1 Introduction of Normal Pressure Changer Technology …

(1) It is highly secure. The whole tool change work is under normal pressure. Compared with the pressurized air intake cabin to strengthen the soil in front of the excavation surface, the normal pressure entering the cabin is in good working condition and much safer.
(2) The construction period is short. It needs about an average of 2 h to replace a knife under normal pressure, only using 3 d to stop the machine to change the knife, compared to the pressure into the cabin work can only work 1.5 h each time, the cabin decompression needs 3–4 h, each work requires 12–15 d, and the normal pressure tool changing efficiency is 4–5 times higher than the high pressure tool change.
(3) Construction costs are low. High-pressure tool change requires 300,000 yuan for each high-concentration mud preparation and 850,000 yuan for professional diving personnel. The pressure of the constant pressure tool is less than 5,000 yuan per mud. The special operation allowance for the tool changer is 10,000 yuan per tool change operation, and the construction cost is obviously reduced.
(4) The environmental impact is small. The normal pressure tool change does not occupy the surface and does not affect the ground transportation and the surrounding environment.

4.2 Normal Pressure Replaceable Cutting Plate Structure Design and Layout

The atmospheric pressure replaceable cutter head comprises a main cutter beam, a secondary cutter beam and a normal pressure replacement cutter, the circular cutter structure is a spoke box type, the main cutter beam is a box type, the auxiliary cutter beam is a spoke type; the atmospheric pressure replacement cutter is both Placed on the main sill beam, the tool holder for changing the tool at atmospheric pressure is embedded in the main knife beam box.

The atmospheric pressure changing tool is fixed on the main beam by the welding seat, and the normal pressure changing tool is a normal pressure changing hob or a replaceable tearing knife; the both sides of the normal pressure changing hob are provided with the guiding slag along the rotation direction of the circular cutter head. The plate, the atmospheric pressure changing hob includes two gates, and the two gate joints are provided with a sealing strip, and the size of the atmospheric pressure changing hob increases as the diameter of the circular cutter head increases.

The cutter arm arms are in the form of hollow bodies. According to the different positions of the cutter cutters, some of the cutter seats are back- mounted, so that the cutting path of the cutters on this part of the cutter covers the entire cutter face. The ram is set in the back-mounted tool knives. The personnel can directly enter the spoke arm, extract the tool from the knives, and then close the sluice gate to separate the high-pressure compartment in front of the cutterhead from the normal-pressure compartment of the cutter arm. After replacing the new tool, open the shutter and replace the tool to replace the tool at atmospheric pressure.

A strip-shaped panel wear detecting device, a real-time flushing device, a hob wear detecting device, and a hob real-time rotation detecting device shall be provided on the cutter head; a ladder is installed in the main cutter beam box, and the cutter head structure and the cutter arrangement are as shown in Figs. 4.2–4.6 is shown.

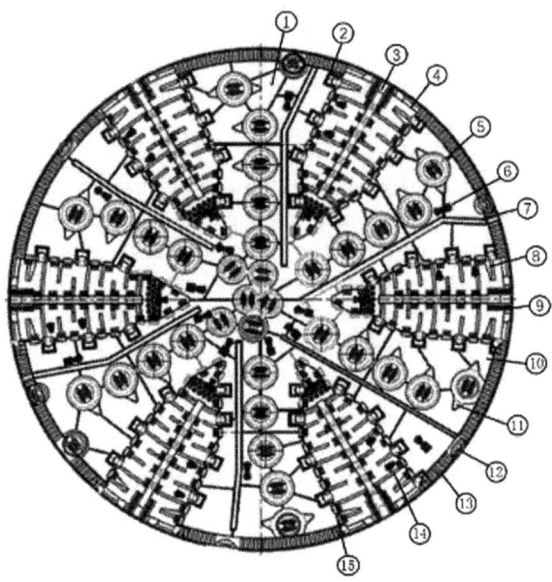

Fig. 4.2 Front view of the normal pressure tool changer

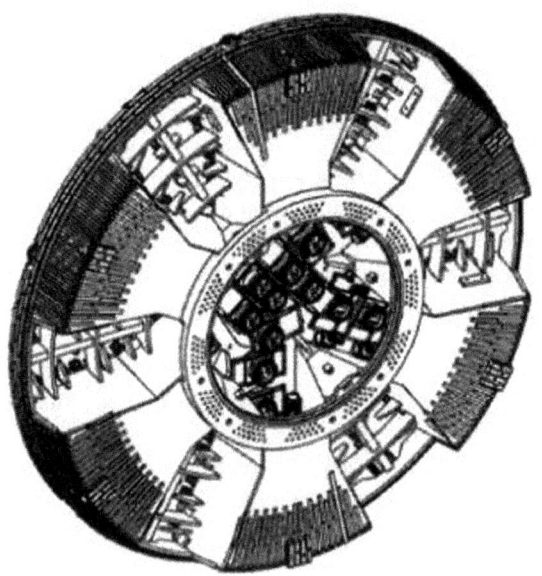

Fig. 4.3 Rear view of the normal pressure tool changer

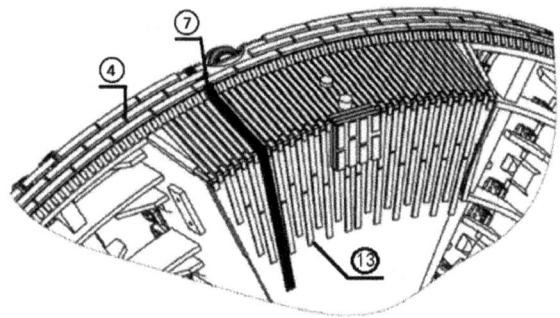

Fig. 4.4 Partial enlarged view of the back of the cutter head

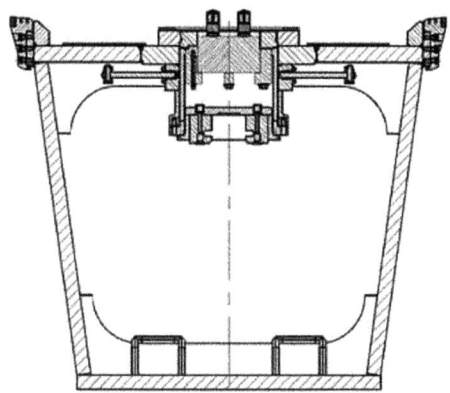

Fig. 4.5 Tool holder installation and replacement tearing knife

The structural components and the numbers correspond to the following: main cutter beam 1, atmospheric pressure replacement scraper 2, sub-knife beam 3, large ring 4, atmospheric pressure replacement hob 5, flushing spout 6, strip-shaped panel wear detecting device 7, slag Mouth size restriction grille 8, replaceable arc scraper 9, wear-resistant composite steel plate 10, slag-plating plate 1, normal-pressure replaceable single-blade hob 12, wear-resistant alloy block 13, Ring 14, with pressure replacement scraper 15, ladder 16.

Taking the shield used in the Nanjing Yangtze River Tunnel as an example, the cutter disc is in the form of a spoke panel with a diameter of 14.93 m, consisting of 6 spokes and 6 triangular panels. The spokes are box- shaped and the interior is a cavity. During the construction period, the construction personnel can enter the cavity inspection under normal pressure to check and replace the tool. The cutter head structure is shown in Fig. 4.7. The tool consists of a leading knife, a tooth cutter and a scraper. The number of the first knife is 16 and the tooth cutter is 189. Among them, the replaceable tooth cutter under normal pressure. The tool excavation track has a tooth knife distribution that can be replaced by normal pressure. The first knife and the tooth cutter are both safe. It is mounted on 6 spokes; the scraper is divided into two small scrapers and a large scraper; the small scraper is 3 pairs, mounted on the spokes, and 6 pairs of large scrapers are mounted on the triangular panel. The

Fig. 4.6 Cutter side view half Cutaway schematic

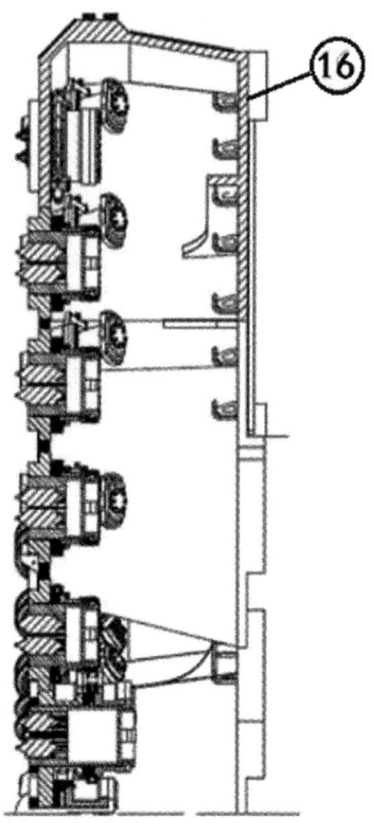

structure of the replaceable tool (the tooth cutter) under normal pressure is shown in Fig. 2.13.

The replaceable tool under normal pressure is mainly composed of a cutter tooth, a knife seat, a fixing bolt, a knife cavity and a gate. The cutter tooth and the tool holder are integrally connected by a fixing bolt, and are fixed and fixed in the cutter cavity, and the cutter cavity is welded to the cutter head on. A special screw can be used to move the cutter teeth and the tool holder back and forth in the tool cavity in the tool axis direction. During the tunneling, the cutter teeth and the seat are pushed to the front end and fixed. The cutter teeth extend out of the cutter head to cut the face of the cutter; when inspecting and replacing, the cutter teeth and the cutter seat are integrally retracted to the rear of the gate at the front end of the cutter. After the gate is closed, the front part of the gate is connected with the excavation chamber, which is a high-pressure chamber, and the rear part of the gate is connected with the spokes of the cutter head, and is an atmospheric pressure chamber. At this time, the inspection and replacement of the cutter teeth can be performed under normal pressure.

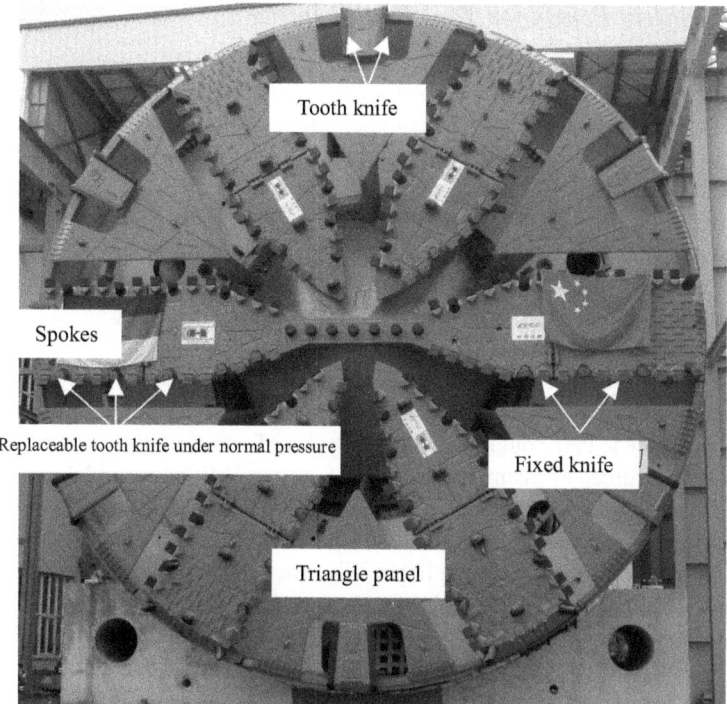

Fig. 4.7 Cutter structure

4.3 Removal and Installation of Normal Pressure Replaceable Tools

When the tool is in the process of excavation, the tool must be replaced due to wear exceeding the limit or falling off, missing or eccentric grinding. Tool can be divided into cutters, scrapers, tearing knives and hobs, and is suitable for different geological conditions. When the local conditions change, In order to ensure the safety of the shield construction and speed up the construction progress, the tools suitable for the formation conditions should also be replaced.

The tool change method adopted by the current shield is still very dangerous with pressure into the cabin. Professional construction personnel are required to enter the shield mud tank for manual replacement. The pressure in the shield mud tank is relatively high and the line of sight is poor, and the replacement speed is slow. This method can not only fully guarantee the personal safety of the construction personnel, but also has low efficiency and high construction cost. In order to protect the staff's safety and improve construction efficiency. For shields which are larger than 14 m, the cutterhead can be designed separately, which means the interior is designed as a cavity, and the constructor can enter the cavity under normal pressure to inspect and replace the tool.

4.3.1 Safety Points for Tool Replacement

(1) The inspection and replacement of the tool must be carried out with safety. Tool replacement is a more complex process, first remove the muddy water and residual soil in the pressure chamber, remove the sediment adhering to the cutter, confirm the tool to be replaced, and clearly transport tools, and set up scaffolding, then remove the old tool and replace it with a new one. The replacement tool has a long downtime, which is likely to cause the overall settlement of the shield, which will cause settlement of the formation and the surface, damage the surface and bury the building (construction), and endanger the safety of the project. To do this, prepare for replacement before minimizing downtime. The replacement work should be carried out in the middle shaft or in a section with good geological conditions and stable formation. If it is necessary to carry out in a geologically poor formation, the formation must be pre-reinforced to ensure the stability of the excavation face and the foundation.

(2) The safety of the operator must be ensured when the tool is replaced. The person who replaced the tool must wear a seat belt. Lifting tools must be used for loading and positioning. Especially when replacing the hob, use the clamp and lifting tool. All used tools of the hoisting must undergo rigorous inspection to ensure the safety of personnel and equipment. When the cutter head needs to be turned, the intake personnel must be evacuated to a safe area and operated by a special person. No one can start it without authorization.

(3) Develop detailed knife change program, emergency plan, and do a good job of technical delivery and personnel training. Tool replacement must be implemented civil and mechanical and electrical engineers on duty system; pressure loading into the cabin must have a strict pressure into the cabin; taking pressure into the cabin must to take Safety measures; the tool changer is used according to the relevant machine operating procedures; the tool transportation and replacement are safe, self-defense, mutual defense and joint defense; the waste used in tool change should be recycled uniformly to avoid environmental pollution. The replacement record must be made when changing the tool. The replacement record mainly includes the tool number, the original tool type, the tool wear amount, the operation record of the repair tool, the replacement reason, the replacement tool type, the replacement time and the operator name.

4.3.2 Tool Inspection and Replacement Plan

From the tool design of the shield and the adaptability of the wear resistance to the geological conditions of the gravel layer, the form of tool wear is: One is normal wear and the other is abnormal bumping and disintegrating. The above two types of wear, the latter accounted for a higher proportion, and the damage to the tool is greater, especially the cutter at the edge of the cutter head. In the section of the stratum, shield tunneling construction, in order to effectively avoid excessive tool wear and damage the cutter-head, the inspection and replacement of the cutter will be a regular work.

According to the wear analysis and shield cutter tool design, it is planned to pass regular tool inspection and replacement and temporary tool inspection, monitoring of the health of the tool.

1) Regular opening tool inspection and replacement plan
   Regularly carry out the inspection of the opening tool and set a certain inspection frequency and replacement plan according to the estimated amount of tool wear. Inspection and replacement of objects are mainly based on edge cutters and tools in the gravel layer.
2) Temporary opening inspection and replacement
   (1) Regular open-out tool inspection, according to the estimated amount of tool wear to set a certain inspection frequency and replacement plan, inspection, replacement objects mainly edge tools and tools in the gravel layer.
   (2) The shield should strengthen the detection of the tool equipped with the tool wear alarm device during the excavation of the gravel layer. If the alarm device is alarmed, the temporary opening of the module on the corresponding digging track line on the replaceable tool open the cabin inspection.
   (3) When a foreign object appears in the mud treatment equipment, it is also carefully analyzed and decided whether to open the cabin for inspection.
3) Tool replacement standard
   (1) Edge tool. Since the edge knife is only 15 mm higher than the knife plate, so when the edge knife wear 10 mm, should replace the head to protect the knife plate, to ensure the diameter of excavation.
   (2) Front tool. The front tool is 200 mm higher than the knife plate, but due to the tool wear after the adjacent tool shape is not protected, will lead to the accelerated wear of adjacent tools, so the amount of wear should not be too large, should be replaced when wear ingresed 15 mm.

4.3.3 *The Process of Changing the Knife Under Constant Pressure*

When the shield structure in hard rock or self-stabilizing capacity of the strong section (the overall better stroke, breezed formation) digging, do not need to carry pressure into the cabin, in this case can enter the knife plate operation under normal pressure conditions; The tool replacement process is shown in Fig. 4.8.

The basic steps of atmospheric pressure change can be summarized as follows: Preparation for cleaning → Cylinder lifting → Loosening the barrel bolts → Retracting the cylinder, lifting the knife → Flushing → closing the gate → balancing

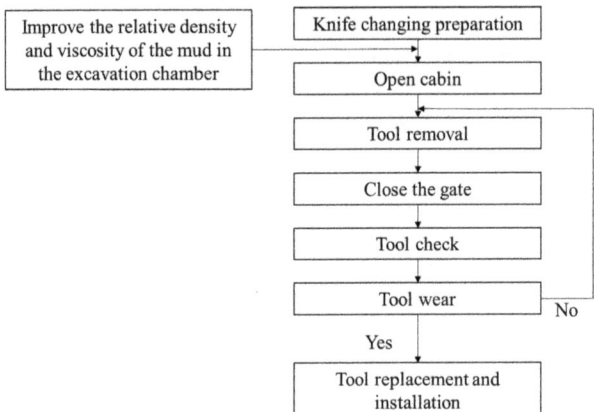

Fig. 4.8 Tool change process flow chart

the internal and external pressure difference → cylinder retraction → dismantling. In addition to the cylinder and temporary storage → the knife is completely extracted. The installation operation of the knife barrel is basically the reverse operation of the removal operation.

4.3.3.1 Preparations Before Changing the Knife

(1) Master plan. In the daily work of the engineer responsible for the knife plate and civil engineers close communication, to strengthen the understanding of the geological conditions of the construction section, the geological data reflect edgy focus and difficulties to pay special attention to this. When planning with accessories, fully estimate the damage degree of the special section to the tool. At the same time, when formulating the tool change plan, communicate with the civil engineer and the driving driver in time and effectively to determine the best opening location. At the same time as the tool replacement plan was proposed, the material preparation and personnel training were prepared in advance.

(2) Equipment and supplies. The preparation of equipment and materials is the fundamental guarantee for rapid knife change. In the case of ensuring that common equipment (implements) and materials are in place, use more advanced tools, such as pneumatic cranes, chain hoists, pneumatic wrenches, etc. Check the various systems of the shield before changing the knives, and coordinate the wind, water, electricity and other aspects to ensure a good working environment during the tool change.

(3) Personnel training. Specially trained personnel are required to enter the cabin for tool replacement.

4.3 Removal and Installation of Normal Pressure Replaceable Tools

(4) Establishment of emergency rescue teams. Knife change is a very dangerous operation, must be set up emergency rescue team, and strict lying "emergency preparedness and response control procedures" to prevent accidents.

(5) Open-out approval. The opening technical plan after the Ministry of Engineering and the Ministry of Construction, by the mechanical and electrical engineers and the chief engineer of civil engineers confirmed, reported to the project manager issued, the owner and the supervisor agreed to open the cabin, the responsibility to implement to the person, in strict accordance with the opening procedures.

The main principle of the mud-water balance shield structure is the balance between the mud-water pressure and the groundwater soil pressure in the excavation chamber, and the mud penetrates into the formation to form a impermeable cement film, thus maintaining the mud pressure in the excavation chamber and supporting the stability of the excavated surface soil. Normal pressure change knife generally requires 2–3d downtime, before changing the knife should be appropriate to improve the viscosity and relative density of mud water in the excavation chamber, to ensure that the excavation surface is stable. The main principle of the mud-water balance shield is that the mud-water pressure in the excavation chamber is balanced with the groundwater pressure. The mud penetrates into the formation to form an impervious cement film, thereby maintaining the mud pressure in the excavation chamber and supporting the stability of the excavation surface. Normal pressure change is generally required for 2–3d downtime, the viscosity and relative density of the mud in the excavation tank should be properly increased before the tool change to ensure the stability of the excavation surface.

Rotate the cutter head to the main arm of the tool to be inspected perpendicular to the horizontal axis of the shield, open the human gate hole on the center cone. Ventilation, connecting shuttle gates and air pipes, water pipes, cables, etc. According to the general rule of tool wear, the edge cutter of the outermost ring wears the fastest. First rotate the cutter head to the position where the main arm of the tool to be replaced is at the bottom, so that the center shuttle is just right. Located in the center of the center cone, and then open the door of the center cone, replace the exhaust gas in the gate through the air pipe installed in the center gate.

After the human gate is filled with fresh air, the cutter checks the replacement personnel to enter the human gate and installs the 2t heavy pneumatic hoist. Vertically perpendicular to the lifting ears above the main arm. Open the main arm cover and replace the exhaust gas in the main arm with compressed air. Two workers stayed at the entrance of the main arm of the center gate, one of them stayed on the partition in the middle of the main arm to transfer the working tools, and the other two entered the position of the tool to perform the specific work of dismantling and installing the tool.

4.3.3.2 Precautions for Normal Pressure Tool Change

When checking and changing tools under normal pressure, in order to ensure construction safety, the following points should be noted:

(1) Before assembling the shield at the site, the pressure-resistant test is required for the knife cavity and gate of the normal pressure replacement tool, and the test pressure is generally 1.5 times the working pressure, ensuring that the tool cavity and gate seal well under this pressure.

(2) Regularly check the condition of the guide rod and the disassembly screw of the disassembling tool, if cracks, stretching or thread damage are found to be replaced in a timely manner, to prevent the failure of the guide rod or disassembly screw when the tool is disassembled.

(3) To formulate and improve the construction emergency plan to ensure that in the event of an accident, the construction personnel can be evacuated from the spokes of the knife plate in a timely manner, and timely close the knife disk spoke gate.

4.3.3.3 Several Specific Schemes for Normal Pressure Knives Change

The following describes the specific scheme of the atmospheric pressure changing tool, including the hob changing knife, the scraper changing blade, the hob and the tooth cutter interchange.

1) Hob changing hob

   (1) Roller assembly (Fig. 4.9). Rollers are mainly used to break rock formations, which are mainly composed of closed units (Fig. 4.10, including gate, top flushing ball valve, bottom flushing ball valve), double-edged roller, replacement unit (Fig. 4.11, including roller, knife, bolt), replacement shell bolt, replacement housing cover, replacement fixed bolt sympathising assembly.

   (2) Install the aids (Fig. 4.12). The installation aids required for the knife change process are: telescopic cylinders, pneumatic hoists, clamps, hoops, cylinder connection bolts, pressure gauges, brackets, cylinder bases.

   (3) Basic operating procedures.
      1. Connect the water pipe (hydraulic pipe) to the flushing ball valve on the knife barrel, open the ball valve for flushing, and close the ball valve after flushing, remove the water pipe afterwards.
      2. Use the tool changer to lift the clamp on the cylinder ferrule so that the cylinder and the cylinder holder are at the same height and use the oil. The cylinder clamp fixes the cylinder and the cylinder holder and finally locks it with a pin.
      3. Connect the control valve to the tubing of the cylinder, extend the cylinder and lock the lock plate in 1st gear (make sure to use the pin lock Dead), loosen the bolts on the knife barrel.

4.3 Removal and Installation of Normal Pressure Replaceable Tools

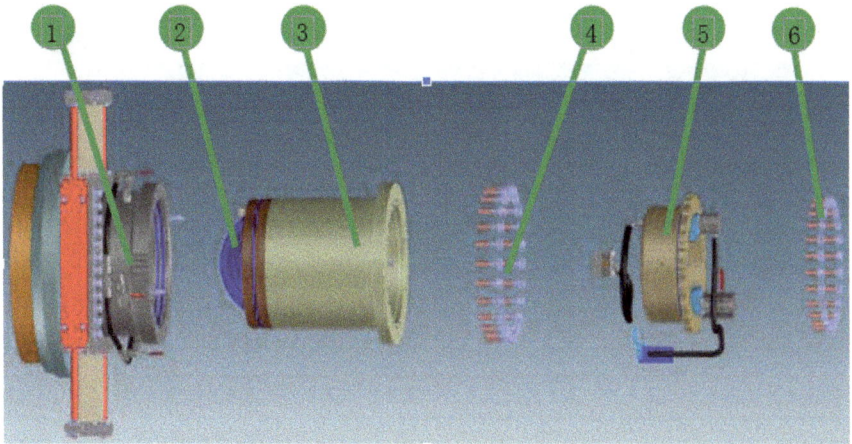

Fig. 4.9 Schematic diagram of the hob assembly. 1—closed unit; 2—double-edged hob; 3—replacement unit; 4—replacement of housing bolts; 5—replacement of housing cover; 6—replacement of fixing bolts

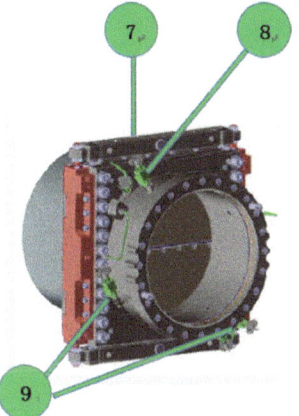

Fig. 4.10 Schematic diagram of the closed unit. 1—gate; 2—top flush ball valve; 3—bottom flush ball valve

4. Completely retract the cylinder, then hang the pneumatic hoist hook on the tool holder and bring the pneumatic hoist to the force (just hang it, bolt the bracket to the bottom of the knife barrel and connect the water pipe (hydraulic pipe) to the flushing line of the gate plate. Open the ball valve for flushing, close the ball valve after flushing, and remove the water pipe.

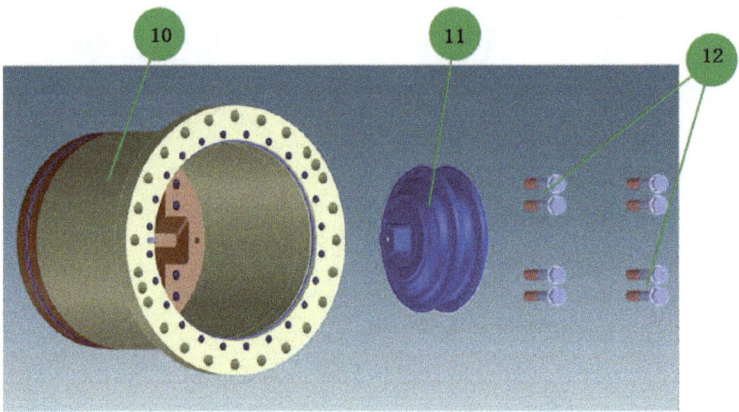

Fig. 4.11 Hob replacement unit schematic. 1—hob; 2—knife; 3—bolt

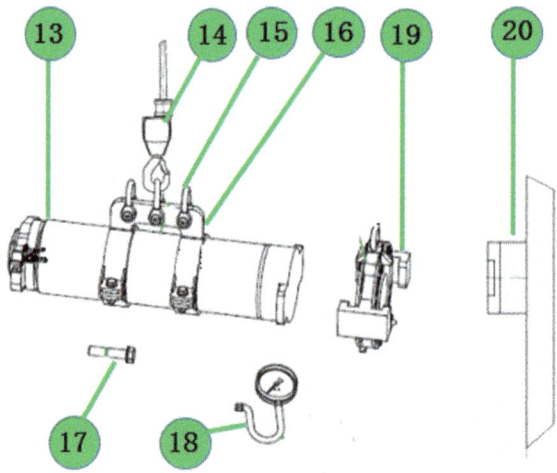

Fig. 4.12 Installation auxiliary tool schematic. 1—Telescopic cylinder; 2—Pneumatic hoist; 3—Clip; 4—Hoop; 5—bracket; 6—cylinder base; 7—pressure gauge; 8—cylinder connection bolt

5. Connect the control valve to the line of the gate plate, remove the yellow protective cover on the piston rod of the gate plate cylinder, and close the gate. And lock with the lock bolt (safe operation).
6. Open the bottom flushing ball valve to balance the internal and external pressure difference. Close. The bottom ball valve after the internal and external pressure difference is balanced, and lock the cylinder lock plate in 2nd gear.
7. The cylinder is fully retracted.

4.3 Removal and Installation of Normal Pressure Replaceable Tools 171

8. Remove the cylinder clamp clamp and use the second pneumatic hoist to lift the cylinder and temporarily store it to a suitable one position.
9. Pull the knife out and remove the worn hob from the knife and replace it with a new double-edged hob. Then the knife is in accordance with the steps from step 8 to step 2 reverse the operation to replace the new hob.

2) Scraper change scraper

Scraper is mounted on the outside of the cutter head to remove the sand from the outermost ring of the face. A schematic view of the tool and the tool holder is shown in Figs. 4.13 and 4.14.

Before disassembling and replacing the tool under normal pressure, first, prepare the relevant tools, including the hoist, the lead screw, the disassembly screw, the wrench, the special cylinder for opening and closing the gate, and the manual hydraulic pump. The specific steps are as follows:

(1) Remove the knife cover
(2) Insert the telescopic cylinder manually into the cartridge, the hydraulic cylinder is bolted to the barrel, and bolts the pendulum sleeve (Fig. 4.16) on the replaceable knife box (Fig. 4.15) as the force fulcrum of pulling out the barrel.
(3) Through the hydraulic line will telescopic cylinder hydraulic valve block and cylinder connection, slowly extend out of the cylinder cylinder, when the

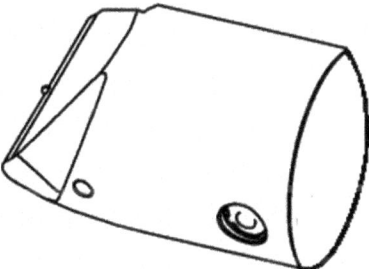

Fig. 4.13 Tool schematic

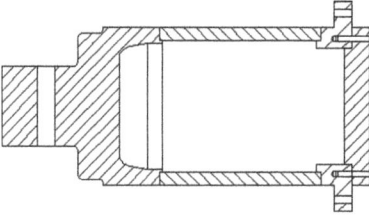

Fig. 4.14 Schematic diagram of the knife barrel

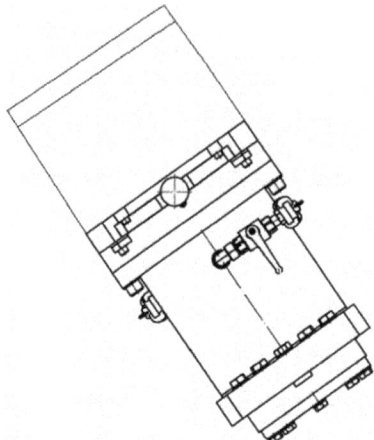

Fig. 4.15 Schematic diagram of the replaceable knife box

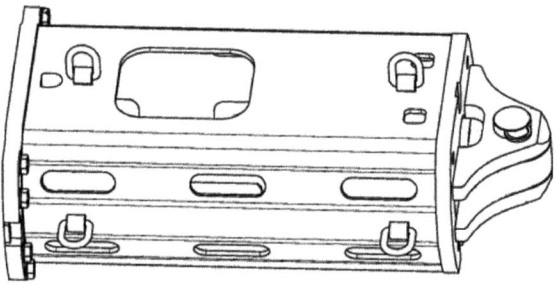

Fig. 4.16 Schematic diagram of the regular pendulum sleeve

      cylinder lifting ear and the sleeve in the same position when stopped, check the vertical position of the cylinder lifting ear, ensure that the cylinder is fully introduced to the sleeve, the cylinder hydraulic pressure remains unchanged, the insert pin inserted into the two lifting ears, and protected with a splint.

(4) Release the bolts attached to the cartridge on the replaceable knife box, slowly and completely retract the cylinder and the knife barrel, connect the flushing line to the flushing ball valve on the knife box, open the flushing ball valve to flush the mud inside the knife tube, the water pressure inside the knife tube is greater than the mud pressure of the palm surface, to ensure that the mud does not flow out, at this time with the gate cylinder to close the knife box (Fig. 4.17), And stop flushing, with bolts and safety rebar to tighten the brake plate cylinder, to prevent the operation of workers due to errors in the knife box spray and other accidents.

4.3 Removal and Installation of Normal Pressure Replaceable Tools

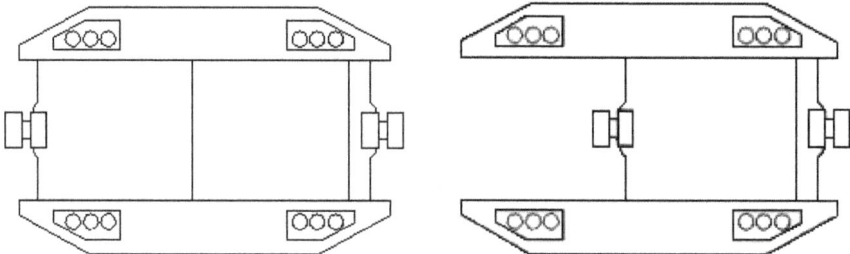

Fig. 4.17 Schematic diagram of the knife box gate

(5) Open the ball valve, release the excess moisture in the barrel, remove the pendulum sleeve, replace the tool.
(6) Install the sleeve, by flushing the ball valve to full, close the ball valve, pull out the pin, push the barrel into the designated position, Make sure that the knife box gate is opened when the seal is sealed, push the knife barrel all the way in, connect the knife barrel to the replaceable knife box with bolts, remove the sleeve and the hydraulic cylinder, and install the knife tube cover.

3) Hob and tooth cutter interchange

According to different cutting trajectories of each tool, different tool changing methods are summarized. The cutting trajectory of the hob can be divided into three categories: track 1–18, track 25–26; track 19–24; track 27/28a, b. The distribution of the cutter hob is shown in Fig. 4.18.

1) Track 1–18, track 25–26 operation steps.

　　1. Install the aids. The installation aids required are the same as in Fig. 4.19, which will not be repeated here.

(2) Basic steps:

　　A. Connect the water pipe (hydraulic pipe) to the flushing ball valve on the knife barrel, open the ball valve for flushing, close the ball valve after flushing, then remove the water pipe, close the wear detection ball valve, and then remove the wear detection line.
　　B. Use a knife-changer to lift the clip on the cylinder, so that the cylinder and cylinder fixed seat at the same height, the cylinder clip will be fixed with the cylinder fixed seat, and finally with the pin lock.
　　C. Make sure that the locking plate of the end step of the cylinder is connected to the chain in a horizontal direction, connect the control valve to the cylinder line, extend out the cylinder and lock the lock plate in 1 stop (ensure lock with a pin), ensure that after the above step, give a little pressure to the cylinder (the cylinder with pressure), and then loosen the bolt on the knife.

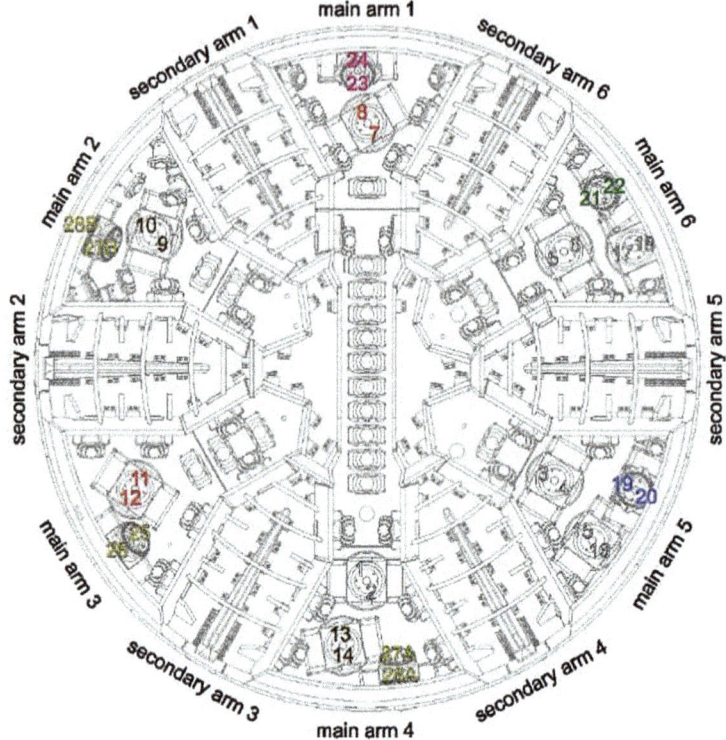

Fig. 4.18 Cutter hob distribution

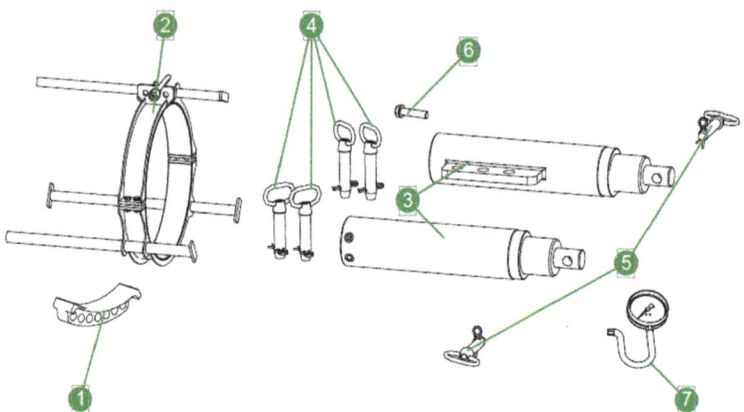

Fig. 4.19 Schematic diagram of the auxiliary tool. 1—tube rack; 2—installation hook; 3—retractable cylinder; 4, 5— bolt with spring plug; 6—hexagon bolts; 7—pressure gauge

4.3 Removal and Installation of Normal Pressure Replaceable Tools

    D. Completely retract the cylinder (#1 to #18 knife about 435 mm, #25/26 knife about 695 mm), fix the knife barrel clamp to the knife barrel, hang the pneumatic hoist hook on the knife barrel clamp, and make the pneumatic hoist band force (just good hanging), the barrel carrier fixed at the bottom of the knife barrel () Using a minimum of 3 M27 × 100 bolts), connect the water pipe (hydraulic pipe) to the flushing line of the gate plate, open the ball valve for flushing, rinse and close the ball valve, and then remove the pipe.

    E. Indirectly handle the control valve with the line of the gate plate, remove the yellow protective sleeve on the piston rod of the gate plate cylinder, and close the gate to ensure that the gate locks the bolt M24 (safe operation).

    F. Open the bottom scouring ball valve to balance the internal and external pressure difference, close the bottom scouring ball valve after the internal and external pressure differential, and lock the cylinder lock plate in 2 stops.

    G. Fully retract the cylinder (92 mm).

    H. Remove the cylinder clips, use a second pneumatic hoist to lift the cylinder sours and temporarily store them in a suitable position (be careful not to place the cylinders and other heavy objects on the removable platform of the changer).

    I. Pull the cartridge out completely (note that the cartridge will shake or swing when drawn out) and transfer it to the material transport channel.

3. Precautions. The installation of the cartridge is basically reverse operation, it is important to note that the following additional steps must be carried out in place:

    A. When the barrel is installed into the gate valve body, do not have to put the lock plate to 2 stops.

    B. Before placing the cartridge into the gate valve body, completely clean the inside (accessible) inside the gate plate and apply the outer surface of the cartridge completely with oil or butter.

    C. Before opening the gate plate, fill the small space between the knife barrel and the gate plate through the flushing ball valve on the knife barrel, and the pressure should be consistent with the pressure on the palm surface.

    D. During the period of watering, a ball valve on the flushing line on the barrel needs to be opened to drain the gas. When the space between the knife and the gate plate is filled with water and water overflows from the flushing ball valve, close the flushing line ball valve again.

    E. When pushing the barrel into the gate valve body, pay attention to the position pin on the knife barrel and the positioning hole on the gate valve body must be consistent (note the position of the position of the positioning pin).

(2) Track 19–24 operation steps.
   1. Installation of aids (Fig. 4.19). Auxiliary tools consist of pipe racks, mounting hooks, telescopic cylinders, bolted spring plugs, inner hexagonal bolts M24 × 135 mm, pressure gauges and other components.
   2. Basic steps:
      A. Control the pressure of the palm surface, connect the water pipe (hydraulic pipe) to the flushing ball valve on the knife barrel, open the ball valve for flushing, close the ball valve after flushing, then remove the pipe, close the wear detection ball valve, and then remove the wear detection line.
      B. Use a changer to lift the cylinder to the position of the barrel, the cylinder is fixed to the barrel of the barrel, to ensure that 4 pins are fixed.
      C. Connect the control valve to the cylinder line, fully extend the cylinder, use 2 pins to lock the cylinder with the seat, and release the threaded hoop of the outer ring of the cylinder piston rod.
      D. Fully retract the cylinders.
      E. Install the knife barrel clamp on the knife barrel, hang the pneumatic hoist hook on the knife barrel clamp, and make the pneumatic hoist band force (just good hanging), secure the barrel carrier at the bottom of the knife barrel, using a minimum of 3 M27 bolts.
      F. Connect the water pipe (hydraulic pipe) to the flushing line of the gate plate, open the ball valve for flushing, rinse and close the ball valve, and then remove the water pipe.
      G. Indirectly good control valve and gate plate line, remove the yellow protective sleeve on the piston rod of the gate plate cylinder, close the gate, and ensure that the gate locks the bolt M24 lock (safe operation).
      H. Open the bottom of the flushing ball valve to balance the internal and external pressure difference, in the internal and external pressure difference balance, the bottom of the scouring ball valve closed, the positioning pin from the barrel hanging ear to pull out, and then the cylinder out 100 mm to the next pin hole, again plug the pin, to ensure that the spring pin lock.
      I. Fully retract the cylinders.
      J. Remove the cylinders and temporarily store them in a suitable location.
      K. Pull the cartridge out completely (note that the cartridge will shake or swing when drawn out) and transfer it to the material transport channel (be sure not to place cylinders and other heavy objects on the removable platform of the tool changer).
   3. Precautions. The installation of the cartridge is essentially a reverse operation, and it is important to note that the following additional steps must be carried out:

4.3 Removal and Installation of Normal Pressure Replaceable Tools

A. Before placing the cartridge into the gate plate, the inner step of the gate plate (accessible) must be completely cleaned and the outer surface of the cartridge is completely applied with oil or butter.
B. Before opening the gate plate, fill the small space between the knife barrel and the gate plate through the flushing ball valve on the knife barrel, and the pressure should be consistent with the pressure on the palm surface.
C. During the period of water ingress, a ball valve on the flushing line on the barrel needs to be opened in order to drain the gas.
D. When the space between the knife barrel and the gate plate is filled with water and water overflows from the flushing ball valve, close the flushing line ball valve again.
E. Make pressure compensation before opening the gate plate.
F. When pushing the barrel into the gate valve body, note that the positioning pin on the knife barrel and the positioning hole on the gate valve body must be consistent (note the position of the position of the position of the position pin).

(3) Track 27/28A, B operation steps.

1. Installation of aids (Fig. 4.20). The mounting aids consist of pipe racks, mounting hooks, telescopic cylinders, brackets, hexagonal bolts M24 × 110 mm, hexagonal bolts M24 × 75 mm, hand-dressed hoists, cylinder control valves, pressure gauges and other components.

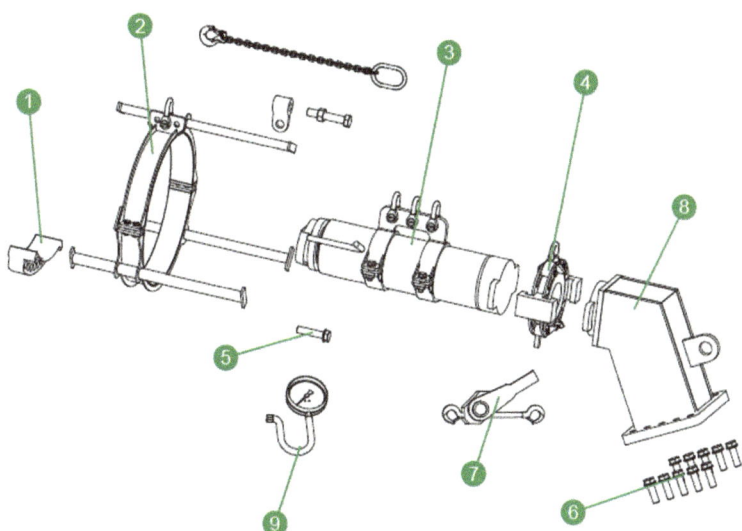

Fig. 4.20 Installation aids required for the trajectory 27/28A, B steps. tuberack; 2—instalationhok;3—retractablecylinder;4—bracket;5-cylindercontrolvalve;6—hexbolt M24 × 75 mm; 7—handdresinghoist;8—presuregauge;9—hexbolt M24 × 10 mm

2. Basic steps:
   A. Control the pressure of the palm surface, connect the water pipe (hydraulic pipe) to the flushing ball valve on the knife barrel, open the ball valve for flushing, close the ball valve after flushing, then remove the pipe, close the wear detection ball valve, and then remove the wear detection line.
   B. Secure the cylinder base to the back of the knife plate, use the knife crane to lift the clip on the cylinder, secure the cylinder holder to the appropriate position of the cylinder body, use the cylinder clip to secure the cylinder base, and lock it with a pin.
   C. Connect the control valve to the cylinder line, extend out the cylinder and lock the locking plate in 1 stop (ensure lock with pin), ensure that after the above step, give a little pressure to the cylinder (cylinder belt pressure), the bolt on the tool tube is loosened.
   D. Fully retract the cylinder (about 695 mm), secure the knife barrel clamp to the knife barrel, hang the buffer hanging chain on the knife barrel clamp, hang a section of the cushioning suspension chain on the barrel fixture on the pneumatic hoist hook, and make the pneumatic hoist force (just good hanging), the knife barrel carrier fixed at the bottom of the knife barrel, at least 3 M27 x x 100 bolts.
   E. Secure the hanging ear to the flange face of the knife barrel, connect the inverted chain to the buffer chain and the hanging ear and tighten, connect the water pipe (hydraulic pipe) to all flushing lines on the gate valve body, open all flushing ball valves for flushing, rinse and close the ball valve, and then remove the water pipe.
   F. Indirectly the control valve and the gate plate line, remove the yellow protective sleeve on the piston rod of the gate plate cylinder, close the gate, and ensure that the gate locks the bolt M24 lock (safe operation).
   G. Open the bottom scouring ball valve to balance the internal and external pressure difference, after balancing the internal and external pressure difference, close the bottom flushing ball valve and lock the cylinder lock plate in 2 stops.
   H. Fully retract the cylinder (92 mm).
   I. Remove the cylinder clips, use a second pneumatic hoist to lift the cylinder steam and temporarily store it in a suitable position.
   J. Remove the cylinder base, pull out the cartridge completely (note that the cartridge will shake or swing when drawn out) and transfer it to the material transport channel.

(4) Precautions. The installation of the cartridge is basically reverse operation, it is important to note that the following additional steps must be carried out in place:

   A. When the barrel is installed into the gate valve body, do not have to hang the lock plate to 2 stops.
   B. Before placing the cartridge into the gate valve body, completely clean the inside (accessible) inside the gate valve body and apply the outer surface of the cartridge completely with oil or butter.
   C. Before opening the gate plate, fill the small space between the knife barrel and the gate plate through the flushing ball valve on the knife barrel, and the pressure should be consistent with the pressure on the palm surface.
   D. During the period of water ingressing, a ball valve on the flushing line on the knife tube must be opened so that the gas is discharged, and when the space between the knife and the gate plate is filled with water and water overflows from the flushing ball valve, close the flushing line ball valve again.
   E. Make pressure compensation before opening the gate plate.
   F. When pushing the barrel into the gate valve body, note that the positioning pin on the knife barrel and the positioning hole on the gate valve body must be consistent (note the position of the position of the position of the position pin).

4.4 Example Swords-Off of Normal Pressure Opening in the Cross-river Section of Wuhan Metro Line 8

4.4.1 Engineering Overview and Geological Conditions

1) Project Overview

Section 3 of the bt project of the civil works section of the first phase of Wuhan Metro Line 8, which mainly consists of one station and one section, namely Xujia Shed station and Huangpu Road Station—Xujiapeng Station Cross-river shield section. The Yuejiang section starts from Xujiapeng Station and crosses the river into Hankou Huang. Pulu Station receives. The administrative division of the station and the originating section is located in Wuchang District of Wuhan City, and the receiving section is located in Jiang'an District of Hankou.

The project runs through the Yangtze River at about 450 m upstream of the Second Yangtze River Bridge. The length of the cross-river tunnel is about 3185.545 m (the design mileage range is right DK9+754.073~ right DK12+939.618, including the left short-chain 1.945), tunnel across the river. The width is about 150 m and the outer diameter of the tunnel is 12.1 m. The project uses a mud-water pressurized shield with a diameter of 12.5 m German Herrenknecht. The layout of the interval tunnel is shown in Fig. 4.21.

2) Engineering geological conditions

The stratum crossing the tunnel in the Yuejiang section is complex and variable, with various geological forms and uneven distribution. The condition of the tunnel crosses the stratiform in Table 4.1 for details. The cross-river tunnel is about 2100 m on both sides of the Yangtze River. The upper part of the stratum is a soft soil layer, and the lower part is a fine sand layer. The tool wear is light. The upper part of the river is a fine sand layer and the lower part is a composite layer such as weathered rock. Mainly include: Q4 round gravel of about 495 m; (15b-1) strong weathered conglomerate of about 1370 m; weakly cemented by about 705 m (15b-2), Conglomerate; about 430 m (15b-3) medium conglomerate, with three rock layers overlapping vertically.

The tunnel crossing the stratum is soft and hard, and the saturated permeable sand layer is closely related to the Yangtze River. The maximum head pressure can reach 58 m. At the same time, according to the experience of Wuhan cross-river tunnel construction, there are many unidentified underground obstacles in the river and on both sides. The long-distance construction of the shield will encounter many different complicated construction environments, which has great construction difficulty and risk.

3) Hydrogeological conditions

According to the conditions of the water medium and groundwater, the groundwater in the area can be divided into upper layer stagnant water, loose rock pore water. There are three types of bedrock fissure water.

The upper layer of stagnant water: It mainly exists in the artificial fill of the two sides of the strait. There is no unified free surface, and it receives the atmospheric precipitation and the vertical infiltration of the leakage of the supply and drainage pipelines. The water volume is limited.

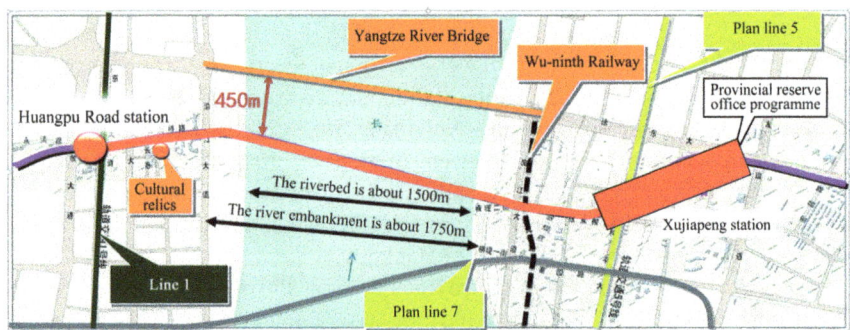

Fig. 4.21 Huangpu Road Station-Xujiapeng Station Section of Wuhan Rail Transit

Table 4.1 List of engineering geological conditions in the cross-river area

Stratum name	Shape	Feature	Thickness
Miscellaneous fill		It consists of gravel, bricks, domestic garbage, etc., and the surface of Wuchang sees a piece of stone, and the structure is dense and dense; the layer is generally distributed along the shore	0.5–4.1
Prime fill	Plastic	The main components are malleable powder clay, powder soil, dense and uneven	0–6
Muddy silty clay	Flow molding	Smelly, local sectionon on shore has distribution, buried depth not more than 4.8 m, high sensitivity, carrying capacity standard value $f_{ak} = 60$ kPa, compression modulus $E_s = 2.8$ MPa	0–3
Silt	Loose	Saturation; the layer is distributed in the surface of the Yangtze River bed	2.3–14.5
Clay	Plastic	Only the Han port has been revealed, for the Yangtze River first-order aboveground hard shell layer, carrying capacity standard value $f_{ak} = 107$ kPa, compression modulus $E_s = 5$ MPa	5.7–10.0
Silty clay	Soft plastic	This layer is mainly distributed on both sides of the Yangtze River, the standard value of carrying capacity $f_{ak} = 104$ kPa, compression modulus $E_s = 4.5$ MPa	~16.50
Silt	Loose ~ Little density	Saturated, local thickness of 5 to 30 cm powderclay, powderth thin layer. The lens is distributed in a body form, with a standard bearing capacity of $f_{ak} = 160$ kPa, compression modulus $E_s = 14$ MPa	0–8.50

(continued)

Table 4.1 (continued)

Stratum name	Shape	Feature	Thickness
Silt clay	Soft plastic	The lens is sporadicly distributed in the (4–2) powdered sand layer, the standard value of the carrying capacity f_{ak} = 105 kPa, the compression modulus E_s = 4 MPa	0.9–6.7
Coarse sand	Medium-density ~ compact	The lens is sporadicly distributed in the (4–2) fine sand layer and at the bottom of the layer, the standard load value f_{ak} = 300 kPa, the deformation modulus E_0 = 25 MPa	
Round gravel	Saturated; medium-density	Pebble content is average about 25%, pebble particle size is generally 2 to 8 cm, up to 15 cm, pebble grinding round medium, composition to microcrystalline gray rock, dolomite and zircon mainly, hard quality; Mainly distributed in the bottom of the riverbed section cover carrying capacity standard value f_{ak} = 350 kPa, deformation modulus E_0 = 28 MPa. Local mezzanine gravel cladding layer, the bond is more solid, looks concrete, thickness 5–10 cm, distribution is not continuous	0–2.9

(continued)

Loose rock pore water: It mainly exists in the Quaternary sand layer, which is the main aquifer of the site. It is closely related to the Yangtze River. The recharge is mainly from the Yangtze River and the water is abundant. Because there are many sand layers in the field. It is lower than the water surface of the Yangtze River, so its pore water is mostly pressure-bearing, and the pressure head is similar to the water level of the Yangtze River.

Bedrock fissure water: mainly occurs in the mid or micro-weathered bedrock fissure, and the recharge mode is mainly the infiltration of the overburden recharge, it has pressure; due to the softness of the bedrock in the field, the bedrock fissures are mostly closed or filled with mud, bedrock fissure water is poor.

4.4 Example Swords-Off of Normal Pressure Opening ...

Table 4.1 (continued)

Stratum name	Shape	Feature	Thickness
Strongly weathered conglomerate	Gravel soil	Contains light gray-green patches, local lycinated cemented rock blocks, can distinguish the structure of gravel-like debris;; More than 2 cm of pebble content is generally more than 20%, up to 60%, the maximum particle size of up to 20 cm or so, the composition of pebbles to gray rock, dolomite-based, hard (pressure strength > 60 MPa)	
Weakly cemented gravel	Gravel structure	Mud pore type or base type weak bondknot, thick layer, crack not developing; 2 cm or more skeleton particle content of about 45%, particle size is more than 2–5 cm, individual up to 20 cm, the composition is mainly gray rock, dolomite and other hard rock, thick particles between the weak connection, rock core easy to break. Carrying capacity standard value $f_{ak} = 1100$ kPa, deformation modulus $E_0 = 46$ MPa	

(continued)

The pore-type confined water in the sand layer of the first-level terraces of the two banks is obviously affected by the river water. During the flood season of the Yangtze River, the river water supply area. When the water is launched, the groundwater is replenished to the river, and the annual variation decreases as the distance from the Yangtze River increases. In January-March, the pressure water level is low. During the dry season (February), the pressure water level is generally 1.6–15.0 m. The pressure level is higher in July–September, and the confined water level in the wet season (August) is generally 20.4–2.8 m. The runoff of groundwater is correspondingly expressed as the flow of groundwater from the Yangtze River to the inside of the terrace during the wet season. On the contrary, groundwater flows from the terrace to the Yangtze River; during the Pingshui period of the Yangtze River, the runoff velocity of groundwater is extremely slow. No mining points in this area. At the point of groundwater exploitation, groundwater is mainly discharged to the Yangtze

Table 4.1 (continued)

Stratum name	Shape	Feature	Thickness
Conglomerate gravel	Gravel structure	Mud calcium pore or base-type bonding, poor bonding, huge thick layer structure, crack not developing, rock body integrity good; 2 cm or more skeleton particle content of about 60%, particle size is more than 2–5 cm, the maximum 10 cm or more, the composition is mainly gray rock, dolomite and other hard rock, rock core more than 15–40 cm column, a small part of 2–5 cm fragmented, soft rock, average natural single-axis pressure strength 14.88 MPa	

River; the annual dry season of the Yangtze River is the main period of groundwater discharge.

The Quaternary pore-bearing aquifer in the riverbed section is closely related to the Yangtze River water, and the groundwater level rises and falls with the rise and fall of the Yangtze River water level.

4.4.2 Shield Basic Parameters and Characteristics

For the cross-river tunnel of Wuhan Metro Line 8, a mud-water pressurized shield with a diameter of 12.5 m Herrenknecht was selected (Fig. 4.22). The basic parameters are shown in Table 4.2. The plate cutter head is selected to have a diameter of 12.5 m and an opening ratio of 28.5%. A total of 243 knives are arranged on the cutter head (Fig. 4.23), and a total of 76 tools are replaced by atmospheric pressure. Among them, 15 can be replaced with normal pressure, 43 can be replaced with normal pressure, and 8 can be replaced with normal pressure. 10 removable center knives, 6 sets of peripheral shovel (12), fixed scraper knives 123 Put and fix the first knives 20.

In order to give full play to the different cutting effects of various types of tools and prevent excessive wear, a height difference is set between various types of tools. The first knife acts to scrape the soil in advance and protect other tools. Therefore, the protrusion height of the first knife is higher than that of other tools. The replaceable

4.4 Example Swords-Off of Normal Pressure Opening …

Fig. 4.22 ""Chutian" shield machine

Table 4.2 Basic parameters of the "Chutianhao" shield machine

Item head	Shield equipment parameters	Need of project
Minimum turning radius (m)	650	700
Maximum withstand pressure (bar)	8	6.74
Maximum slope (%)	5	2.7
Excavation diameter (mm)	12,540	12,540
Maximum tunneling speed (mm/min)	60	60
Maximum thrust (kN)	156,753@350 bar	140,000
Maximum torque (kN m)	18,373–24,620	17,194–20,312
Maximum inflow and discharge flow rate (m^3/h)	2000/2500	2000/2500
Rear matching arrangement	Can be installed box culvert	Can be installed box culvert

Fig. 4.23 Shield cutter tool

advance knife protrudes 25 mm from the cutter head, and the fixed advance cutter is higher than the cutter head. 205 mm, because it can be replaced under normal pressure, allowing the wear value to be greater than the fixed advance cutter, so the replaceable leading knife is more protruding than the fixed forepass.

The hob protrudes 25 mm from the cutter head and is flush with the replaceable advance cutter. Since the hob destroys the hard rock and soil by the pressing force and friction between the hob blade and the rock and soil body, if the hob setting height is too small, firstly, The hard rock mass will be cut by the replaceable cutting knife first, which will cause a large amount of wear of the replaceable cutting knife, which will affect the smooth tunneling of the shield. Secondly, the rock mass that has been scraped cannot give the hob sufficient friction, resulting in the hob, causing the roller to produce a rolling knife bias phenomenon without turning. The scraper mainly plays the role of scraping the scraped soil into the earthen tank, so the setting height is the lowest, which is 185 mm.

In order to ensure the correct number of tools on the cutter head and to make full use of the independent cutting action of the tool, different tool spacings are set between the various types of tools: the distance between the reels is 10 mm, the spacing between the knives is 195 mm, the first knife The spacing between the two is 20 mm.

4.4.3 Shutdown Position and Knife Change Plan

1) Geological conditions at the stop position

In the cross-river tunnel project of Wuhan Metro Line 8, the shield was opened due to the severe wear of the tool which is uncommon. The position of the shield stop tool change is mainly divided into two types: The first type is a fine sand stratum, and the excess cutters, tooth cutters and scrapers on the cutter head are replaced; the second type is conglomerate area, the cutter head replaced the excess knives, knives, scrapers and hobs with excessive wear.

The shield passes through the upper soft-hard composite stratum over long distances (the upper part is a full-fine fine sand stratum, the middle is 1365 m strong weathered conglomerate layer, 750 m weakly cemented conglomerate and 430 m medium cemented conglomerate, as shown in Figs. 4.24, 4.25, and Table 4.3), Faced with problems such as slow tunneling speed, offset direction, and serious cutter wear.

2) Tool change plan

According to the engineering geological conditions and tool configuration, drawing on the experience of similar engineering construction, it is expected that a total of 8 times of tool inspection and tool change work, 6 times of atmospheric pressure river bottom tool change in the river's composite stratum, as shown in Fig. 4.26. The first tool inspection and replacement before crossing the river. The second is

4.4 Example Swords-Off of Normal Pressure Opening … 187

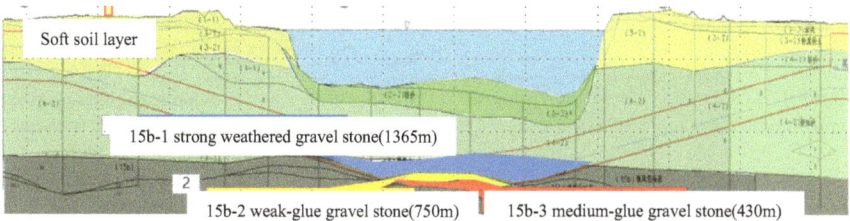

Soft soil layer

15b-1 strong weathered gravel stone(1365m)

15b-2 weak-glue gravel stone(750m)   15b-3 medium-glue gravel stone(430m)

Fig. 4.24 Schematic diagram of the upper soft and hard composite stratum crossing the shield

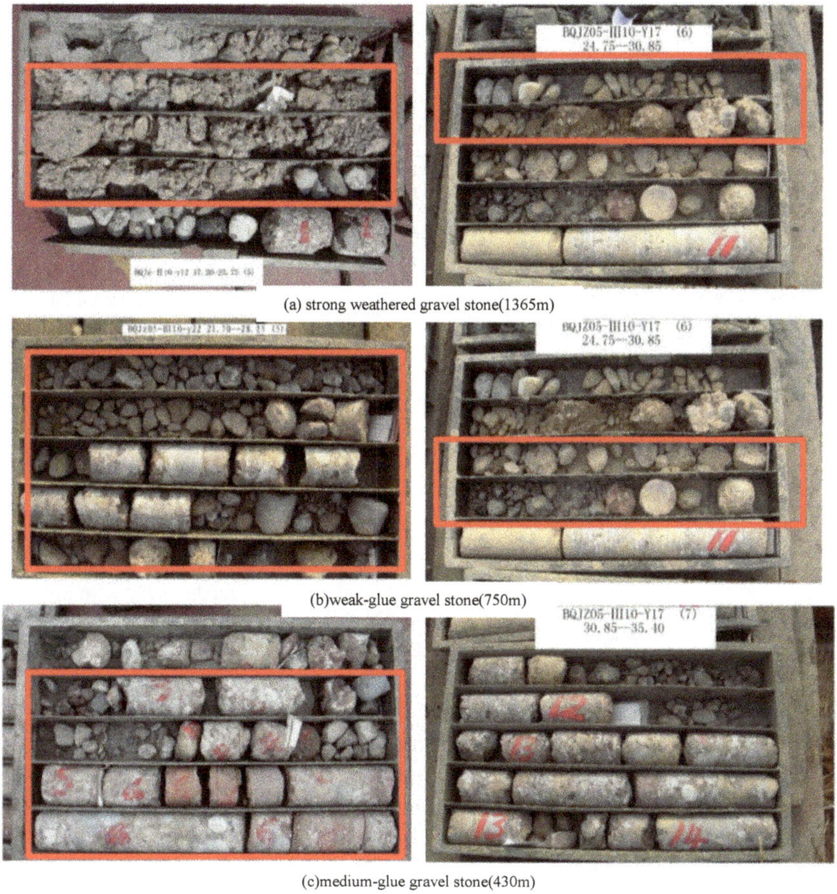

(a) strong weathered gravel stone(1365m)

(b) weak-glue gravel stone(750m)

(c) medium-glue gravel stone(430m)

Fig. 4.25 Geological survey map of the upper soft and hard composite stratum

Table 4.3 Geological conditions of the upper soft and hard composite stratum

Stratum name	color	Shape state	Feature description
Strongly weathered conglomerate	Purple red, grayish yellow	Gravel soil	Containing light gray-green patches, partially weakened cement blocks, and discriminating gravel-like debris structures; The content of pebbles above 2 cm is generally more than 20%, up to 60%, The particle size can reach about 20 cm, the composition of the gravel is mainly limestone and dolomite, and the hardness is hard (compressive strength > 60 MPa)
Weakly cemented conglomerate	Gray, maroon	Gravel structure	Mud pore type or base type weak bondknot, thick layer, crack not developing; 2 cm or more skeleton particle content of about 45%, particle size is more than 2 to 5 cm, individual up to 20 cm, the composition is mainly gray rock, dolomite and other hard rock, thick particles between the weak connection, rock core easy to break. Carrying capacity standard value $f_{ak} = 1100$ kPa, deformation modulus $E_0 = 46$ MPa

(continued)

the inspection and replacement of the tool before the river, the second is the hob replacement before the cemented conglomerate, the third to the sixth is the tool inspection and replacement for the soft and hard formation, and the seventh is the cementation. The hob after the conglomerate is replaced with a tooth cutter, and the eighth time is the tool check replacement before the arrival of the shore. The tool change is replaced with a normal pressure tool.

4.4 Example Swords-Off of Normal Pressure Opening …

Table 4.3 (continued)

Stratum name	color	Shape state	Feature description
Medium cemented conglomerate	gray, grayish purple, And maroon	Gravel structure	Mud calcium pore or base-type bonding, poor bonding, huge thick layer structure, crack not developing, rock body integrity good; 2 cm or more skeleton particle content of about 60%, particle size is more than 2 to 5 cm, the maximum 10 cm or more, the composition is mainly gray rock, dolomite and other hard rock, rock core more than 15 to 40 cm column, a small part of 2 to a few 5 cm fragmented, soft rock, average natural single-axis pressure strength 14.88 MPa; In case of water easy softening, the rock softening coefficient is about 0.70. After alternating wet and dry, the core is disintegrated

Fig. 4.26 Schematic diagram of tool change and tool inspection position

4.4.4 Atmospheric Pressure Changing Process and Program

4.4.4.1 Atmospheric Pressure Change Process

The basic steps of normal pressure tool change can be summarized as follows: cleaning preparation → cylinder lifting → loosening the barrel bolt → cylinder retraction, knife lifting → flushing → closing the gate → balancing the internal and external pressure difference → cylinder retraction → removing the cylinder and

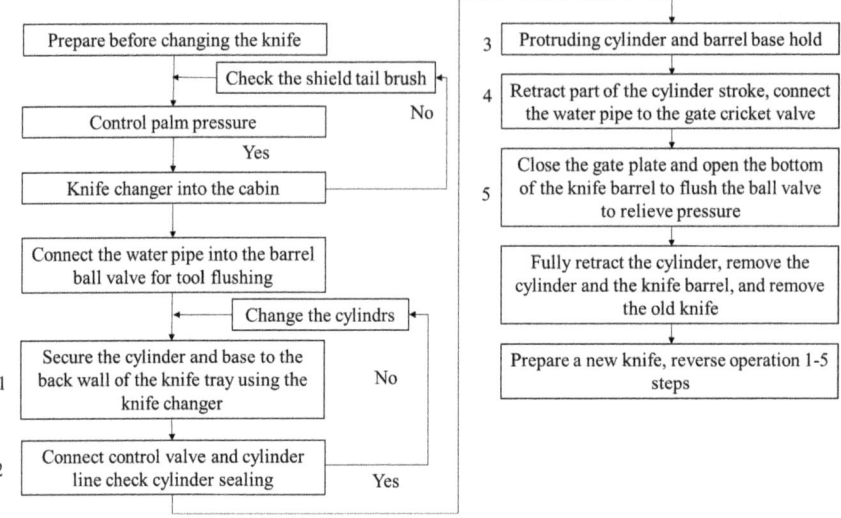

Fig. 4.27 Atmospheric pressure tool change flow chart

temporary storage → The knife tube is completely extracted. The installation operation of the knife barrel is basically the reverse operation of the removal operation. The normal pressure tool change process is shown in Fig. 4.27.

4.4.4.2 Preparations Before Changing the Knife

(1) According to the tool change inspection plan, the technology and safety of the replacement tool are prepared and submitted to the site, the operator, make sure that content is comprehensive and accurate.
(2) Check the replacement tool before doing a good job of synchronous slurry tube blocking, and carry out 1–2 h mud water large cycle, to mud pipe in and out of the mud parameters consistent, mud station separation equipment does not slag as the standard, as much as possible to remove the soil in the soil tank to prevent fine sand and the gravel enters the knife cavity during the tool change, which makes the tool disassembled and difficult to install. After that, the high-concentration bentonite slurry is injected into the excavation chamber to form a stable mud film to ensure the stability of the face.
(3) By the captain of all tool change (pneumatic hoist guide chain, hook and hanging hoist wire rope and other tools) to carry out a comprehensive inspection and record, there is a safety hazard of the implement equipment is strictly prohibited, and timely sent to the ground repair workshop replacement.
(4) The knife is circled to the main arm where the tool needs to be checked perpendicular to the shield horizontal axis, opening the person in the center body of the knife plate to ventilate through the hole, connecting the trachea, water pipes,

hydraulic pipes, lighting cables, etc. Before changing the knife work, first turn the knife to the main arm in the bottom position, open the center cone door, connect the trachea, install the ventilator, replace the air inside the center of the knife plate, reduce the temperature inside the knife change cabin.

4.4.4.3 Atmospheric Pressure Change Program

1) Normal pressure hob scheme

   (1) The line connects the flushing wear tool and lifts the cylinder, as shown in Fig. 4.28a.
   (2) The hydraulic line connecting the cartridge bolts is loosened, as shown in Fig. 4.28b.
   (3) Cylinder recovery, drag cartridge recovery, as shown in Fig. 4.28c.
   (4) Close the gate, as shown in Fig. 4.28d.
   (5) Release the pressure, balance the internal and external pressure difference, lock the cylinder lock plate, as shown in Fig. 4.28e.
   (6) The cylinder sleeve is fully retracted and fixed, as shown in Fig. 4.28f.
   (7) Remove the cylinder, as shown in Fig. 4.28g.
   (8) Sleeve stoist and reverse replacement, as shown in Fig. 4.28h.

2) Atmospheric pressure changing knife program

   (1) Open the ball valve on the cover to release (the ball valve on the housing must be closed), as shown in Fig. 4.29a.
   (2) Remove the bolts and washers from the rear cover of the knife box and remove the cover as shown in Fig. 4.29b.

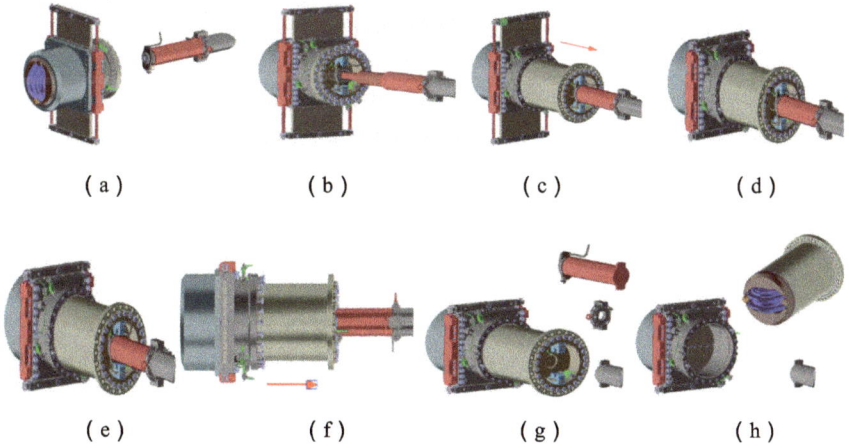

Fig. 4.28 Schematic diagram of the atmospheric pressure changing hob

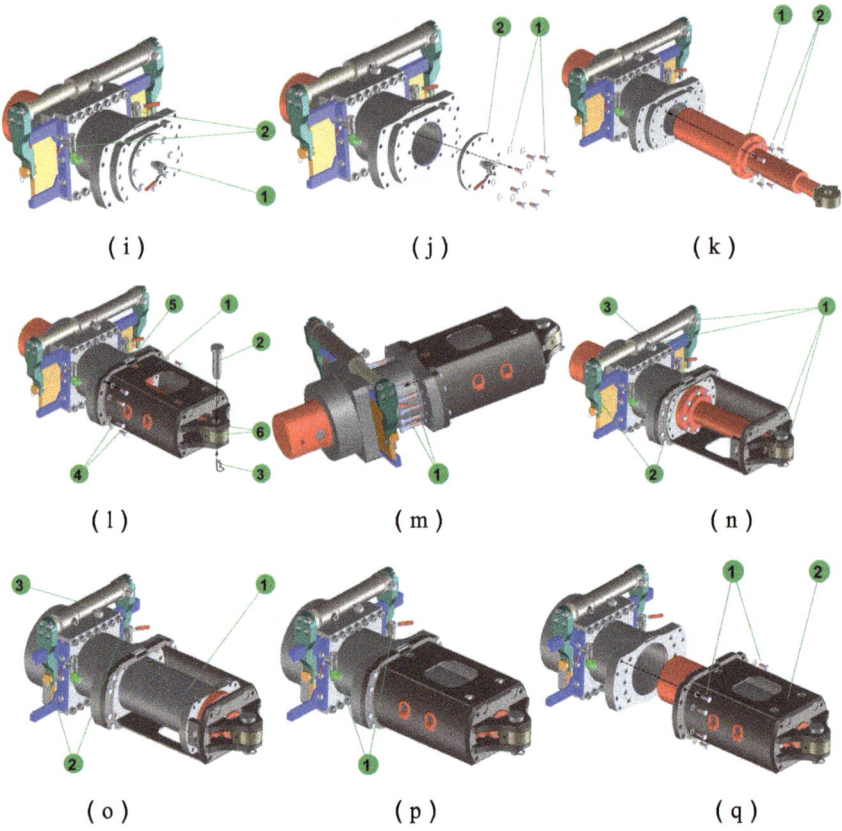

Fig. 4.29 Schematic diagram of the atmospheric pressure changing knife. 1, 2, 20—balvalve; 3—coverplate; 4—bolt, washer; 5—dragcylinder; 6, 12, 13, 21—bolt; 7—mechanicalcode; 8, 2—tolchanger; 9—plug; 10—hydraulic pipe; 1—spring pin; 14, 17—hydraulic cylinder; 15—hydraulic pipe; 16—gate clamp; 18—shell; 19—slider

(3) Place the drag cylinder into the knife case and tighten the bolts, as shown in Fig. 4.29c.

(4) Secure the position of the tool replacement equipment, observe the mechanical coding, insert the pin and secure it with a spring pin, screw it into the bolt and tighten it (to the specified torque); and connect the hydraulic tube, as shown in Fig. 4.29d.

(5) Release the bolt and remove the tool, as shown in Fig. 4.29e.

(6) Connect the hydraulic pipe and install the gate clamps and the hydraulic cylinder, as shown in Fig. 4.29f.

(7) Retract the telescopic cylinder (shell and tool retract inside); retract the slider and hydraulic cylinder shrivelled, as shown in Fig. 4.29g.

(8) When the gate is closed, the pressure compensation is made using the ball valve (remove the pipe, observe whether there is pressure and mud

4.4 Example Swords-Off of Normal Pressure Opening … 193

flowing out of the knife cavity, if there is a statement that the gate is not completely closed, then safely pull out the tool); When the pressure is compensated, close the ball valve, as shown in Fig. 4.29h

(9) Release the bolt, remove the tool with the tool to replace the equipment, the tool is drawn out to observe the wear of the head, replace the head, as shown in Fig. 4.29i.

(10) The installer of the new tool is basically the opposite of the removal procedure.

The atmospheric pressure changing scene is shown in Figs. 4.30–4.35.

Fig. 4.30 Installing a multi-stage cylinder

Fig. 4.31 Installing the limit tube

Fig. 4.32 Shrink cylinder closing gate

Fig. 4.33 Shrink cylinder extraction tool

Fig. 4.34 Retaining the tool together with the limit cylinder fixing bolt

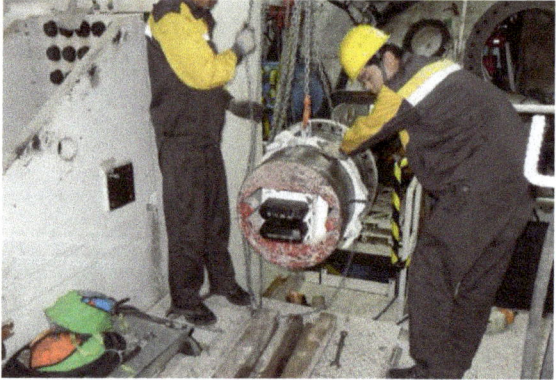

Fig. 4.35 Connecting the tool

4.4.4.4 Precautions in the Process of Changing the Knife

(1) Before the new tool is loaded into the tool holder, the seal ring must be checked by the person in charge of the change of knife, the tool model and the installation direction of the head is correct, confirmed correctly after the corresponding cylinder on the application of butter before loading.

(2) Move the replaced head out of the center cone in time to prevent the head from stacking in the center cone and affecting other tool change work.

(3) For the replaced tool, first clean up, clean up after the label, measure the wear value, record and take a picture archive.

(4) The pressure of the manual oil pump should not exceed 20 MPa to prevent damage to the gate or head. Although some knives are equipped with sensors, when these knives wear to the extent of the design, will automatically alarm prompt, but due to geological unevenness, still need to dig into a certain distance to check the tool once, in order to prevent local wear of the knife plate, damage the safety of the knife plate.

Check the tool can also choose to operate with pressure, but because the pressure operator must have pressure operation qualification and familiar with the tool change process, in addition to the pressure operation has a certain time limit, so often choose the normal pressure replacement tool.

The frequent pressure change knife has a certain risk, the knife changer must work in strict accordance with the operating process, to avoid the security risks. Before the start of the knife.

Chapter 5
Stabilization Technology and Tool Change Technology for Open Face with Pressure Open

In the shield construction method, the pressure of the compressed gas is used to replace the mud pressure or the mud pressure in the pressure chamber, and the air pressure is used to support the excavation surface according to the principle of the gas pressure construction. In the case of encountering highly permeable ground such as sand, it is not good to maintain the stability of the air pressure. In this case, it is necessary to carry out the auxiliary construction of the formation to increase the gas-closing performance. The method of the operator entering the pressure chamber to repair the tool under such a certain pressure environment is a conventional compressed air method with a pressure opening and opening tool change technique.

Compared with the normal pressure open-cavity tool change technology, the advantages of the conventional compressed air method with pressure-opening tool change technology are: The technology has a wide application range, saves the construction period, is more economical, and has less impact on the surrounding environment; the preparatory work with the pressure opening and opening technology method is less, more convenient and flexible to use; the pressure opening and opening technology is highly versatile.

Although conventional compressed air pressure opening technology has many of the advantages mentioned above compared to the atmospheric pressure opening method, many problems are still encountered during the opening process, and the existence of these problems also limits the pressure opening method. For example, the location of the forced stop of the shield is often in the soil with deep depth, complicated hydrogeological conditions or the bottom of the river, etc. Under the action of pressure and water pressure, the construction measures and improper construction management during the opening of the cabin often cause accidents such as collapse of the excavation surface and groundwater breakdown, which may cause personal safety and even the entire tunnel to be scrapped. At this time, you need to consider the pressure, stability problem of the excavation face when opening the cabin. On the issue of personal safety involving the personnel entering the cabin, there may also be improperly, when using the compressed air to maintain the stability of the

Fig. 5.1 Living and living cabin control room

excavation surface, the workers in the tunnel will have symptoms of decompression sickness, such as eardrum rupture, hearing impairment, nitrogen poisoning, hypertrophy, decompression sickness and the like. Incoming operator when using the conventional compressed air belt opening method. Before entering the cabin and before leaving the cabin, it is necessary to adapt to the pressure in the cabin and the external pressure. It is necessary to carry out the pressurization and decompression process in the bubble chamber. This process takes a long time, occupies the effective working time of the operator in the cabin, and the frequent increase and decompression process is unfavorable to the health of the operator.

How to ensure the stability of the excavation face when opening the cabin in various ground conditions, to ensure the safety of the passengers entering the cabin, and to improve the safety Industry efficiency has become a major factor limiting the use of pressure-opening methods. The saturation method of pressure-opening and tool change has better solved this problem. The so-called saturation method of pressure-opening and tool change refers to the saturation of the body of the cabin personnel, that is, through balancing the gas and the pressure to balance the pressure in the cabin crew with the ambient pressure. The hoisting personnel experience a pressurization process and live in a saturated life in a set pressure for a long time (15–30d) (Fig. 5.1). Every day, take the shuttle cabin and transport it to the shield mud tank for 4–6 h. After the day's work, return to the living cabin by the shuttle cabin. After the task is completed, the entry personnel perform tool inspection, repair, replacement, etc. in the state of saturated body. As a person, a person is a uniformly pressurized carrier. In this process, as long as the body's constant oxygen partial pressure is controlled, it will not cause various symptoms.

In addition to the advantages of conventional compressed air pressure opening and opening technology, the saturated method of pressure-opening and tool change avoidance is also avoided. The regulation of compressed air is required to pressurize and decompress each time, which makes the working time in the mud tank greatly prolonged, greatly improving the working efficiency of the operators in the cabin. Especially in rock formations, the tool wear is severe, frequent, and maintenance. When the workload is large and the pressure operation lasts for a long time, the saturation method with pressure opening and opening can greatly reduce the operation time of the shutdown maintenance tool, speed up the project progress, and the economic and social benefits are obvious. In addition, work with conventional compressed air belt pressure opening method the person breathes compressed air, and the compressed oxygen-enhanced mixture for the operator's breathing when using the saturation method to open the cabin. Gas (replacement of nitrogen in the air with helium) not only reduces the breathing resistance of the workers, but also avoids the occurrence of nitrogen anesthesia, avoids the defect that the decompression must be decompressed per shift operation, and greatly reduces the occurrence of decompression sickness. Saturating method is often used in the case of large opening work and large opening pressure. It is a high efficiency and high safety method of pressure opening. The gas required for the saturation method is mainly composed of three types: (1) 15% of helium–oxygen mixed gas, and the gas components mainly include 5% oxygen, 75% helium and 20% nitrogen are the main types of gas used in the saturation process. The main gas that the person breathes during the whole process of opening; (2) 10% of the helium–oxygen mixed gas, the gas composition mainly includes 10% Includes 10% oxygen, 55% helium and 35% nitrogen; (3) Medical oxygen, 10% helium oxygen mixture and medical oxygen are mainly used to regulate the oxygen content of the living and shuttle compartments and the treatment process of decompression sickness.

5.1 Pressure Opening and Opening Face Stability Technology

5.1.1 Mud Preparation Technology

1) Use of mud

The mud retaining wall forming technology was firstly applied to oil drilling engineering. The application in underground continuous wall, bored pile, non-excavation horizontal directional drilling and mud water pressure shield engineering also originated from oil drilling engineering. Mud plays an important role in oil drilling projects. Drilling works play an important role, and people are likened to the blood of drilling. It can be seen that mud plays a decisive role in the construction of such projects. Drilling works play an important role, and people are likened to the blood of drilling. It can be seen that mud plays a decisive role in the construction of such projects.

During the construction of the mud-water shield, mud is formed on the excavation surface by mud penetration (Fig. 5.2). Because the permeability coefficient of the mud membrane is small, it not only reduces the penetration of water in the formation to the excavation surface or the mud into the formation.

Fig. 5.2 Mud film formation at the excavation

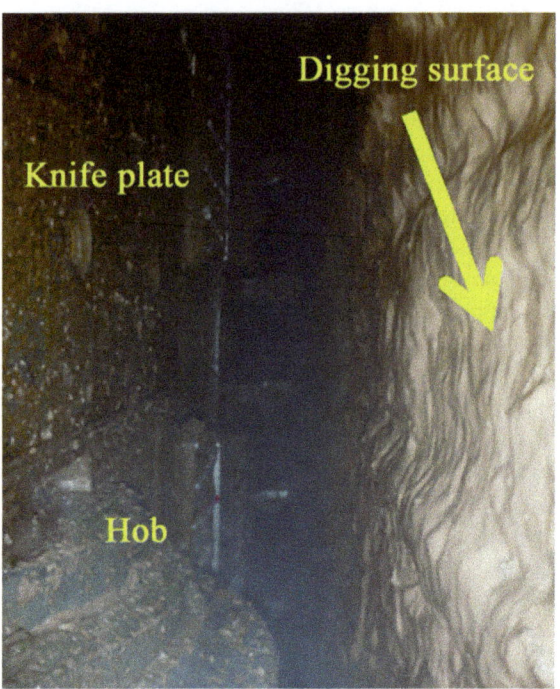

The fluid loss prevents excessive pressure fluctuations. At the same time, the mud membrane is used as a conversion medium to convert the mud pressure into effective stress during normal excavation to balance the soil stress at the front end of the excavation face, thereby maintaining the stability of the excavation surface. When it is necessary to use the pressure opening method to carry out the inspection and cutting of the cutting tool, the mud film plays two roles: one is to maintain the stability of the air pressure in the working chamber; the other is to convert the air pressure into effective pressure to open Excavate the surface to maintain the excavation surface The stability creates safe and stable construction conditions for the entry personnel.

It is another important role of mud to suspend the underground soil and carry it out. Will be broken or cut down the formation soil is discharged to the ground in time to maintain the construction operation space at the excavation surface, which requires the mud to have better fluidity, so that the fractured or cut formation soil still has good fluidity after mixing with it. Using the mud pump through the pipeline. Discharge it to the ground; secondly, it also requires mud to have better fluid properties such as viscosity to ensure formation soil or cuttings, etc. It is suspended in the mud to prevent the danger of a large amount of deposition, blockage of holes or grit, and also reduce the wear of the pipe due to particle deposition during transportation.

In addition, the mud also has the function of cooling, lubricating the cutting equipment and reducing the wear of the pipeline. Due to excavation equipment, etc. during construction.

Long-term work heat is not conducive to the effective service life of the equipment. The mud can effectively cool the cutter cutters and the like in the mud tank, and the mud cake or mud film formed on the excavation surface has the characteristics of slipperiness. Helps reduce tool cutters and ground. The friction between the layers extends the life of the tool cutter.

2) Mud materials and types

The mud shield mud material contains clay, bentonite, sodium carboxymethyl cellulose (CMC) and other additives.

Clay is one or more water-aluminosilicate minerals that are loose or colloidal, and are a mixture of various fine minerals and impurities. Compound. The clay in nature is a mixture of various hydrous aluminosilicate minerals. The appearance is mostly earthy or dense. It is white, yellow, red, black, gray and other colors. The particles are fine, most of which are not more than 2 μm, the crystal is in the form of a sheet, a tube, a sphere, a hexagonal scale, and the like. The main body of clay is composed of various types of clay minerals, usually containing impurity minerals, such as unweathered rock fragments. Chips, quartz sand, feldspar, mica, calcite, pyrite, organic impurities, etc. The natural clay is formed by the weathering, hydrolysis and hydrothermal alteration of rocks rich in aluminosilicate minerals such as feldspar and pegmatite. According to the similarities and differences of clay mineral content, clay is divided into three types: kaolinite, montmorillonite and illite.

Bentonite is a layered aluminosilicate clay mineral with montmorillonite as its main component. Its unit cell is composed of two layers of siloxane tetrahedron. A layer of aluminum oxide octahedron is added, and the oxygen atoms are shared between the tetrahedron and the octahedron to form a highly ordered quasi-two-dimensional sheet with a large specific surface area, and the silicon-aluminum ions therein can be magnesium, iron, Substituting low-priced ions such as lithium, the electricity price between the bentonite layers is not saturated, resulting in a permanent negative charge between the layers. This negative charge is usually balanced by hydration cations that are exchangeable between the layers. Therefore, bentonite has a good series of ion exchange, water absorption, swelling, adhesion, and adsorption. The nature of value. Using its excellent ion exchange performance, the natural calcium-based swelling is subjected to sodium treatment, and its hydration performance is improved, and it can be used as a material for drilling mud and mud-water shield mud.

CMC, fullname sodium methyl cellulose, is the largest production, the most widely used and most convenient products in the cellulose ether category, commonly known as "industrial MSG". CMC is an important cellulose ether, is a kind of water-soluble good anion cellulose compound obtained by natural fiber after chemical modification, which is easily soluble in hot and cold water. CMC is a macromolecule chemical that can absorb water to expand, when dissolved in water, can form a transparent viscous adhesive, in acidity and alkalinity of the performance of neutral. Solid CMC is stable for light and room temperature and can be preserved for a long time in a dry environment. CMC is white or yellow powder, granular or fibrous solid, odorless, non-toxic. The important characteristics of CMC are the formation of high viscosity colloids, solutions, there are sticky, thickening, flow, water conservation, protection colloid, film molding, acid resistance, salt resistance, suspension and other characteristics, physiologically harmless. It has been widely used in the production of construction.

The sodium salt of cellulose methyl ether (Fig. 5.3) is an anion-type cellulose ether, which is white or milky fiber-like powder or particles, 0.5–0.7 g/cm^3, almost odorless, and hygroscopic. It is easily dispersed in water into a transparent gel-like solution, insoluble in organic solins such as ethanol. 1% aqueous solution pH is 6.5–8.5, when pH 10 or pH 5, the viscosity of the glue is significantly reduced, the best performance at pH 7. For thermal stability, the viscosity of less than 20 °C rose rapidly, the change was slow at 45 °C, and the long time heating above 80 °C, which reduced the viscosity and performance of its colloidal denaturation, soluble in water and transparent solution. In the alkaline solution is very stable, acid is easy to hydrolysis, pH of 2–3 when precipitation will occur, in the case of polyvalent metal salts will also react to precipitation. Mainly used in water-based drilling fluid, mud shield mud adhesive, has a certain amount of filter loss, especially salt resistance, temperature resistance is strong.

In order to increase the stability of the mud, sodium carbonate will also be added to the mud, playing a scattered, uniform role. The principle can be called salt immersion principle, the silica aluminum magnesium calcium in the montiline can be chemically reacted with alkaline substances, alkali is designed to destroy the lattice structure of

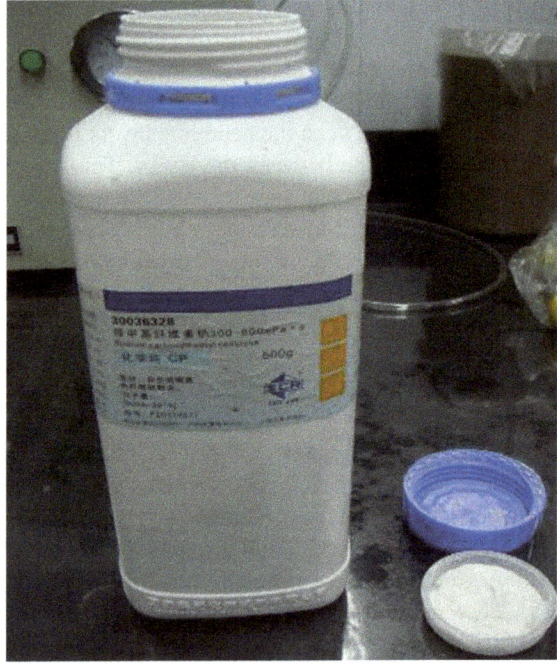

Fig. 5.3 Sodium methyl cellulose

Fig. 5.4 Na$_2$CO$_3$

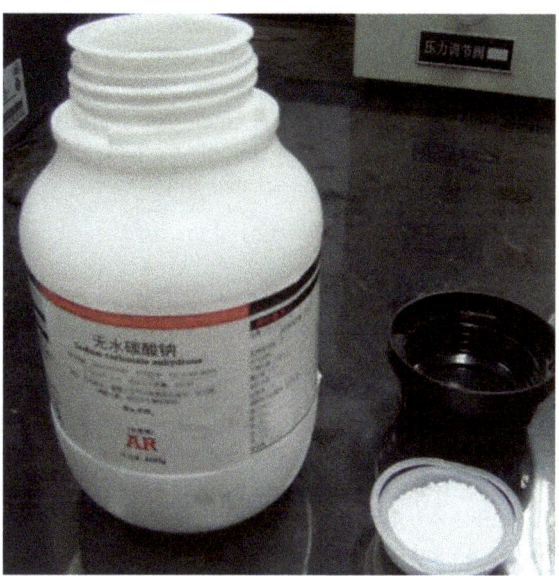

the monoshid, so that the main component of the amorphous state of the monoshidate and NaOH or Na2CO3 solution is the polysilicone sodium gel and sodium alumina. Because the monosal stone contains the gender substance Al$_2$O$_3$, the lattice structure is often destroyed when the Al$_2$O$_3$ dissolution reaches 75–85% when the melting state of NaOH or Na$_2$CO$_3$ (Fig. 5.4) reacts at high temperature conditions and destroys the cell Coupled with the solubility of other ions, the amorphous silicon in the lattice structure is converted into active silicon, thus achieving the function of activation.

Sodium carbonate additives can electrolysis a certain amount of Na$^+$ plus and polymerized anions. The ion exchange between the Ca^{2+} plus and the clay surface increases the thickness of the diffusion layer and increases the potential of the clay particles due to the intervention of the sodium-hydrated ions. In addition, polymerized anion can be partially adsorbed by the positively charged clay particles end surface, so that the end-face charge modification, breaking the side of the clay particles combined to form a card structure, releasing part of the free water, so can play a certain dilution, dispersion role.

Different formation composition, the permeability of the formation is not the same, the corresponding mud shield used mud is also different. In general, mud used in mud shields is generally used in the following types: pure bentonite mud, pure clay mud, mixed mud, polymer mud (added methylphenidate cellulose, sodium carbonate, adhesive, etc.). Each mud can be further subdivided according to its composition. Pure bentonite mud can be divided into sodium-based bentonite and calcium-based bentonite according to the different composition of bentonite. The pulp rate of sodium-based bentonite is higher than that of calcium-based bentonite. Pure bentonite mud and pure clay mud are suitable for the formations with smaller permeable coefficients such as clay layer and fine sand layer. Mixed mud according to the mixing material, can be divided into: bentonite and clay mixed mud, bentonite and slurry mixed mud, clay and slurry mixed mud. This type of mud is suitable for areas with large penetration coefficients, such as coarse sand and gravel layer. Polymer mud is mainly made up of several kinds of product composite products, the structure is more complex, suitable for leak blocking, reinforcement, with pressure open cabin film and other special conditions.

3) Basic physical properties of the mud

According to the composition of the formation, the used mud is determined and the mud is formulated. The mud is formulated according to the basic nature of the mud. The basic properties of mud include the density, viscosity, particle level, and physical stability and colloidality of the mud.

The mud density characterizes the content of solid-phase particles in the mud, which has an important influence on the penetration of mud in the formation, is one of the most important nature parameters of mud, and is also an important index of mud management in mud shield construction. The instrument commonly used to measure mud density is an easy-to-operate mud density scale, which is tested as a ratio of mud density to a pure water density of 4 °C, and the reading on the scale is the mud density. The density scale used is a 1002 type mud density scale, shown in Fig. 5.5, consisting of a mud cup, beam, swimming weight and bracket, with a regulation tube and horizontal bubbles on the beam. Mud density also has other methods of determination, such as weight tolerance, gamma-ray attenuation and so on.

The mud viscosity is an indicator of the size of the viscosity of the mud, which reflects the rheological properties of the mud, and is also an important index of mud management in the construction of mud and water shield. The greater the viscosity of the mud, the better the physical stability, the mud is not easy to separate. The smaller the viscosity of the mud, the worse the physical stability, the more easy the mud to produce separation. The combination force of particles and water in the mud is weak, although the grain silt in the mud blocks the pores of the formation, but the water in the mud is easily pressed out under the influence of mud pressure, leaving the mud in the grain siltation surface, which makes the formation of mud film is

Fig. 5.5 Type 1002 mud density scale

thick but loose, is not conducive to the stability of the excavation surface. The mud viscosity commonly used in the project is basically funnel viscosity, funnel viscosity instrument is cheap, compact, easy to carry, and easy to operate, easy to operate in the field, its measurement is the viscosity of the mud in the low flow rate state. The funnel viscometer is divided into a Su-style funnel viscosity meter (Fig. 5.6) and a Mars funnel viscosity meter (Fig. 5.7). Take the Sioux funnel viscosity meter as an example (the Mars funnel viscosity meter is used in a similar way). The funnel viscosity determination method is to hold the funnel vertically, hold it tightly and block the tube mouth with its index finger, and then use the two ends of the barrel to pour 200 and 500 mL mud into the funnel. Place the 500 mL section upward under the funnel, release the index finger, and start the stopwatch clock to record the time required to flow 500 mL mud, i.e. the viscosity of the measured mud.

The grade matching of mud reflects the size and distribution range of solid-phase particles in the mud, and is an important parameter that affects the penetration of mud in the formation. The mud grade is different from the same formation, and the permeation form of the mud is different. The requirements for mud performance in the construction of mud shield also include the grade of particles in the mud to ensure that the mud can form a stable and dense mud film on the excavation surface. Traditional soil sieves are mainly suitable for testing the grade of sand (-75 m) and cannot be determined for grade settings with a particle size of less than 75 m. Because the mud contains a lot of fine particles, so for the mud grade, the current commonly used experimental instrument is the laser granularizer. The particle-level distribution curve of the mud is measured using the MS2000 laser particle size analyzer (Fig. 5.8).

Fig. 5.6 Su-funnel viscometer

Fig. 5.7 Markov funnel viscometer

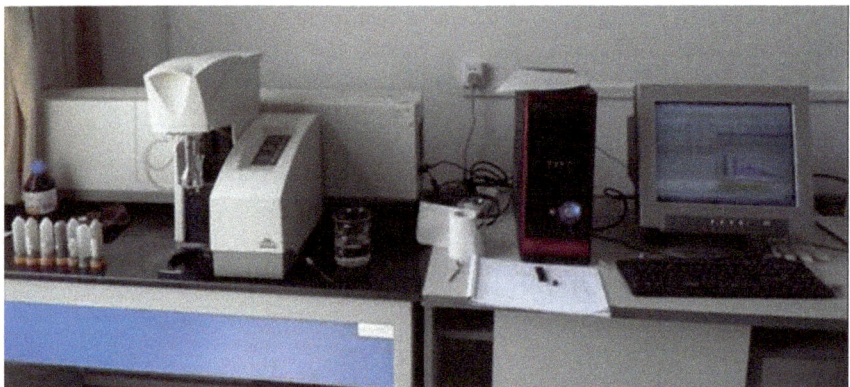

Fig. 5.8 MS200 laser particle size analyzer

The mud colloidal rate is a rough indicator of the degree of dispersion of clay particles in mud, characterized by the hydration capacity of mineral components in the mud, mainly the absorption of mucosa, glue particles and other particles, molecular key elongation, the combination of water film thickening around the mucous grain process, macro-performance of the particles absorb water, scattered into a more uniform colloid. The higher the colloidal rate of mud, its physical properties are more stable, mud is not easy to occur separation, water analysis and other phenomena, the formation of mud is more uniform. Therefore, colloidal rate is the basic index of the physical stability of the mud. Test method: 1000 mL mud poured into the 1000 mL barrel, covered with glass static 2 h or 24 h, observation of the volume of the upper clarification liquid and the precipitation volume of the lower sand particles, if the upper clarification rate of 3%, the lower precipitation rate of 2%, then the mud colloid rate of 95%.

Physical stability refers to the ability of solid-phase particles in mud to maintain the physical state of floating dispersion, usually described by interface height. A certain amount of mud water static placed in the measuring tube for a period of time, some soil particles will lose the suspension characteristics of precipitation, mud water surface appeared clear water, the bottom of the soil particles, the middle is still mud water, water and mud water appear edgy interface. By observing the boundary height of water and mud water, the sedimentation degree of soil particles in mud water can be identified. The physical stability of mud is measured by the usual mud colloidal rate in engineering. As shown in Fig. 5.9, pour 1000 mL mud into the 1000 mL cylinder, set aside 24 h, observe the volume of the liquid in the upper part of the measuring tube. If the clarification liquid is 50 mL, the mud colloidal rate is 95%. The mud colloid rate should generally be greater than 90%.

In addition, the basic properties of the mud also include the sand content of the mud, water loss and pH.

The sand content of mud refers to the volume percentage of sand in the mud that cannot pass the 200 screen (equivalent to the diameter greater than 0.075 mm), which reflects the amount of coarse particles contained in the mud, and also reflects the grinding process level of bentonite ore.

In the field of shield mud, the amount of mud loss is used from the filter loss of drilling mud. Drilling filter loss can be divided into low temperature low-pressure

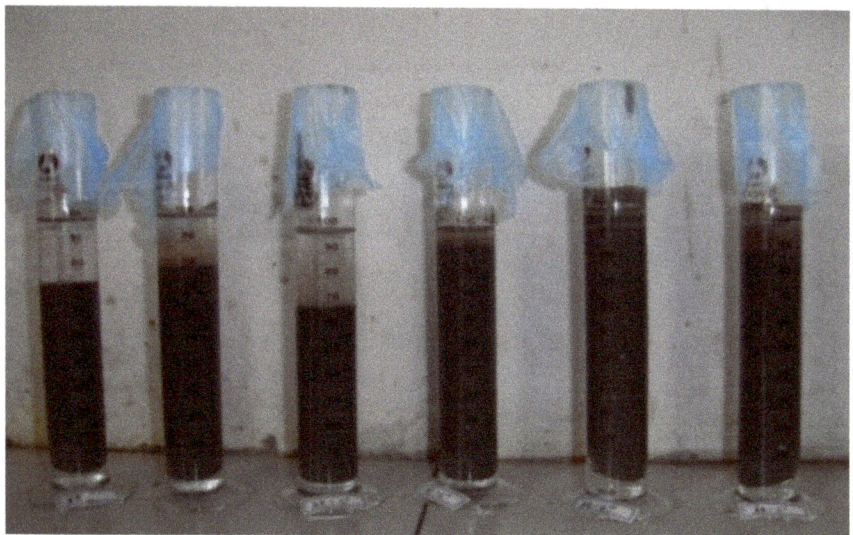

Fig. 5.9 Mud colloid rate test

filter loss (API filter loss) and high temperature and high-pressure filter loss, usually used API filter loss at home and abroad to evaluate the filter loss of drilling mud. Because API filter loss is measured under the same pressure and the same osmosis medium, and the formation is variable and the mud pressure is different, the same mud appears different types of osmosis in different formations, forming different forms of particle accumulation. Therefore, the API filter loss of mud only reflects the quality of the mud itself and cannot be used as the standard for evaluating the permeation process of the mud in the formation.

The pH of the mud is mainly used to reflect the acid–base degree of the mud, and the method is pH test paper color. The chemical stability of the mud is usually used in pH test paper to identify the mud, usually the chemical stability of the mud with pH value of 8–10 is good.

5.1.2 Mud Film Forming Technology

Shield tunnel engineering and academia generally believe that the penetration of mud will form a layer of "mud film" on the excavation surface, in the pressure chamber. The mud penetrates into the excavation surface under pressure, and the coarser particles are blocked on the surface of the formation, while the finer particles infiltrate into the pores of the formation to reduce the permeability coefficient of the formation, and the permeability coefficient formed in a certain penetration range of the formation surface area is small. The particle aggregation layer is a mud film.

There are different penetration modes when the mud penetrates in different formations, and the phenomenon is quite different. In various penetration modes, Mud infiltration will form different types of mud membranes. The type of mud film depends on the matching between the particle size distribution of the mud particles and the pore distribution of the formation. The indoor simulation of different muds in different layers penetrates, the mud state after the test is basically. There are three types of mud film formation in the formation after mud infiltration, as shown in Fig. 5.10.

(a)type I——mud-type mud film   (b)type II——mud +penetration belt mud   (c)type II I——no mud film

Fig. 5.10 Typical mud type

Type I: Almost all of the soil particles in the mud are silted on the surface of the formation, forming a thin and dense mud-type mud film on the surface of the formation. Type II: Some fine particles in the mud penetrate into the formation, due to the filtering of the formation and stay in the formation pore, forming a mud penetration zone, at the same time, the formation surface also formed a thin layer of mud, this type can be called mud skin and permeable belt mud film. Type III: Mud penetration through the formation, the formation can observe the trapped fine particles of mud, but no obvious mud or dense mud penetration zone, can be considered unable to form a mud film. It should be noted that only type I and type II count mud film, type III can not count mud film, here it is classified as a mud film, in order to include all forms of mud penetration.

If the membrane formation or not of different mud in different formations and the type of film forming are marked in the coordinate system of the average aperture D_0 of the formation and the mud particle d_{85} (Fig. 5.11), it can be found that the three forms of the mud film can be separated by two dividing lines a and b. A and b can be similarly represented by two straight lines, $d_{85} = D_0$ and $d_{85} = 0.5D_0$, respectively. Therefore, the formation type of mud film can be approximated by $d_{85} \geq D_0$ and $d_{85} = 0.5D_0$. When the mud particles can not penetrate into the formation pores, then the formation of mud-type mud film, when the mud $0.5D_0 < d_{85} < D_0$, part of the mud particles can enter the formation pore blocked in the formation, forming the mud mesh and penetration belt mud film, when the mud $d_{85} \leq 0.5D_0$, Almost all of the mud particles can flow through the formation pores, at which point it is not possible to form a mud film or a stable osmosis zone.

The average aperture of the five groups of formations, $D_0 = 0.18$–$0.2D_{15}$, was taken from the average aperture $D_0 \approx 0.19D_{15}$, indicating that the small particle size in the formation controlled its aperture. Therefore, the formation form of mud film can also be divided by the relatively simple ratio of the formation of the formation of the layer D_{15} and the mud d_{85}. Mud-type mud film is basically formed in the $d_{15}/d_{85} \leq 5.26$, while the mud-skin + penetrating-belt mud film is mainly formed

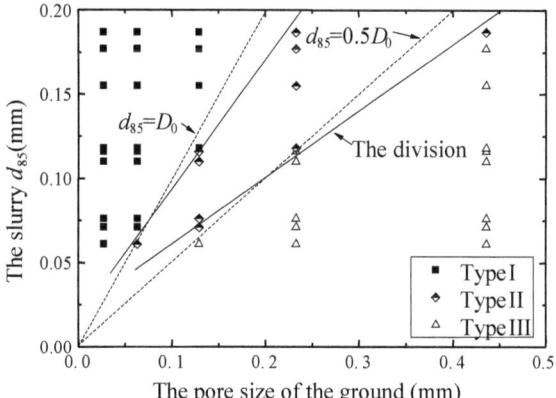

Fig. 5.11 Correspondence between stratum D_0 and mud d_{85}

in the range of 5.26 D_{15}/d_{85} 10.53, when $D_{15}/d_{85} \geq 10.53$, the mud is through the formation outflow, can not form the mud film and stable penetration band.

When the mud-water shield is constructed in a high-permeability formation, the mud easily penetrates into the formation under the action of the pressure difference between the mud pressure and the groundwater pressure, causing a large fluid loss or even complete fluid loss, which cannot be formed on the excavation surface. A good quality mud film is not good for maintaining the stability of the excavation surface. Even if the same type of mud penetrates into different formations, the type of mud film formed may not in the same way, this requires a certain matching between the mud and the permeable formation. The formation type of the mud Shield mud film and the formation quality of the mud film affected by different formations are as follows types:

(1) The formation type of mud film can be divided into two types, namely mud skin type mud film and mud skin + permeable belt type mud film. They have their own characteristics: The formation speed of the mud-type mud film is fast, the fluid loss of the mud is small during the formation process, and the permeability coefficient of the mud film is small after formation; the mud skin + permeable belt type mud film formation speed is slightly slow, and the mud filtration during the formation process The loss is slightly larger, but because of its mud and osmotic belt, its ability to resist interference is strong.

(2) When the permeability coefficient of the local layer is less than 1×10^{-2} cm/s, the film formation of the mud is easier, and a muddy mud film is formed. The main factor in the quality of the mud film formation is the density of the mud, followed by the particle size of the solid phase particles in the mud and the viscosity of the mud. To form a dense mud film in such formations, the mud must meet the following two conditions: There are enough powdery particles in the mud to clogging the surface pores of the formation, and at the same time have a lower sand content; the mud has a higher The viscosity and good physical stability make the solid phase particles and water have strong binding force, and the mud is not easy to isolate and lose water under the action of pressure.

(3) When the formation penetration coefficient is between 1×10^{-2} cm/s and 1×10^{-1} cm/s, the mud in the formation is also relatively easy to form a film, and can not only form a mud-type mud film, but also the formation of mud-skin-penetration-belt mud film. When the mud has a lower density and viscosity, as well as a small effective particle size, it can form a mud skin and osmosis belt mud film on the excavation surface, when the mud contains appropriate powder or sand to make the effective particle size of the mud larger, or increase the density or viscosity of the mud, it can form a mud-type mud film on the excavation surface. In such formations, any change in the density, viscosity or effective particle size of the mud in any of any indicators, will have a greater impact on the type and quality of the mud film. Only mud should not contain too much large particles such as sand, otherwise the formation of good mud film is not conducive.

5.1 Pressure Opening and Opening Face Stability Technology

(4) When the local layer penetration coefficient is between 1×10^{-1} cm/s and 1 cm/s, the formation of mud film is not easy, mud and formation mismatch will probably occur mud in the film ingestion process filter loss is too large, or even can not be filmed phenomenon. In such formations, the most important impact on the formation type and quality of mud film is the particle size of solid-phase particles in the mud, followed by the density of the mud, and finally the viscosity of the mud. In order to ensure that the mud can form a good quality mud film on the excavation surface, the mud must meet the following two requirements: mud must contain a certain amount of powder or sand particles, to ensure that the effective particle size of the mud is large enough to block the formation pores; To ensure that the larger solid-phase particles in the mud can be evenly dispersed in the mud for a long time, no precipitation phenomenon.

(5) When the local layer penetration coefficient is greater than 1 cm/s, the mud in the formation of a membrane is very difficult, the general mud will form a large amount of water filter (or even completely filtered), but can not effectively silt the formation of pores or silt on the surface of the formation to form a mud film. In this type of formation, the size of solid phase particle size in the mud and how much determines whether the mud can form a membrane and the quality of the film, the viscosity of the mud is only to ensure that the larger solid-phase particles in the mud can be evenly dispersed in the mud, will not precipitate, separation, water loss and so on. In order to ensure that the mud can form a mud film on the excavation surface, the mud must meet the following three conditions: mud must contain a certain amount of particles with a large particle size as a leak-blocking particles to block the larger pores in the formation; It is necessary to ensure that the mud has a large viscosity, so that the particle-blocking particles with large particle size are suspended evenly in the mud for a long time and do not precipitate.

(6) When the mud $d_{85} \geq D_0$, the mud particles can not penetrate into the formation pores, when the formation of mud-type mud film, when the mud $0.5D_0 < d_{85} < D_0$, part of the mud particles can enter the formation pore blocked in the formation, forming the mud mesh and penetration band mud film, when the mud $d_{85} \leq 0.5D_0$, Almost all of the mud particles can flow through the formation pores, at which point it is not possible to form a mud film or a stable osmosis zone. Using the representative particle size of the formation and mud to divide the formation type of the mud film, the mud-skin-type mud film is basically formed in the range when $D_{15}/d_{85} \leq 5.26$, while the mud-skin-penetrating-belt mud film is mainly formed in the range of $5.26 < D_{15}/d_{85} < 10.53$, when $D_{15}/d_{85} \geq 10.53$ Mud flows through the formation, unable to form a mud film and a stable osmosis zone (d_{85} is the representative particle size of the mud; D_0 is the average aperture of the formation; D_{15} is the representative particle size of the formation).

5.1.3 Stable Excavation Face Technology Based on Mud Membrane Closed Gas

1) The importance of mud membrane gas tightness

At present, the pressure-opening cabin mainly uses mud film forming and air pressure support technology to ensure the stability of the excavation surface. When the cabin is opened, In order to ensure the stability of the mudstone shield excavation surface, the pressure and pressure in the pressure chamber must balance the water and earth pressure at the excavation surface, that is, to ensure the stability of the mudstone shield excavation surface, the pressure and pressure must balance the water pressure and soil at the excavation surface pressure. The balance between water pressure and earth pressure requires that a muddy membrane that is impervious to water or slightly permeable to water must be formed on the excavation surface. The intermediate medium can be used to convert the atmospheric pressure into effective stress to achieve the balance of soil water pressure at the excavation surface.

When the mud-water pressurized shield is pressed open, a high-quality mud film is first formed on the excavation surface, and then on the mud tank. The part is filled with a certain volume of compressed gas to reduce the mud level and replace the space available for the construction personnel to enter the front of the cutter head. The mud film is directly exposed to the compressed air. At this time, the role of the mud film on the excavation surface should be considered from two aspects:

(1) The mud film is the guarantee of stability of the excavation surface under the air pressure support. When the mud-water pressurized shield is held apart, the balance of force on the excavation surface is shown in Fig. 5.12. With the pressure open cabin, through the action of air pressure on the mud film to balance the soil and water pressure in the formation, support the excavation surface, theoretically the working chamber of the air pressure p_a should not be less than the water pressure on the excavation surface p_w and soil pressure p_{soil} sum, that is:

$$p_a \geq p_w + p_{soil} \tag{5.1}$$

Among this, air pressure and water pressure belongs to pore pressure, change equation (5.1) to

$$p = p_a - p_w \geq p_{soil} \tag{5.2}$$

The difference between air pressure and groundwater pressure should be greater than or equal to soil pressure, and the hole pressure is not balanced soil pressure, so the pressure difference needs to be converted by mud film to effective stress before the soil pressure can be balanced.

(2) The closed air time of the mud film determines the working time of the cabin with pressure. The closed gas time refers to the mud film under the action of

5.1 Pressure Opening and Opening Face Stability Technology

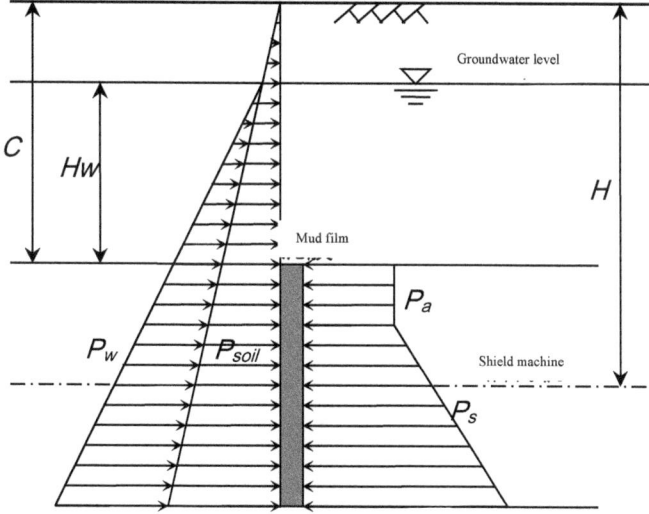

Fig. 5.12 Balance of force on the excavation face when the tank is opened

air pressure can maintain the closed gas for a certain period of time, beyond this time, the mud film will be gas penetrated. With the pressure open cabin needs construction personnel under high pressure to complete the maintenance work, according to different work needs, a time into the cabin 0.5–4 h, which requires the mud film under the action of high pressure has a larger air closure time.

2) Mud membrane closed gas destruction form

In the process of the mud shield, the main function of the mud membrane is to block the pores of the formation. When the mud membrane loses gas tightness, It is believed that it has ventilated damage. There are several typical closed-gas destruction phenomena in the process of mud film from air-tight to gas-permeable damage.

(1) Large holes are breathable (Fig. 5.13). When the air pressure is greater than the mud membrane closed gas value, it is easy to appear large hole breathable destruction phenomenon. At the moment of air pressure, the mud film is pierced and several large holes 2 mm in diameter can be observed on the surface of the mud film. In the bad form, the mud film consolidation drainage process has not been fully carried out, the mud film has been destroyed, and the mud film hardly produces compression. When the pressure is slightly smaller than the closed gas value of the mud film, the form of large pores will also appear. In the phenomenon of large pore gas permeability, the mud membrane has a short gas-closing time or is instantaneously broken by air pressure, and the moisture content of the mud film before and after the air-tightness does not

Fig. 5.13 Mud membrane large hole ventilation

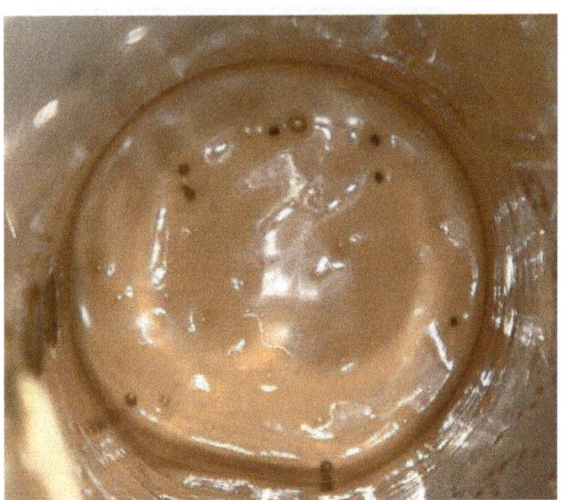

change significantly. Large-pore venting generally occurs when the pore size of the formation is large and the mud and the formation are not matched. Because the mud particles are difficult to completely block the pores, the formed mud film is not dense, and the air pressure easily pushes out the moisture in some large pores to form a gas leak. Channels, such damage should be controlled by adjusting the mud index to form a mud film with a smaller and denser permeability coefficient.

(2) Small holes are breathable. When the air pressure is much smaller than the closed gas value of the mud film, the damage pattern of the small hole ventilation is likely to occur. In this failure mode, the moisture content of the mud film is significantly reduced, the compression amount is large, and there is a clear consolidation drainage process. Because the air pressure is much smaller than the mud film closed gas value, air pressure is difficult to push the pore water to discharge outward, and the pores of the mud membrane are further compressed during the gas-closing process, which leads to an increase in the gas-closing time of the mud membrane, which has positive significance for practical engineering. Small hole ventilation generally occurs under the condition that the mud and the formation are well matched, and the formed mud film is thick and dense. It is difficult for the air pressure to quickly break through the moisture constraint in the mud film, forming a long-term gas-closing time, and after a long time of compression. After that, the moisture in the surface of the mud film gradually migrates to the bottom, resulting in a difference in moisture content between the surface layer and the formation. The surface mud film is cracked due to water shrinkage (Fig. 5.14).

Shear damage. When the surface of the local layer is extremely uneven, it is prone to the shearing of the mud film that is instantaneously broken down to form a large hole. The particle size of the formation is poorly matched, the surface is not flat, and

5.1 Pressure Opening and Opening Face Stability Technology

Fig. 5.14 Mud cracking damage

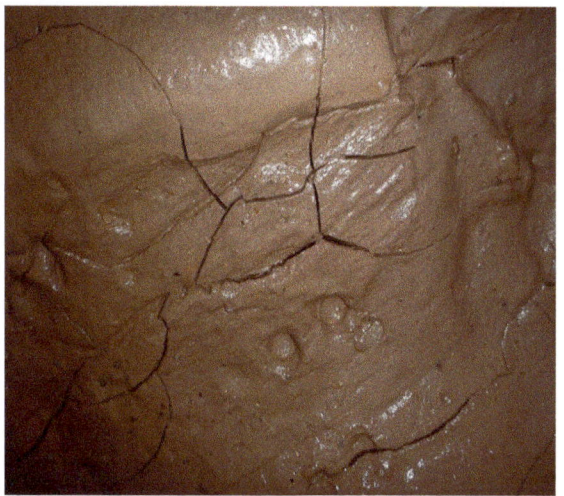

there may be uneven thickness on the surface of the formation after the mud film is formed. Like, under the action of air pressure, uneven deformation is easy to occur. Because the shear resistance of the mud film is small, the mud film is easy to follow along the air pressure. Shear damage occurs around the hole. The fracture mode of shear failure is mainly caused by the unevenness of the formation. When the attitude of the shield is adjusted and over-excavated, it is easy to cause collapse and holes in some areas of the excavation surface, which is extremely difficult for mud formation and gas shut-off. The appearance of this failure mode is independent of the quality of the mud film and should be avoided by fine control of the shield tunneling.

3) Method for improving the airtightness of the mud film and the stability of the excavation face

When the permeability of the formation with the pressure open compartment is large, in order to improve the airtightness of the mud membrane and enhance the stability of the excavation surface, Use the following method:

(1) First, an osmotic beltre-filming technique is formed. Considering that the formation also has a certain ability to close the air, the formation should be fully utilized. The ability to close the air, that is, a certain amount of penetration of the mud in a certain range in front of the excavation face, and densely block the formation. The pores between the particles form a closed gas layer. At the same time, the mud particles penetrate into the pores of the formation for a certain distance, which improves the cohesive force of the formation and is beneficial to improve the stability of the excavation surface. This type of mud film is the permeable belt + mud skin type mud film. According to the gradation of the formation particles, the pure bentonite slurry with low density and low viscosity can be infiltrated into the excavation surface in the pressure chamber first. Then, the low-density mud in the pressure chamber is replaced

by a mixed mud of bentonite clay with a large viscosity and density, so that a dense mud-type mud film is formed on the excavation surface, and then the belt opening operation is carried out. After this adjustment, the fluid loss of the mud can be effectively reduced, the gas permeability of the formation can be reduced, and the closed gas value and the closed gas time of the mud film can be improved.

(2) Lifting mud level method. When the tank is opened, the mud film will be exposed to air pressure for a long time, and the mud film will crack and leak. Gas and other phenomena, when severe, will result in poor pressure holding effect and instability of the excavation surface. When the cabin is being repaired, the working time of the human body under high pressure is generally not more than 4 h. When the shifting operation is carried out, the mud slurry can be repaired to face the mud film, so that the closed time of the mud membrane can reach the closed time of the fresh mud membrane again. Repeated rise of mud level can repair mud film well and form thicker, denser mud film. More importantly, it is possible to repair the defects that may have formed the mud film in time, and the gas sealing ability of the mud film can be greatly improved.

(3) Additive improvement method. When the pressure conditions are constant, the nature of the mud is the main factor determining the quality of the mud film. The engineering properties of the slurry include mud viscosity, density, particle gradation, etc., so there are many factors that affect the gas tightness of the mud film. In terms of the effect of viscosity on the gas-closing time of the mud film, the test shows that the increase of mud viscosity can increase the gas-closing time of the mud film. After adding a small amount of CMC to the mud, the macromolecules in the CMC bond with the clay particles to form a space grid structure, thereby reducing The contact of the soil particles makes the coalescence of the soil particles more stable, the thickness of the mud film increases, and the airtightness of the mud film is enhanced. But adding too much CMC will result in a high-viscosity and high-elasticity hydration layer is formed around the clay particles, and the mud viscosity is high due to the high viscosity of the CMC solution itself. The rapid rise makes it difficult for the water in the mud to penetrate into the formation, causing the mud film to be loose and the thickness of the mud film to become very thin. It should have a certain thickness to ensure a long closed gas time, the mud viscosity should be controlled within a certain range, and the high viscosity mud will have an adverse effect on the mud film formation and gas shut-off. In addition, adding appropriate amount of additives such as vermiculite, wood chips and sand to the mud to enhance the plugging function of the mud in the formation can also enhance the airtightness of the mud film.

4) Mud membrane closed gas pressure into the cabin operation

The key to the operation of the pressurized cabin is to maintain the stability of the air pressure in the earthen tank. If the airtightness of the mud membrane is not good, the air leakage will lead to sudden changes in air pressure, groundwater infiltration, and a large amount of drifting sand, resulting in instability and collapse of the excavation surface, will endanger the safety of construction workers. Therefore, before the construction personnel enter the cabin, in order to ensure the stability of the air pressure during the pressure into the cabin, the airtightness of the pressure chamber must be checked first.

The airtightness of the test chamber is divided into the following two parts:

(1) Indoor simulation test. In the self-made mud film-forming device, air tightness testing on the formed mud film, test whether the quality of the mud film can effectively prevent gas leakage, meet the airtightness requirements of the pressure chamber, and then determine the mud formula.

(2) Field test. On the basis of the indoor test, the air tightness of the pressure chamber after grouting and sealing is directly checked on the spot. Under the condition of simulated working condition, observe the air supply interval time of the air compressor, the gas leakage speed in the pressure chamber, and the pressure. The change of cabin air pressure, the amount of slag washed after the pressure is maintained and the ground monitoring, etc., to analyze whether the excavation surface is unstable and whether the air pressure in the cabin can be kept stable, so as to determine whether it can enter the cabin. During the test, if the pressure in the pressure chamber is stable at the set value, the starting frequency of the air compressor is stable, the air supply is normal, and the ground is not deformed or leaked, which proves that the formation can achieve gas pressure regulation; if there is no extra after washing The muck was taken out, indicating that the soil layer on the excavation surface was stable and there was no collapse. After verifying that the airtightness of the pressure tank mud membrane is good, the air pressure is replaced by a part of the mud pressure in the upper part of the pressure chamber. If the gas permeability of the mud film is obvious, the air pressure is dissipated from the pores of the formation by the pore pressure, the air pressure in the tank is difficult to maintain constant, and the earth pressure on the excavation surface is difficult to balance.

5.2 Pressure-Opening Tool Changer

5.2.1 Diving Operation

Diving is a compressed air or artificial mixed gas that is equal to the pressure of the human being under water or high pressure. The process of returning to the surface or atmospheric environment. The shield cutterhead is in a high-pressure environment during the process of opening the cabin. The overhaul construction is a kind of diving operation. The high-pressure diving operation in the tunnel belongs to the diving operation mode, and the pressure and decompression are consistent with the diving operation. High-pressure operation in the tunnel is equal to or lower than the diving operation regardless of construction difficulty or safety risk. In the shield pressure maintenance work, the contents of the diver's work mainly include: replacement of the hob under high pressure, daily inspection, cutting and welding.

In the closed high-pressure environment, the following three problems should be solved: First, whether the welding strength can meet the construction requirements; second, whether the welding oxygen partial pressure is too high in a high-pressure environment; and third, whether toxic gases are generated during the welding process and How to deal with it.

Divers should be familiar with high pressure operating procedures. Check all gates and systems before pressurization to understand the source Reserve volume and pressure, check whether the pipeline, valve and oxygen device are good, adjust the oxygen pressure reducer output pressure (close the oxygen gas source after adjustment), check the door, valve, instrument, communication device, gas detection instrument and video monitoring device Is it good, check whether the various therapeutic supplies are complete, and whether the performance of each accessory in the cabin is good. Check for equipment and materials (lighters, closed tanks, etc.) that are prohibited from being brought into a high-pressure environment. The diver enters the human lock cabin and closes the hatch to maintain communication and video connection during the pressurization system. The pressurization rate is controlled to not exceed 0.15 MPa/min. Try to avoid single pressurization, if special circumstances require single person plus. When pressing, the operating specifications should be strictly observed. The work location should be clear and the call should be kept in contact.

Do not perform high voltage operation in the event of a system failure or an emergency. When there is not enough oxygen (including standby oxygen), it is not possible to pressurize. If you are not qualified for professional training and medical examination, you cannot enter the cabin for pressurization.

During the construction, the diver must obey the diving supervision arrangement until the operation is completed. The hyperbaric chamber operator should arrange for shift duty and there must be at least two operators on site. The diver must not leave the site within 24 h of the completion of the high pressure operation. Gas sampling analysis should be carried out regularly during the operation, and it is found that harmful gases should be discharged and treated in time. Have to work on the job during the job. The process monitors and finds hidden dangers to stop the operation and eliminate them in time.

Provide comprehensive health and safety protection for each operator during construction. All operators must have at least 12 h of unpressurized time within 24 h. In the case of continuous operation, the pressure is 0.1 MPa or higher, and up to 5 continuous belt press operations are allowed, and 48 h of unpressurized time is ensured between continuous operations; when the pressure is lower than 0.1 MPa, continuous belt pressure is 6 times. The time without pressure is 24 h. During the high pressure operation of the shield, the human brake chamber should maintain working pressure and facilitate the work. When the cabin is evacuated, it is isolated from the other compartments. Leave a cabin to maintain normal pressure to ensure access to the emergency. If the diver suffers from decompression sickness during construction, he should be treated promptly. After the physical rehabilitation, it needs to be evaluated before the high pressure operation can be resumed.

Safety analysis and evaluation should be carried out before high-pressure welding and cutting operations, safety precautions should be taken, and the operation should be monitored. Cutting and welding should be approved by the diving supervisor. Divers involved in welding and cutting should be familiar with high pressure sealing. Space work regulations, and professional training to obtain welding and cutting certificates. All staff should practice using a protective mask respirator before starting work.

Wear protective mask respirators for welding and cutting operations to prevent inhalation of harmful gases. When cutting and welding work, special personnel should be assigned to take care of the high pressure gas cylinders and monitor the vicinity of the high pressure gas cylinders. Ventilation and ventilation are used to discharge harmful gases generated by cutting and welding operations in time to ensure the quality of the inhaled air.

After the high-pressure welding and cutting operations are completed, the oxygen valve and power supply should be turned off in time. If cough, asthma, or other respiratory illnesses occur after inhaling welding fumes, treat them immediately.

The first-aid kit should be equipped at the job site, and the operator should understand the function and usage of the emergency equipment. Wear protective clothing such as dry clothing (protective clothing), rubber insulated shoes, welding protective gloves, goggles, and protective mask respirators (Fig. 5.15).

High-pressure fire extinguishers should be provided at the job site. All textiles and flammable items should be removed from the work area. If conditions do not allow, cover them with sand or fireproof materials. The escape routes in the cabin are kept clean and smooth at all times. The number and location of fire extinguishers and faucets should be preset and the fire extinguisher should meet the maximum working pressure.

Fig. 5.15 Welding under high pressure environment in shield machine

The decompression procedure should be performed in accordance with the decompression protocol established by the diver. All door lock chambers and systems should be inspected prior to decompression to ensure proper system use.

After the diver has reached the specified working time, he should return to the cockpit immediately. The diver cleans the exposed skin, removes the clothes and shoes, then enters the lock chamber, puts on the clean jacket and shoes, closes the door of the human lock compartment and the work area, starts decompression according to the decompression scheme, and works with the cabin operator. Keep communication and video contact. After the diver enters the cockpit, the built-in door lock valve can only be adjusted under the guidance of the hyperbaric operator.

The hyperbaric chamber operator should record the decompression condition and report it to the diving supervisor and the diver in time after the abnormal condition is found. The oxygen concentration in the door compartment should be controlled at around 21%, the highest is not higher than 23%, and the lowest is not lower than 19%.

The diver must not be away from the high pressure chamber within 24 h after decompression.

5.2.2 With Pressure into the Cabin

Before the implementation of the cabin, the various tools, mechanical equipment, personnel equipment, etc. required for the whole process of the cabin. shall be inspected and counted. It plays an important role in the smooth implementation of the opening and closing operation.

In order to ensure the staff from the danger of water immersion and collapse in the excavation cabin, according to the geological conditions of the stratum, the excavation surface is supported by high-quality bentonite mud after the excavation stops. Firstly, the bentonite slurry with a concentration of 8–12% is used as the infiltration mud to form a certain distance of the infiltration zone in the excavation surface to increase the cohesive force of the formation; then the mud with a relative density of 1.15–1.20 and a viscosity of 25 s or more is opened. Mud-type micropermeable mud film is formed on the excavation surface, and the mud pressure is effectively applied to the excavation surface, reduce the loss of mud pressure.

Changing the tool in the pressure chamber, repairing the cutter head, etc. requires that the part of the cutter head to be repaired be transferred to the upper part, while the mud level in the excavation chamber is appropriately reduced to support the excavation surface with compressed gas. In order to ensure safe opening, the excavation face and mud level must be closely monitored at all times.

In order to pass the pressurized work area, the person and the tool need to first enter the sub-cabin through the human gate, where the air pressure is raised first. Now introduce the pressurization process of the shield opening of the Nanjing Yangtze River Tunnel in the form of a schematic diagram.

Fig. 5.16 Initial state before the shield is opened

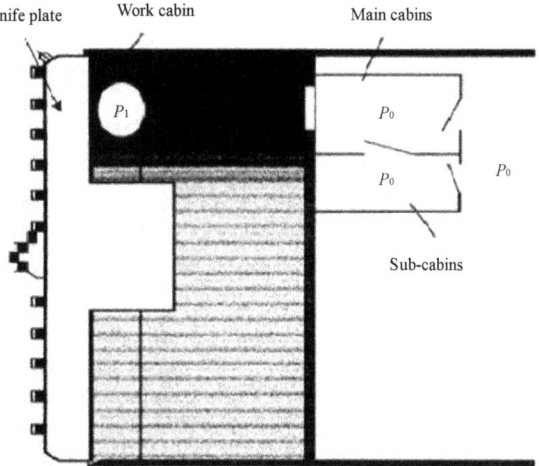

The shield diagram is shown in Fig. 5.16, defined as: p_0 is atmospheric pressure, p_1 is the notch pressure, P2 is the pressure value between P0 and P1. In the initial state, only the working chamber (excavation and pressure chamber) is under pressure and the pressure is p_1. If all the conditions of pressurization and decompression are available, the pressurization can be started.

First step, the main cabin is pressurized. As shown in Fig. 5.17, close the main compartment gate and open the main compartment air valve to increase the main compartment pressure from p_0 to p_2 and then to the notch pressure, i.e.the working chamber pressure p_1.

In the second step, the auxiliary compartment is pressurized. The subchamber pressurization procedure is shown in Fig. 5.18 and is similar to the main cabin

Fig. 5.17 Schematic diagram of the main cabin of the shield

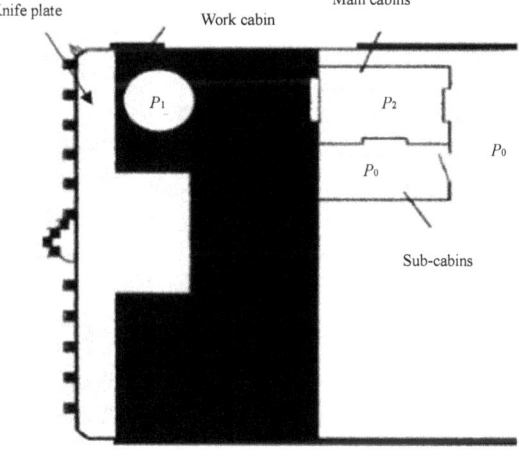

Fig. 5.18 Schematic diagram of the shield sub-cabin pressure

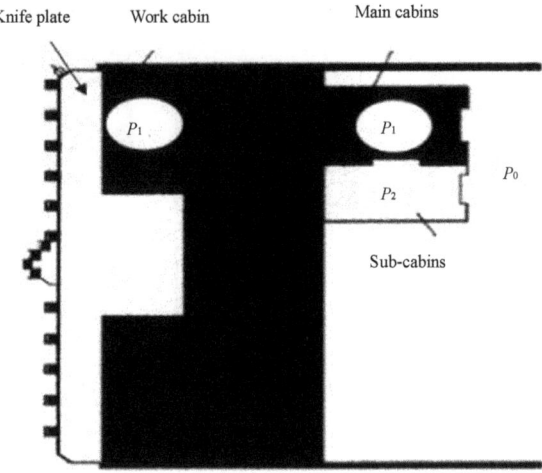

pressurization procedure. When the main cabin pressure and the cabin pressure are balanced, the personnel enter the sub-cabin and begin to pressurize the sub-cabin. Generally, the main cabin is always pressurized first, and the sub-cabin always maintains a pressure relief state. The incoming personnel are pressurized in the auxiliary compartment to accommodate the pressure in the main and working compartments. The operation of adding a person to the pressure working area can be performed by pressurizing the sub-chamber.

The third step is to enter the cabin. When the pressure of the sub-cabin is balanced with the pressure of the main cabin and the working chamber, as shown in Fig. 5.19, the cabin personnel can enter the work chamber to repair the cutterhead and other equipment without any adverse reaction.

Fig. 5.19 Shield sub-cabin pressurization completed

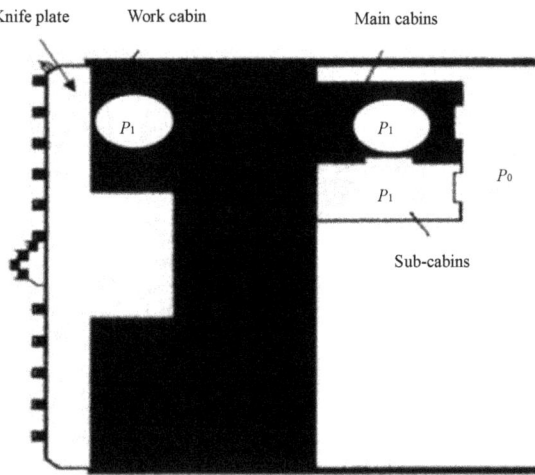

5.2 Pressure-Opening Tool Changer

The incoming crew must return within the specified time. The pressurization time depends on the person who arrives at the pressure compensation in the sub-tank. The time required to enter the sub-cabinet is also calculated within the working hours of the site. If the pressure drops slowly during the decompression process, the released gas can pass through the blood circulation and the lungs; if the pressure is quickly reduced, the bubbles will form a carbonated water effect in the body fluids and tissues. These can cause a series of decompression symptoms, the release of gas can cause temporary or permanent tissue damage, causing serious physical harm to the staff. Therefore, the decompression process and precautions must be strictly observed during decompression.

In the initial state, as shown in Fig. 5.20, the working chamber and the main cabin are under pressure, the pressure is p_1, and the auxiliary cabin is under normal pressure. When all conditions of the staff and decompression are ready, the decompression operation can be started. Normally, the decompression is only carried out through the main compartment; at the same time, the sub-cabin is always in a normal pressure state.

As shown in Figs. 5.21 and 5.22, the main cabin pressure is reduced from p_1 to p_2, then to p_0, the decompression operation ends.

The above opening procedure is for the conventional compressed air method. The opening step of the knife technology is the main development direction for the opening of the saturation method with the open-face tool change technology, the stability principle of the excavation face and how to better improve the efficiency of the work under high pressure environment.

Fig. 5.20 State before the shield decompression operation

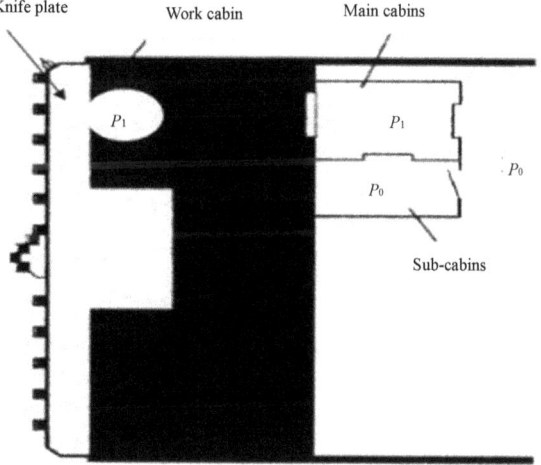

Fig. 5.21 Schematic diagram of the decompression process of the shield main cabin

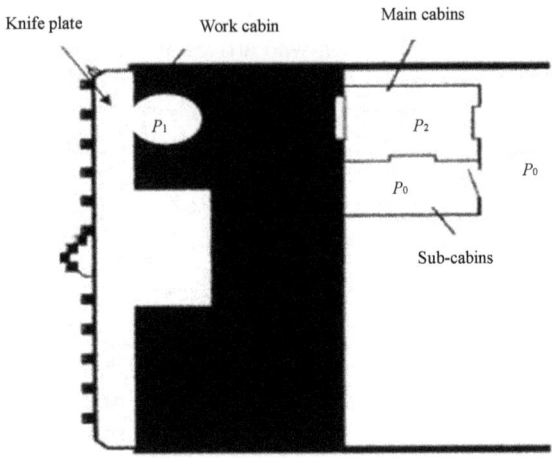

Fig. 5.22 Shield decompression process ends

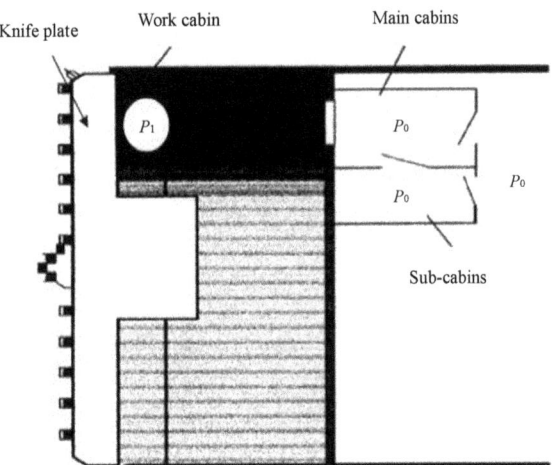

5.2.3 Pressure Change Tool Technical Produces

When carrying out the pressure-opening tool change operation, the operator enters the rear end of the cutter head from the human cabin and the bubble chamber after the pressurization procedure, and the cutter head and the cutter are inspected (Fig. 5.23). If the fixed tool on the cutter head needs to be replaced after inspection, it needs to be welded and welded. For replaceable knives on the cutter head, use tools such as hoists and wrenches for disassembly and installation (Fig. 5.24).

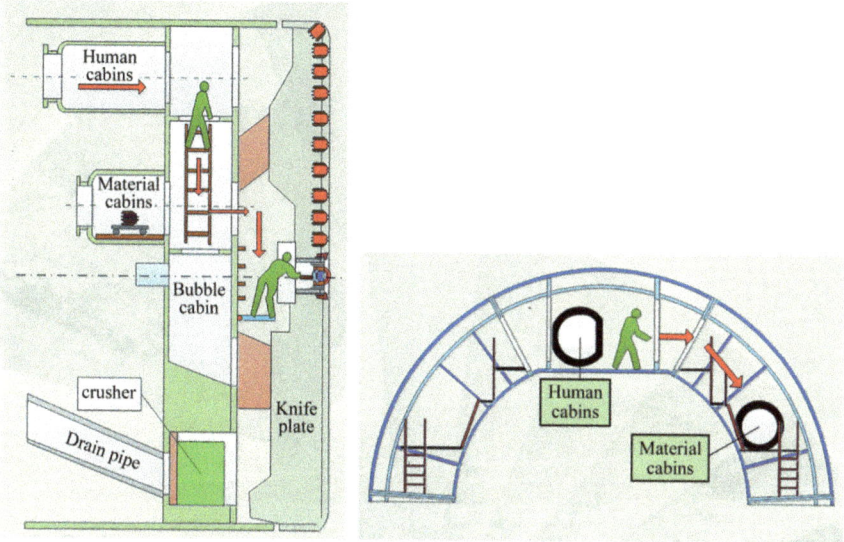

Fig. 5.23 Inlet process of the operator with pressure

5.3 Nanjing Yangtze River Tunnel (Wei7Road) Example of Conventional Compressed Air Pressure Opening and Changing

5.3.1 Overview of the Nanjing Yangtze River Tunnel Project

1) Project overview

Nanjing is located in the Yangtze River Delta. The Yangtze River passes from the southwest to the northeast in the north of Nanjing, and the south of the Yangtze River is south. In the key areas of the development of Beijing, there are currently three bridges to communicate with the main city and Jiangbei. The Nanjing Yangtze River Tunnel is located between the Nanjing Yangtze River Bridge and the Third Bridge (Fig. 5.25). It is 9 km above the Third Bridge and 10 km below the bridge. It is connected to Nanjing City and Pu. One of the most direct fast lanes in the mouth area is also the first road tunnel in Nanjing that connects the Yangtze River to the north and south. Tunnel adopts "tunnel on the left and bridge on the right" is 5853 m in length, of which the length of the shield tunnel is 3020 m. After reaching the Jiangxin Island from north to south, it is connected to the urban road through the bridge, as shown in Fig. 5.26.

The Nanjing Yangtze River Tunnel is designed as a two-way six-lane tunnel with a design speed of 80 km/h and consists of two shield tunnels. The construction of two mud-water pressurized shields with a diameter of 14.93 m was started from the

(a) Roller wear check (b) Old roller removal

(c) Old roller handing (d) New roller installation

Fig. 5.24 Process of changing the chamber with pressure

jiangbei starting well and carried out in the same direction, with a tunnel excavation diameter of 14.96 m. The circular tunnel section is basically located in the water bottom of the middle of the tunnel, which is the main tunnel section form, with a pipe lining of 14.5 m and an internal diameter of 13.3 m. The pipe sheet is 60 cm thick and the ring width is 2 m, each ring consists of 10 pieces, divided into 7 standard blocks, 2 adjacent blocks and 1 cap block, and the ring and block are connected by oblique bolts, 1/3 of the stitches are misassembled. The pipe plate design strength rating is 60 MPa, the anti-penetration grade is P12, the concrete design has a service life of 100 years, and the ratio is designed for high-performance concrete.

The minimum plane curve radius of the shield tunnel is 2500 m, the minimum center distance of the left and right line is 23.33 m, and the general lot line spacing is 35 m. The tunnel cover thickness is up to 30 m and the minimum is 6.0 m (starting section). The thickness of the mulch in the middle of the river is controlled by the diameter of no less than 1 × shield, but subject to environmental conditions, the minimum depth of the tunnel at the bottom of the river at the bottom of the river on

5.3 Nanjing Yangtze River Tunnel (Wei7Road) Example ...

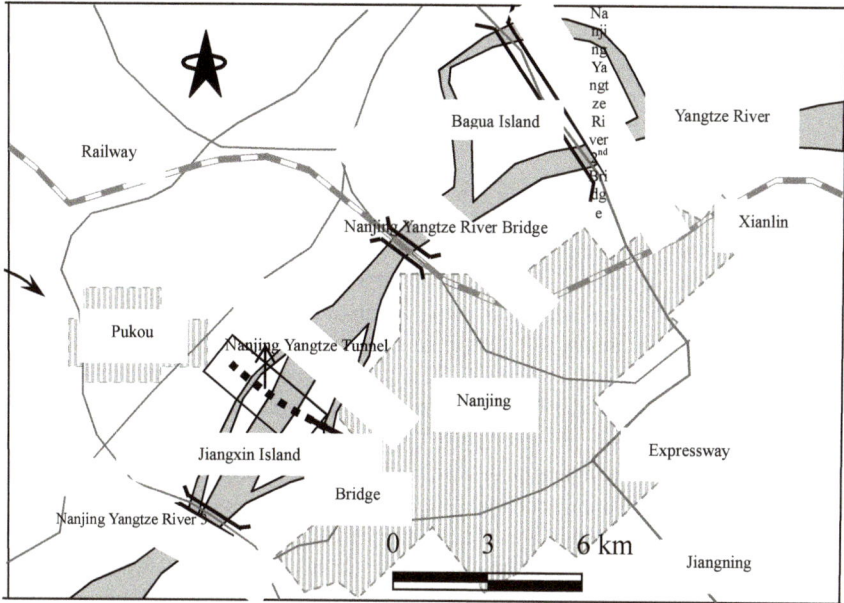

Fig. 5.25 Location map of Nanjing Yangtze River Tunnel

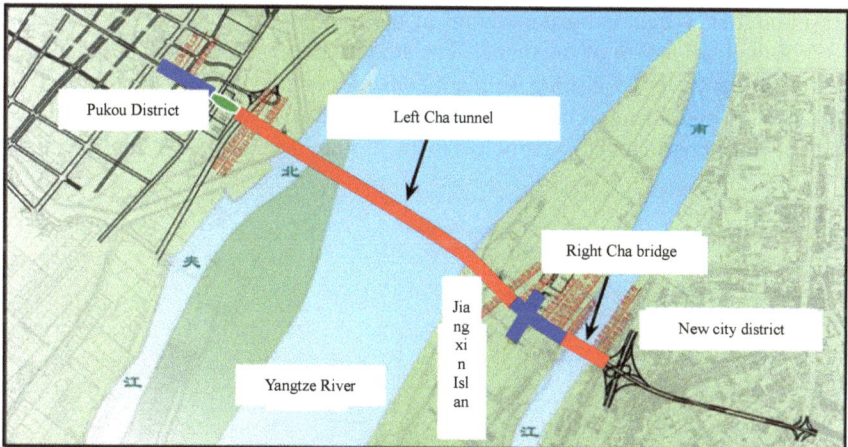

Fig. 5.26 Schematic diagram of Nanjing Yangtze River Tunnel Project

the side of the river is only 11.2 m. The maximum vertical slope of the tunnel line is 4.5%, the minimum slope is 0.6%, the maximum slope is 1130 m long, the minimum slope is 850 m long, and the tunnel section has 3 vertical curves, and the minimum vertical curve radius is $R = 750$ m.

With a total investment of about 3.3 billion yuan, the Nanjing Yangtze River Tunnel Project was laid in March 2005 and civil construction began in September 2005; On May 28, 2010, the Nanjing Yangtze River Tunnel opened to traffic.

2) Engineering geological conditions

The shield section of the Nanjing Yangtze River Tunnel Project is located at the bottom of the Yangtze River riverbed and the Yangtze River rushing and silting low floodplain. The shield tunnel crosses the river surface with a width of about 260 m. The high water level has a multi-year average of 8.37 m and the maximum water depth is about 28.8 m. The tunnel mainly crosses the Quaternary and white. In the lanthanide stratum, there is no fracture or fracture zone at the passage of the tunnel site. The upper Quaternary strata mainly include: The upper part of the underwater strata of the Yangtze River is composed of the Quaternary Holocene newly deposited loose fine sand, the middle part is composed of the Quaternary medium-density-compact fine sand, and the lower part is composed of the Upper Pleistocene dense gravel. Sand, gravel, etc.; the basal strata underlying bedrock is Cretaceous calcareous mudstone with fine calcium sandstone. The main strata of the tunnel crossing are: silty clay strata, fine sand strata, gravel strata, gravel strata and few strongly weathered-mud stones. The geological section is shown in Fig. 5.27. The formation grading curve is shown in Fig. 5.28. The basic physical properties of each layer are shown in Table 5.1.

The stratum crossing the tunnel is mainly composed of fine sand, gravel sand and gravel stratum. The composite stratum of the three strata accounts for more than 43% of the total length of the tunnel. The stratum has soft and hard, soft and hard, and water permeability. Strong, difficult construction and so on. Among them, the permeability coefficient of gravel and gravel stratum is as high as 10^{-2} cm/s, and the content of clay particles (fine particles) in the stratum is less than 5%, and the muddy

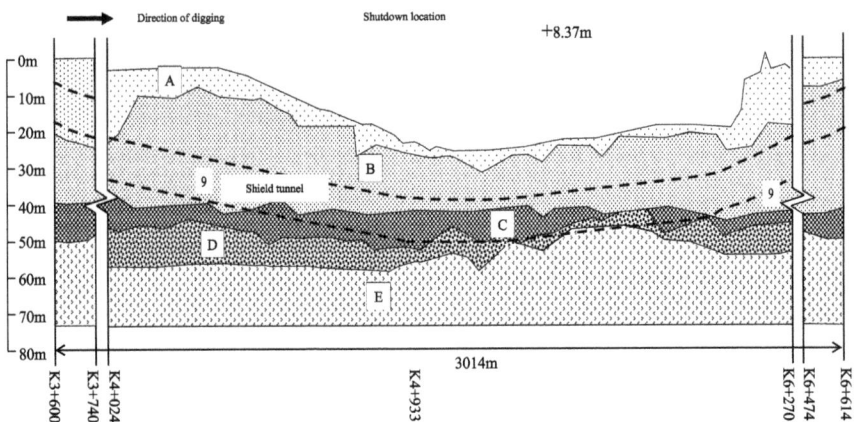

Fig. 5.27 Main geological section of Nanjing Yangtze River Tunnel. A-filled and silty clay; B-powder fine sand; C-gravel sand; D-boulder and pebble stratum; E-strong weathered mudstone

5.3 Nanjing Yangtze River Tunnel (Wei7Road) Example ...

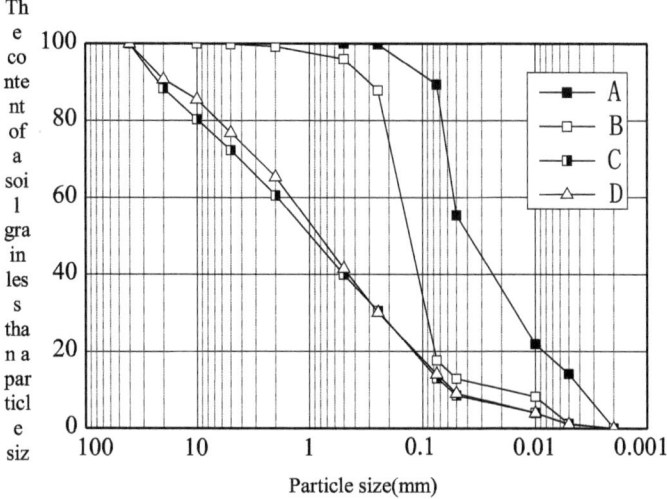

Fig. 5.28 Main stratigraphic grading curve of Nanjing Yangtze River Tunnel

Table 5.1 Basic physical properties of main strata in Nanjing Yangtze River Tunnel

Formation number	Stratum name	γ (kN/m³)	k (cm/s)	Quartz content (%)	C (kPa)	φ (°)	Clay content <5 μm (%)	d50 (mm)
A	Muddy silty clay	17.4	3 × 10⁻⁶	14.5	10	10	14.1	0.05
B	Fine sand	19	6 × 10⁻³	39.43	4	31.4	1.2	0.15
C	Gravel	20.1	3 × 10⁻²	67.09	6	34	1.1	0.8
D	Pebble and round gravel	20.7	5 × 10⁻²	72.67	–	–	1.0	1.0
E	Strong weathered mud rock	22.55	2 × 10⁻⁶	15.6	–	–	–	–

water is added. It is very unfavorable for the pressure shield to form a mud film on the excavation surface; the quartz content of the sandstone in the various layers that the shield passes through is generally higher, especially the quartz content of the gravel and gravel formation is as high as 70% or more, and the selection of the cutter and Wear resistance puts higher demands.

3) Shield section

The correctness of the shield selection is a key issue related to the success or failure of the project. The key to shield selection is based on engineering geology.

Conditions, hydrological conditions and environmental conditions determine its performance parameters, making it more adaptable to the project, higher reliability and lower risk. That is to say, the matching of shield selection and geological conditions is the basis for the successful completion of the project. The selection of the shield machine should be highly valued. The two mud-water pressurized shields used in the Nanjing Yangtze River Tunnel are based on the geological conditions of the German Herrenknecht Company. A mud-water pressurized shield specially designed and manufactured for use in permeable structures with high water pressure. The basic structure of the shield is shown in Fig. 5.29. The main parameters of the shield are shown in Table 5.2.

The cutterhead of the mud-water pressurized shield used in this project is the center support and spoke panel type, as shown in Fig. 5.30, the cutterhead opening rate is 30%, the cutterhead is set with 6 spokes, and a total of 18 fixed scrapers are arranged. There are 16 first-hand knives, 12 edge knives, and 7 prototype knives. In addition, in order to avoid the damage caused by the damage of the cutter head, 71 normal pressure replaceable scrapers are also arranged. The knife, that is, the person can enter the spoke arm under normal pressure to replace it, which is convenient for checking the wear of the tool. At the same time, at8 blade scrapers and 2 blades are equipped with a tool wear warning device, and the detecting device uses a wire buried inside the blade body. Monitor the current buildup when the wire is worn and

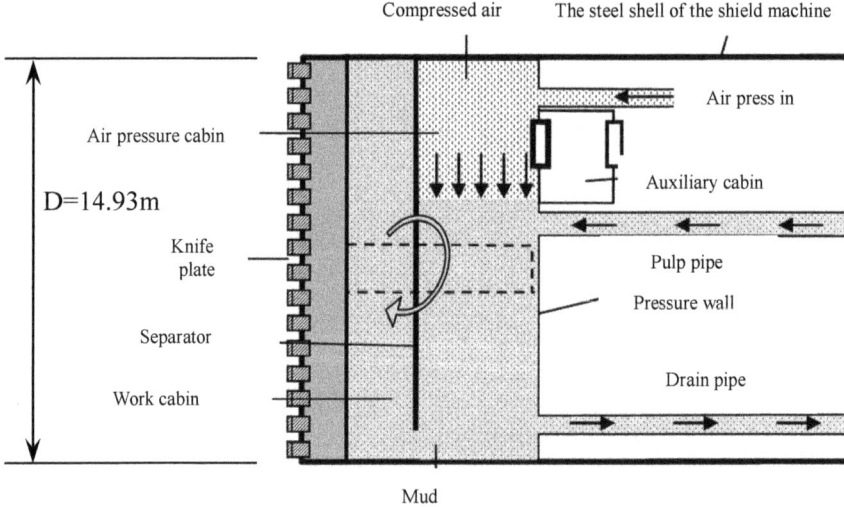

Fig. 5.29 Schematic diagram of the mud-water pressurized shield structure of the Nanjing Yangtze River Tunnel

Table 5.2 Main parameters of shield

Name		Specs	Note
The body of the shield	Front shield diameter (mm)	14,930	Without weld
	Rear shield diameter (mm)	14,900	
	Minimum horizontal turn radius of TBM (m)	750	
	Working water pressure (MPa)	0.75	Shield axis position
	Shield front and rear structure	Designed as 10 blocks	
	Platform and Walkway and Staircase (set)	1	
	Pressure sensor (a)	5	
	Pressure tank (a)	1	
	Partition (a)	1	
	Air pressure adjusting device (a)	2	For air pressure adjustment. adjustment accuracy is 0.002 MPa
	Front gate (a)	1	Hydraulic drive DN1200mm, maximum strength 200 MPa
	Stone crusher (a)	1	
	Shield lubrication point (a)	2×15	
The tail of the shield	Diameter	14,870	
	Shield tail type	Fixed	
	Shield tail seal system	3 sealing brushes and 1 spring plate	
	Emergency shield tail sealing system (a)	1	Inflatable emergency sealing system
	Grease tube (root)	20×3	
	Grouting pipe (root)	6 + 6 (spare)	Integrated in the tail of the shield, quick
	Grouting wash system (root)	1	
Cutter plate	Diameter (mm)	14,960	According to the drawing A-1855-03
	Structure	6 star-arranged spokes	
	Types	Soft soil type	

(continued)

Table 5.2 (continued)

Name		Specs	Note
	The cutter head has a tool that can be replaced from the cutter arm	Main soft soil tool	
	Cutter	replace soft soil cutter	
	Soft soil cutter	112 scrapers, 63 replaceable scrapers, 12 blades (4 pieces/handfuls, bolted connections) + 12 blades (1 piece/handful, welding), 32 tooth knives, 2 over-diggers (travel 0–40 mm), 7 fishtail knives	Replaceable tool change from the cutter arm at atmospheric pressure
	Wear monitoring system	6 scrapers + 2 blades	
	Rotating center connector (piece)	1	Used for bentonite scouring, circuits, hydraulic lines, etc.
	The direction of turn around	Left, right	
	Center flushing	2	DN150

Fig. 5.30 Nanjing Yangtze River Tunnel shield cutter

determine the wear level. The maximum design torque of the shield is 39945 kN m, the largest thrust is 199504 N.

The shield's pressure chamber is divided into two compartments by a partition: a mud chamber and a pressure regulating chamber. When the shield is normally excavated, the working compartment is filled with mud, the upper part of the pressure chamber is high-pressure gas, the lower part is filled with mud and communicates with the working chamber, and the pressure control is completed by adjusting the air pressure. The air pressure compartment is provided with a human gate and an auxiliary compartment. Through the human gate, the technician can enter the internal maintenance equipment of the pressure chamber, and the setting of the main auxiliary cabin can realize the adaptation of the personnel to the adding (reducing) pressure before in(out) the cabin. In order to prevent the occurrence of decompression and decompression, to ensure the safety of the personnel entering the cabin. This dual-cabin configuration of the pressure chamber makes it possible to press-fit the cabin access to the equipment. In addition, in the crushing machine is equipped with a crusher to break the large-diameter gravel encountered during the excavation process (Fig. 5.31).

The two unique advantages of the shield—equipped with tools that can be replaced under normal pressure and a powerful rock crusher: With the design of the original atmospheric pressure changing tool, the left-line shield ensures smooth passage through the composite formation without causing serious damage to the cutter head. The total number of atmospheric pressure tools in the construction of the Nanjing Yangtze River Tunnel exceeds 700, the system contributed to the smooth progress of the Nanjing Yangtze River Tunnel. Figure 5.32 shows the engineers changing the tool under normal pressure. In the shield selection stage, the two shields are equipped with sufficient strength due to the large stones in the gravel and gravel formations.

Gravel machine. Practice has proved that the configuration of the crusher is very successful. It breaks up large mudstones and stones into the size allowed by the

Fig. 5.31 Crushed rock crushing rock

Fig. 5.32 Changing the tool under normal pressure

muddy pipeline, eliminating the possibility of accumulation at the bottom of the excavation tank and frequent blockage of the muddy water pipeline.

4) Analysis of engineering characteristics and mud application difficulties

According to the engineering design data and geological report analysis of Nanjing Yangtze River Tunnel, the project mainly has the following difficulties of characteristics and construction:

(1) The shield has a large diameter. Nanjing Yangtze River Tunnel two mud-water pressurized shield diameter of 14.93 m, after completion of the tunnel diameter of 14.5 m, the same as the construction of the Nanjing Wei Three-passing river tunnel, in the country's large-scale Cross River shield tunnel, second only to the completed Shanghai Yangtze River Tunnel and the Construction Of the Qianjiang Tunnel (both shield diameter 15.43 m, The tunnels are 15 m in diameter, ranking second. The size of the shield diameter is very unfavorable to the stability control of the excavation surface, and the higher requirements are put forward for the excavation parameters, the shield attitude, the mud properties, the post-wall slurry properties and the precision of pipe fitting.

(2) High working water pressure. The maximum water pressure of the Greenhart Tunnel in the Netherlands is 0.55 MPa, and the other two shield tunnels on the Yangtze River, the Shanghai Chongsu Tunnel and the Wuhan Crossing Tunnel, and the water pressure is 0.5–0.55 MPa. The Nanjing Yangtze River tunnel shield working pressure of up to 0.65 MPa, at that time the same diameter and larger diameter mud water pressurized shield project is the world's largest. This also greatly increases the risk of the shield crossing the river, the shield itself and the construction process is also a test.

(3) When the shield crosses the Yangtze River levee, it is easy to cause the uneven subsidence of the embankment, which poses a threat to flood control. Shield excavation through the Yangtze River flood control embankment under the

powder fine sand layer, the formation particle level with a single, 70% of the particle particle diameter concentrated in 0.075–0.25 mm, less than 0.075 mm particles accounted for less than 20% of the entire formation particles, greater than 0.25 mm particle proportion of only 15% of the entire formation particles. This type of powder fine sand floor level is poor, slightly disturbed formation is very easy to liquefy, and with the excavation construction of the formation disturbance and the pressure water effect of the formation, it is very likely to cause the tube surge or the uneven subsidence of the embankment, forming the groundwater path, the flood control work poses a great threat.

(4) The gravel formation through the tunnel has a high permeability coefficient and difficulty in filming the mud. The middle section of the Nanjing Yangtze River tunnel river is mainly loose, slightly dense to medium dense fine sand formations, as well as some large sections of gravel, pebble layers, high penetration coefficient, of which the penetration coefficient of gravel formations as high as 10^{-2} cm/s. Under the high water pressure of 0.65 MPa, the mud in the formation is not easy to quickly form a good quality mud film, mud pressure to balance the formation of soil and water pressure, to maintain the stability of the excavation surface is more difficult. At the same time, due to the formation less than 0.075 mm below the fine particle content is very small, less than 15%, mucous grain content of nearly 1%, in the process of excavation mud fine particles loss more, the nature of deterioration is more serious, need to supplement a large number of mud particle filling material. In addition, because the above-layered layers of the tunnel section of the middle section of the river are mostly permeable or strong, groundwater and river water are connected, making the excavation surface mud film directly face the impact of high water pressure, and has an impact on tunnel construction and force. Therefore, the mud parameters should be strictly controlled in the construction to ensure the formation of high-quality mud film, in order to ensure the stability of the excavation surface.

(5) Powder fine sand, gravel composite layer on the soft and hard, high level of formation quartz, shield excavation difficulties. Tunnel through the powder fine sand and gravel mixed formation up to 1500 m, the formation rock up and down soft and hard uneven, the difference is obvious, the tunnel is prone to produce uneven settlement, resulting in the lining structure of the local additional stress and stress concentration; At the same time, the geological report shows that the sand pebble formation contains large grain-diameter stones, the quartz content of the formation is generally higher, the wear resistance of the knife plate and the tool is very high, easy to lead to serious tool wear or pressure chamber blocking and other accidents, shield excavation control technology is difficult, high construction risk.

(6) The shield starting section, the river in the punch section and the shield reach section cover soil thickness is small: the origin depth is only 5.5 m, the thickness of the river chute section cover soil is only 9 m, the thickness of the river bottom cover soil on the side of the shield reaches the well is 9.5 m, the shield

construction control difficulty, the security risk of the shield crossing is very high.

5.3.2 Analysis Process of Shield Tunneling and Shutdown of Nanjing Yangtze Tunnel

In the August of 2008, When the right-hand shield of the Nanjing Yangtze River Tunnel is dug into the 659 ring (mileage K4 + 918), the torque of the knife plate increases sharply and the digging speed is reduced to less than 5 mm/min. When the machine was stopped for inspection, it was found that the cutter head and the tool were seriously worn, and it was impossible to continue the excavation. It was necessary to carry out the pressure-opening and maintenance of the cutter head and the replacement of the cutter. Finally, the pressure-opening operation is carried out in the fine sand and gravel sand stratum at the mileage K4 + 933, the underwater 53.25 m, and the covering soil 30 m (to the center of the tunnel section). This opening The cabin example is one of the most difficult projects to date for shield opening conditions. This section is based on the example of the opening of the project. Investigate and analyze the conditions of the indoor experiment, analyze the various conditions of the shield opening, and start from the basic principle of the balance and stability of the excavation surface, introduce the application of the mud film technology in the pressurized open cabin, and summarize the experience of successful opening.

When the shield machine is tunneling in the silt clay of ring 0–200 (K3 + 600–K4 + 000), torque of cutter head is around 3 MN m, and the tunneling speed is about 30 mm/min. When the shield machine is tunneling in the silt clay of ring 201–500 (K4 + 000–K4 + 600), torque of cutter head is around 5 MN m, and the tunneling speed is about 30 mm/min. When the shield tunneling machine reaches 500 ring (K4 + 600) and enters the silty fine sand and gravel sand strata, the torque of the cutter head is significantly increased to about 7 MN m, and the tunneling speed is significantly reduced to about 20 mm/min, since then, the project will continue to advance. By the beginning of August 2008, when the 650–659 ring (K4 + 900–K4 + 918) had been excavated, the torque of the knife plate had risen sharply to about 15 MN m. the highest time reaches 20 MN m (Fig. 5.33), and the tunneling speed is reduced to below 5 mm/min, at which time the tool wears. The warning device still did not issue an early warning message. Since the torque and thrust have reached the limit state of the equipment, it can only be stopped for inspection.

After the shield enters the gravel sand stratum, more than 20 cm of pebbles appear in the discharged muck. Before the shutdown inspection, it is found that the worn parts of the cutter head are discharged in the muck (Fig. 5.34). First, check the 71 replaceable cutter heads under normal pressure, found that the outermost cutters of each spoke are seriously worn, the blade and the cutter body wear more than half, and the cutter head still has serious chipping phenomenon (Fig. 5.35), some of the tools have been ground to the holder (Fig. 5.36). It can be inferred from the wear

5.3 Nanjing Yangtze River Tunnel (Wei7Road) Example ...

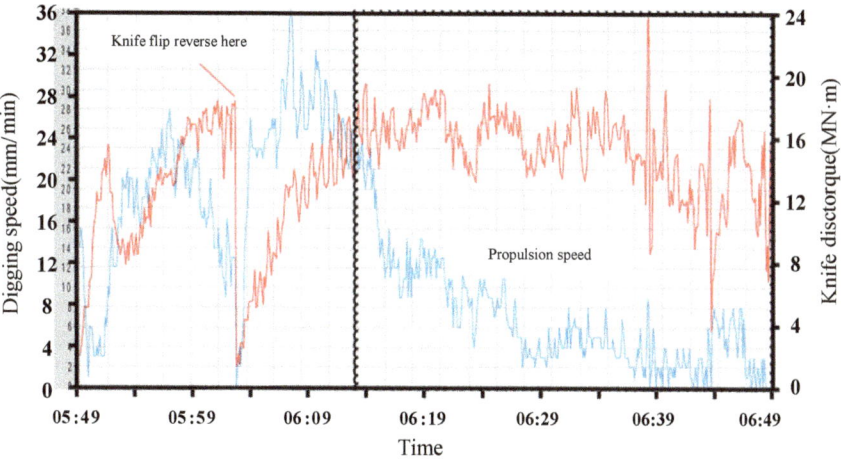

Fig. 5.33 658 ring propulsion speed and cutter torque curve

Fig. 5.34 Massive stones and foreign bodies in the discharged muck

condition of the replaceable cutter head that most of the cutters are seriously worn, and the cutter heads of some outer edges are also worn more seriously. Only by replacing 71, the cutter head can no longer be restored to normal push. Therefore, in order to complete the smooth and safe completion of the project, it is necessary to open the cabin to check the geological conditions in front of the excavation as soon as possible, remove the obstacles, and replace the tools or cutters or large-scale maintenance.

Combined with the above tool inspection results and the analysis of the slag conditions, the reasons for the shield shutdown can be summarized as follows:

(1) The actual wear coefficient of the tool is much higher than the empirical wear coefficient of the tool material, indicating that the tool is abnormal wear. The blade selection material is not suitable for the composite layer of soft and hard fine sand and gravel sand. The performance parameters (fracture resistance

Fig. 5.35 Wear of the replaceable tool

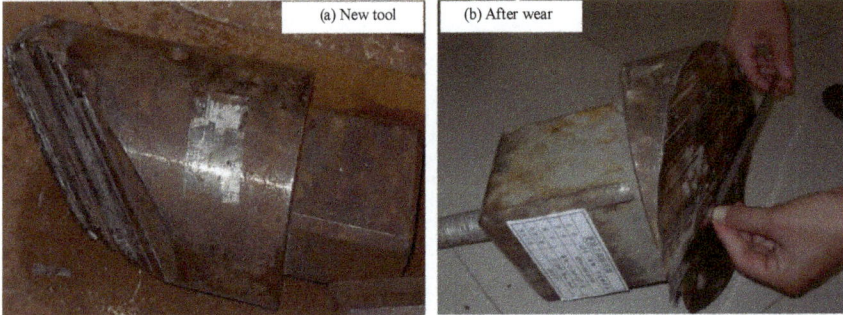

Fig. 5.36 Edge tool wear before and after wear

and hardness) of the blade material need to be tested to determine whether it is suitable for a full-face gravel sand layer.

(2) The phenomenon of blade collapse (leaving a slot) is very serious, indicating that the brazing process of the blade is not suitable for the formation or the depth of the cutting groove design is too shallow (Note: Brazing is a solid phase connection. Unlike the fusion welding method, the base metal does not melt during brazing, and the solder is cooled at a lower temperature than the base material. The heating temperature is lower than the solid phase of the base metal. And a connection method higher than the liquidus of the solder).

5.3.3 Control of Stability of Excavation Face when Opening

1) Selection of the opening plan

The position of the shield stop of the project is located at the bottom of the river at 53.25 m, and about 1/4 of the upper part of the section is a fine sand stratum. The permeability coefficient is about 6×10^{-3} cm/s; about 3/4 of the lower part is a gravel sand formation with a permeability coefficient of about 3×10^{-2} cm/s. Due to the navigation of the river surface, etc. it is impossible to carry out foundation reinforcement on the water surface. At the same time, there is no horizontal drilling device in the shield, nor can it be in the tunnel. Advance reinforcement in the road cannot be carried out by the scheme of normal pressure opening after reinforcement. Therefore, the air pressure support excavation surface and pressure are selected as plan for opening the cabin.

In the case of a pressure-opening compartment, the cabin is filled with mud and the cabin is replaced with a pneumatic cabin. Due to part, the cutter head wear has reached the position of the holder, and the holder must be welded in order to replace the cutter head. Underwater welding is extremely difficult, and the diver's view in the mud is not good, and the operation in the mud cannot be carried out. When the pressure chamber is replaced by air pressure, the stability of the excavation surface. It is difficult to guarantee. Therefore, the pressure of the upper part of the pressure chamber is 3 m, and the lower part of the 1.96 m range is still supported by mud. Since the cutter head wear is mostly distributed on the outer edge of the cutter head, each time the upper 3 m range cutter head is repaired, the next time the adjacent cutter head to be repaired is transferred to the upper 3 m range through the rotary cutter head for repair, so that the rotation is about 6 times, you can complete all the knives' repair of the disk.

2) Principle of stable balance of excavation face under local air pressure support

When the muddy water pressurized shield is normally excavated, the mud forms an impervious or slightly permeable mud film on the excavation surface, passing through the mud. The membrane balances the mud pressure with the effective stress and the pore water stress to ensure the stability of the excavation face. The force analysis at the geological section and the excavation face of the shield stop position is shown in Fig. 5.37. Replace the top 3 m with air pressure, the ideal state is through the action of the mud film, the air pressure can balance the effective stress and pore water stress at the top of the excavation face. At this time, to maintain the stability of the excavation surface, there should be.

$$P_0(P_{sl}) \geq P_w + P_e(P_a) \tag{5.3}$$

$$\Delta P = P_0(P_{sl}) - P_w \geq P_e(P_a) \tag{5.4}$$

Which in the formula:

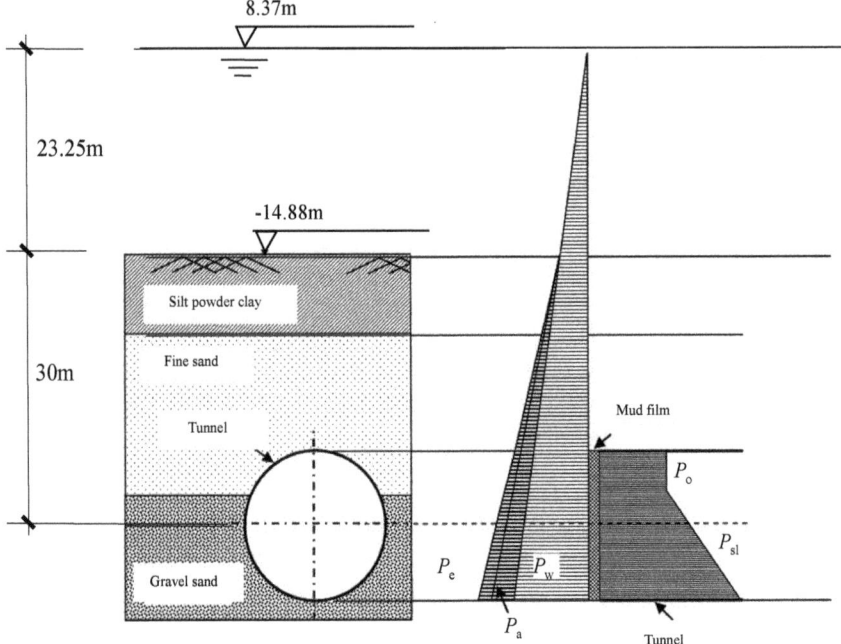

Fig. 5.37 Analysis of the geological section and the excavation face at the stop position of the shield

P_w—water pressure;
$P_e(P_a)$—static soil pressure (active soil pressure);
P_0—air pressure which in the cabin;
P_{sl}—pressure chamber mud pressure.

According to the actual formation conditions, the static soil stress (K_0 is 0.5) calculated from the top 3 m (calculated at the lower end of the air pressure surface at 25.52 m and the water depth of 48.77 m) is 0.1 MPa, and the active soil stress (K_0 is 0.3) for 0.067 MPa, the pore water stress is 0.48 MPa, the total static earth pressure is 0.59 MPa, and the total active earth pressure is 0.547 MPa. General pressure method construction situation. Under the circumstance, the air pressure can balance the pore water stress. In the fine sand stratum with large permeability coefficient, the gas pressure is difficult to balance the soil stress. If the mud film is used to offset the effective stress, the mud film needs a good air-closing effect (or a very small gas permeability coefficient). If the mud film is not gas permeable, the gas pressure acts on the soil particles by means of the mud film to resist the soil stress. Therefore, in such a high permeability and high water pressure formation, the tightness of the dense mud film and the mud film under the action of air pressure becomes a key factor.

5.3 Nanjing Yangtze River Tunnel (Wei7Road) Example …

Table 5.3 Basic properties of mud and summary of film formation experiment

Number of the mud	Density (g/cm³)	d_{85} (μm)	Viscosity (s)	Forming of mud film	Time of forming mud film (s)
C1	1.04	45	18.9	8 cm penetration belt	About 10
C2	1.06	52	21.3	2 mm mud + 3 cm penetration belt	About 2
C3	1.08	55.3	23.8	2 mm mud + 2 cm penetration belt	About 2
C4	1.15	75.8	25	5 mm mud	About 2

3) Mud film formation and air tightness test

The mixed mud of bentonite and clay on site can form a muddy mud film. Table 5.3 is the basic properties of several muds actually used in engineering, where C_1, C_2, C_3 are pure bentonite mud, and c4 is a mixture of bentonite mud and clay mud. The indoor experiment uses the gravel sand stratum with the largest permeability coefficient as the experimental stratum, and the results of the osmotic film formation in the gravel sand stratum are shown in Table 5.3.

From the experimental results of C1–C3, it can be seen that the C1 slurry penetrates into the formation under the pressure difference of 0.3 MPa, forming the osmotic zone where the mud particles block the pores, while C2, C3 The mud will form a mud on the surface after partial penetration, that is, a muddy permeable belt type mud film. In order to increase the thickness of the mud, a film formation experiment was carried out by mixing C4 mud with bentonite mud and natural clay mud, which can form a mud skin with a thickness of 5 mm faster (Fig. 5.38).

Fig. 5.38 Mud-type mud film formed by C_4 mud

Fig. 5.39 Mud film breath test

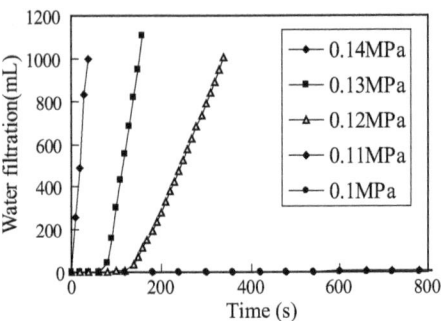

Five sets of parallel pressure gas permeation experiments were carried out after the formation of the mud film, and the displacement curves of the mud film at 0.1–0.14 MPa were respectively tested (Fig. 5.39). It can be seen that at pressures of 0.1 and 0.1 MPa, substantially no moisture oozes out of the mud film, and the experiment still oozes little water for 24 h, indicating that the compressed air cannot penetrate the mud film; when the pressure difference is added to 0.12, 0.13, 0.14 MPa, the filter out of the water suddenly increases and gas is revealing. Experiments show that the critical pressure difference (the closed gas value of the mud film) can be closed gas is 0.11–0.12 MPa, and when the air pressure is less than the closed gas value of the mud film, the mud film can meet the requirements of the closed gas. The limit pressure value that can be used in the actual project should be the stratigraphic water pressure plus the mud membrane closed gas value, i.e. 0.48 MPa + 0.12 MPa, which is 0.60 MPa. In addition to offsetting the pore water stress, the part of the closed gas value can cancel the effective stress on the excavation surface in this pressure range. This limit pressure is slightly greater than the total stationary soil pressure on the excavation surface.

4) Mud osmosis film forming scheme during opening

After the indoor experiment, a two-step film formation scheme was determined:

(1) In order to increase the stability of the excavation surface, first with a small density, low viscosity of pure bentonite mud, the use of normal excavation mud pressure to make the mud into the excavation surface a large number of penetration. The purpose is to form a certain range of mud permeability zone in the excavation surface stratum, which can increase the cohesive force of the formation and reduce the permeability. Specifically, a bentonite slurry having a density of 1.08 g/cm^3 is infiltrated at a pressure of 0.6 MPa for 8–12 h.

(2) Form a dense and closed air mud film. After the first step of mud penetration is completed, the pressure chamber is replaced with a C_4 formula for the mud. The slurry was permeated at a pressure of 0.6 MPa for about 12 h to form a mud film having a large thickness.

5.3.4 With Pressure Open Cabin Changer Implementation

Compression and air tightness tests were carried out on site prior to opening the cabin. After the C_3 mud penetrates to form a permeable zone, replace the C_4 mud; after forming the mud film, reduce the mud level by 3 m and replace it with compressed air. Considering the safety factor of the gas breakdown mud film, the pressure less than the ultimate pressure gas pressure value of 0.6 MPa is selected, and the top excavation surface is supported by the pressure of 0.5–0.57 MPa, which is greater than the total active earth pressure of the excavation face. 0.547 MPa. Maintain the pressure of the compressor and check the gas in the pressure chamber. The change in pressure, and the air supply interval between air compressors (to determine the amount of outgassing). The data shows that the pressure is stable under this pressure and the amount of outgassing is small, indicating that the mud membrane plays a role in closing the gas. Then the professional enters the cabin for maintenance work. The working time is within 2–5 h, and the mud level is restored after the personnel exit. Before the next entry, the mud injection and the gas pressure stabilization work are also performed twice, and the operation is performed in this order each time.

The situation observed by the operator in the cabin (Fig. 5.40) shows that a good quality mud film is formed on the excavation surface, don't find the phenomenon of mud film falling off and gas leakage was found; there was no unstable phenomenon such as water seepage and collapse on the excavation surface, and the excavation surface was stable. The formed mud film has good compactness, and the excavation surface is in a stable state, which can meet the requirements of safety for opening and repairing in a short time. Figure 5.41 shows the technician repairing the damaged cutter head by welding at 0.6 MPa.

From December 5, 2008 to February 13, 2009, after 71 days, after 70 times of pressure in the cabin maintenance, all the cutter repair and tool change work was completed. After the maintenance of the shield in the subsequent tunneling (660–680 ring), the torque and speed gradually returned to normal, and the torque value was significantly reduced, always maintained at around 4 MN m, and the tunneling speed

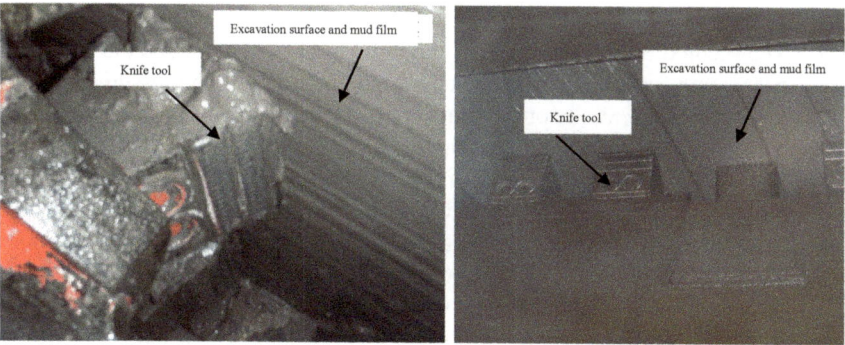

Fig. 5.40 Mud film at the excavation face in front of the pressure chamber at the time of opening

Fig. 5.41 Welding repair damaged cutter-head at 0.6 MPa air pressure

was higher than before the shutdown. There has also been an increase of more than 10 mm/min. It can be seen that the effect of the cabin repair is better, which ensures the smooth construction of the second half of the project.

5.3.5 Open Cabin Experience Summary and Discussion

1) Summary of successful experience in opening the cabin

The Nanjing Yangtze River Tunnel has been successfully opened under large depth and large water pressure, and its experience is summarized as follows:

(1) Through the closed gas and pressure action of the mud film, the water pressure on the excavation surface is completely offset, and part of the soil pressure is offset, which guarantees the stability of the excavation surface during the opening.
(2) The method of local pressure of the top 3 m is used to reduce the risk of the excavation surface stabilization caused by the full section pressure. Since the wear tool is essentially on the outer edge of the knife disc, the most worn tools can be repaired within 3 m by rotating the knife disc.
(3) The design of the pressure chamber can meet the control state of the upper air pressure and the lower mud, while the upper part of the pressure chamber is equipped with a pedestrian gate, which avoids the situation of personnel entering directly into the pressure mud.
(4) Although the open compartment section is in a large permeable powdered sand land layer, but the internal friction angle of the stoic φ reached 31.4°, has the basic strength guarantee, in close to the active soil pressure under the balance conditions of the excavation surface is basically stable.

5.3 Nanjing Yangtze River Tunnel (Wei7Road) Example ...   245

2) Stability problem of the excavation surface

The project uses a compression pressure that is less than the static earth pressure slightly greater than the total active earth pressure to ensure the stability of the excavation face. During normal shield construction, the mud control pressure is generally carried out at a static earth pressure plus an additional pressure of 20 kPa, and the pressure at the time of opening is less than this control pressure. The situation that silty sand lacking cohesive force can be slightly larger than the active earth pressure strip. The reason for the stability of the piece should be that its own strength produces a certain degree of self-standing stability. At the same time, the mud that infiltrated in the early stage may have a certain cohesive effect. In fact, the gas pressure at the top of 3 m also determines the mud pressure of the lower part of 12 m. The increase of the pressure of the gas pressure plays an important role in the stability of the entire excavation surface. On the other hand, before changing to mud before air pressure. The osmotic pressure of the slurry film formation is greater than the pressure of the gas, and the problem of mud peeling will occur after the pressure of the gas is reduced. However, after the actual opening, the mud peeling phenomenon could not be seen. It can be considered that the mud peeling of the excavation surface is caused by the groundwater pressure, and the pressure of the gas is greater than the water pressure, which can ensure the stability of the mud.

3) Regularly check the wear of the tool

The cutter head is the main component of the shield. The cutter head should be inspected regularly to check the wear of the cutter and record the key parts. Wear of blades (such as blades, wear parts, tools, etc.). The integrity and torque of the fixing material also needs to be checked. For sharply worn cutters and tools, they should be replaced and repaired in time, because the use of damaged tools for tunneling will seriously damage the cutter-head and cause more serious consequences. The frequency of inspections should be determined based on the current geological environment. After the start of construction, the best inspection interval should be determined. When the geographical conditions change, the tool should also be checked in time. To avoid the wear of the cutter and the forced opening of the cutter. The concept should be changed, and the wear of the cutter should be actively opened and repaired in time.

In order to avoid the forced shutdown caused by the serious wear of the cutter during construction, measures should be taken from the aspects of shield design and construction management. First of all, according to the geological conditions of the project, the mechanical parameters of the soil and the length of the tunneling, etc., the shape, material, structure and welding process of the tool should be rationally designed. Consider setting the normal pressure replaceable knife to facilitate the tool under high water pressure. Inspection; reasonable design and layout of the wear warning system to effectively detect the wear of the tool cutter and avoid the inspection of the cabin.

In addition, in the construction management should also: to eliminate blindly fast, to ensure the quality of construction; according to the actual changes in geological conditions, timely adjustment of excavation parameters; pay attention to the management and changes of muck, control the amount of excavation within a reasonable range, and pay attention to soil The size of the particles changes, whether there are foreign objects, etc.; regular inspection, to obtain the wear coefficient of each layer, predict the amount of wear, and pay attention to the accumulation of experience and timely summary; focus on the first-hand data collection, sorting, analysis, etc. The change of the excavation parameters is related to the formation of the shield machine and the wear of the tool, and the law is summarized.

5.4 Example of Saturating Method of Pressure-Based Open Caving in Nanjing Wei Three Road Crossing River Tunnel

5.4.1 Nanjing Wei Three Road Crossing River Tunnel Project

1) Project overview

The Nanjing Wei Three Road Crossing River Tunnel Project is located between the Nanjing Yangtze River Bridge and the Nanjing Yangtze River Tunnel of Wei Seven Road. It is about 4.9 km away from the Yangtze River Bridge and 4.7 km above the Nanjing Yangtze River Tunnel. The tunnel section of the engineering shield tunnel is designed as a double-tube tunnel. It runs from the west side of the top mountain turntable in Pukou District of Nanjing, across the planned sand river road and Jiangbei Riverside Avenue. After tunneling through the Yangtze River and Jiangnan Riverside Avenue, the South Tube Tunnel and Dinghuai Street is connected, with a length of 5530 m, of which the shield section is 4134.8 m, which is mainly responsible for the traffic of Wei Three Road and Pukou. Tunnel is connected with Yangtze River Avenue, with a length of 4930 m, of which the shield section is 3537.8 m, which mainly bears the traffic connection between Yangtze River Avenue and Pukou. The project location is shown in Fig. 5.42.

The tunnel section of the Wei Three Road Crossing River Tunnel Project in Nanjing is constructed by two Φ14.93 m mud-water pressurized shield tunnels, which are started from the Jiangbei starting well and tunneled in the same direction. The tunnel excavation diameter is 14.98 m, the tunnel outer diameter is 14.5 m, and the inner diameter is 13.3 m. The completed Nanjing Yangtze River Tunnel (Wei 7 Road) is the same. The tunnel is designed as a double-deck, two-way, eight-lane tunnel. To the four lanes, the upper level is from the north bank to the south bank, and the lower floors are from the south bank to the north bank. The single pipes have independent capacity. From Pukou to Dinghuaim, there will be two tunnels

5.4 Example of Saturating Method of Pressure-Based ...

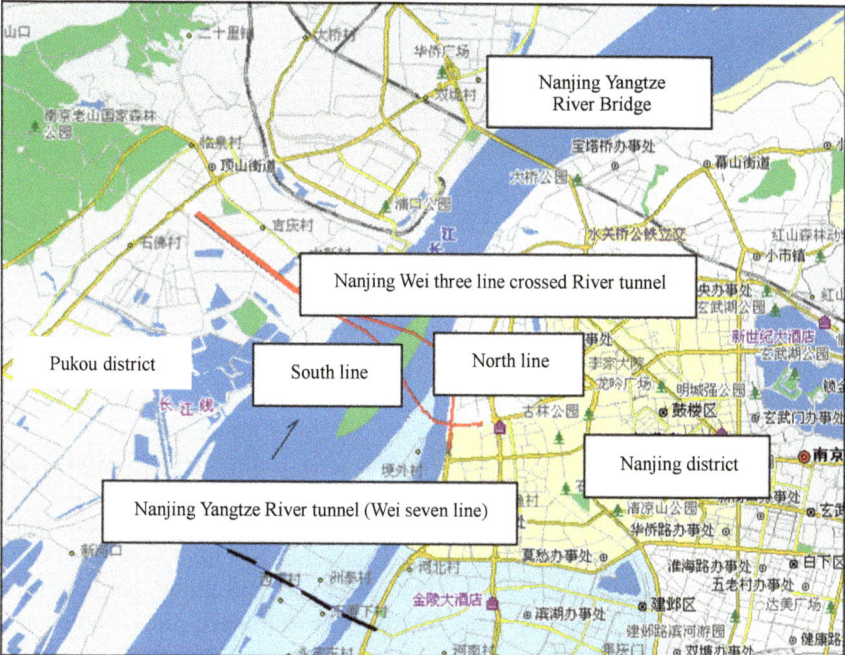

Fig. 5.42 Location of Nanjing Wei Three Road Crossing River Tunnel Project

"X" shaped crossing the river. The lane width is 3.5 m, the height of each floor is 4.5 m, and the design speed is 80 km/h.

2) Geological conditions

The tunnel crossing site is the Yangtze River alluvial plain, mainly for the levee beach, the high floodplain in the levee, the low floodplain in the levee, the Yangtze River waters and the Jiangxin island and Qianzhou. The average annual high water level of the Yangtze River in this region is 8.37 m. The key water conservancy facilities of the tunnel are the Yangtze River embankment in Nanjing. It is a high-grade embankment. The near-water side of the embankment adopts dry block stone slope protection and slurry block stone foot protection, and the bank slope and embankment are stable. The distribution of strata in the section of the project is shown in Fig. 5.43. It can be seen from the geological survey report that there are mainly four types of stratum lithology in the cross-river tunnel section.

(1) powder, fine sand formation: saturated, dense to dense, particle-grade matching, the main mineral composition for quartz, long stone, including shell fragments and a small amount of humus, local laminate clay thin layer, single layer thickness 1–30 mm, the bottom contains a small diameter of 2–10 mm gravel, penetration coefficient is about 7×10^{-3} cm/s.

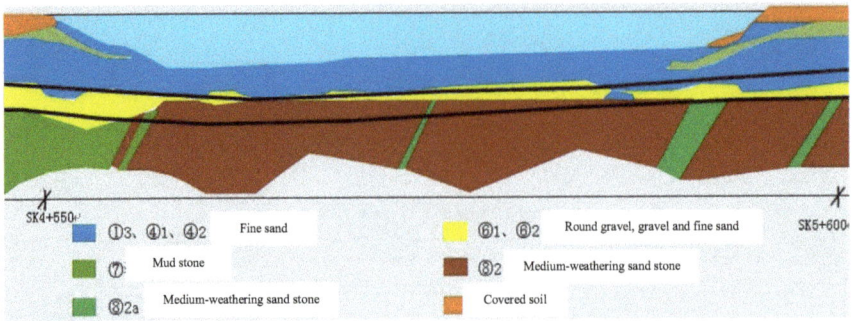

Fig. 5.43 Schematic diagram of the geological section of the middle section of the Weisan Road Crossing River Tunnel

(2) Round gravel, pebbles and powder fine sand composite formation: round gravel, pebble layer saturation, dense, particle-level matching, female rock composition to quartz sandstone, zircon and gray rock-based, sub-circular, particle size 2–100 mm, occasionally large up to 150 mm, content 50–80%, medium coarse sand and a small amount of clay soil filling, partially sandwiched silty clay and fine sand. The layer is distributed more continuously except for local deletions.

(3) Mud rock formation: mud structure, layered structure, composition of clay mineral-based, containing a small amount of long stone and quartz, mud cement, rock integrity, fissure non-development, crack surface see plaster filling, hammering dumb, no rebound, there are deep dents, hands can be crushed, easily dispersed after immersion, air dry and easy to crack. Localised as mud powder sandstone, RQD (rock mass index, mainly reflects the degree of rock integrity, i.e. the degree of development of the fissure in the formation of the site. The higher the RQD value, the more complete the rock, the better its mass) is generally 70–90%, the formation self-stabilization capacity is better.

(4) Medium weathered sandstone: brown red, powdered structure, layered structure, mineral composition for quartz, long stone and other mineral composition, mud calcium bond. Among them, $⑧_2$ medium weathered sandstone layer in the fissure is more developed, the fissure inclination is more in 40°–85°, the fissure is more closed or filled with solution stone veins, the rock body is more complete, the core is column, the local is a block, RQD is generally 50–80%; $⑧_{2a}$ medium weathered sandstone layer affected by the structure, rock fragmentation, rock fragmentation, the core is broken. A small number of short bars, after immersion hands can be opened, RQD is generally 10–30%.

At the same time, the geological report also shows that the quartz content of the sandstone in the various layers that the shield passes through is generally high (about 60%), especially the quartz content of the gravel and pebble formation is as high as 70% or more, and the rock is hard.

5.4 Example of Saturating Method of Pressure-Based …

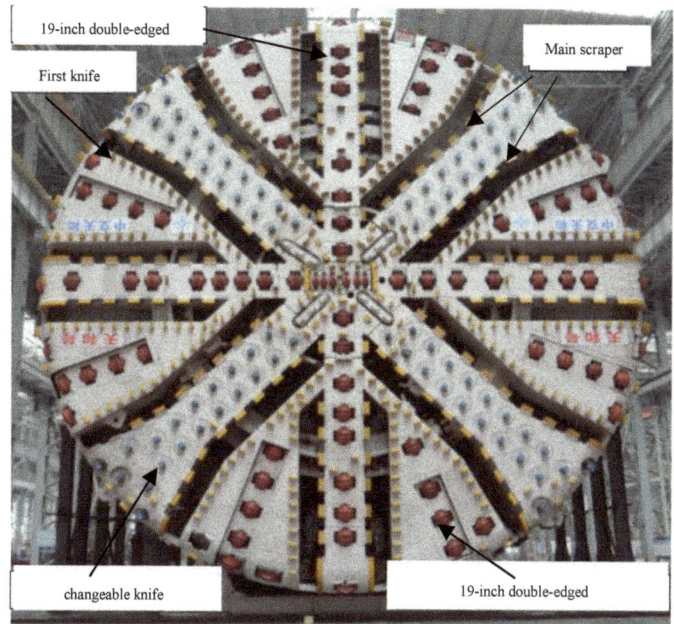

Fig. 5.44 Mud shield shield panel

3) Shield selection

The Nanjing Weisan Road Crossing River Tunnel Project uses a 15.02 m mud-water balance composite shield. The cutter is designed to meet the sandstone pebble formation with a quartz content of up to 65% and the rock layer with a compressive strength of 120 MPa, especially the upper soft and hard composite. Demand for excavation in the stratum. Figure 5.44 shows the shield cutter panel.

The cutter head adopts a spoke panel type with an excavation diameter of 15.02 m and an opening ratio of 25.7%. The cutter head is equipped with a 19in double-edged hob, a leading knife, a scraper, a replaceable cutter and a 19in double-edged roll. Tool configuration design concept: When digging in the clay layer or sand layer, the soft soil is mainly excavated by the first knife and the replaceable cutter, and the cutter can be worn by the wear detection.

Check it. When the cutter needs to be replaced, the normal pressure tool change method can be adopted, and the operator enters the sealed spoke through the normal pressure passage to replace the cutter. Before the shield passes through the sand egg layer and rock formation, the cutter retracts 10 mm, which is lower than the 60 mm of the hob. Afterwards, the excavation process mainly relies on the hob on the spoke panel to excavate the geotechnical soil (the distance from the panel is 160 mm), and the panel arrangement 38 spare push-out hob is flush with the panel height. When the hob detects severe wear, the hob can be replaced by a pressure change tool.

4) Analysis of engineering features and difficulties in mud application

The buried depth and water pressure of the tunnel are large and the geological conditions are complex. The maximum buried depth of the Wei Three Road Tunnel in Nanjing is 7 m, and the top of the tunnel is -62 m. It is a difficult technique to open the ground in the groundwater with large groundwater pressure, complex geological conditions and high permeability. It will cause the excavation surface to collapse, the groundwater to penetrate, and even the entire tunnel will be scrapped when it is bad, so it has great engineering risks. Among the completed cross-river tunnels, the maximum water pressure of the Greenhart tunnel in the Netherlands is 0.5 MPa; the other two shield tunnels on the Yangtze River (Shanghai Hu Chongsu tunnel and Wuhan cross-river tunnel) have a water pressure of 0.5–0.5 MPa; The shield working pressure of the Yangtze River tunnel is 0.65 MPa, and the maximum water pressure of the tunnel passing through the stratum reaches 0.77 MPa, and the stratum crossing is complex and in the composite formation of fine sand, sand pebble and sandstone, and the full-section pebble layer. These are highly permeable formations, muddy forming is a difficulty. Even if a mud film is formed, the stability of the mud film needs to be carefully considered. A schematic diagram of the section of the tunnel crossing the composite stratum is shown in Fig. 5.45.

The shield crosses the gravel sand stratum, and the particle size is single, and the mud film is difficult to form. According to the development direction of the shield

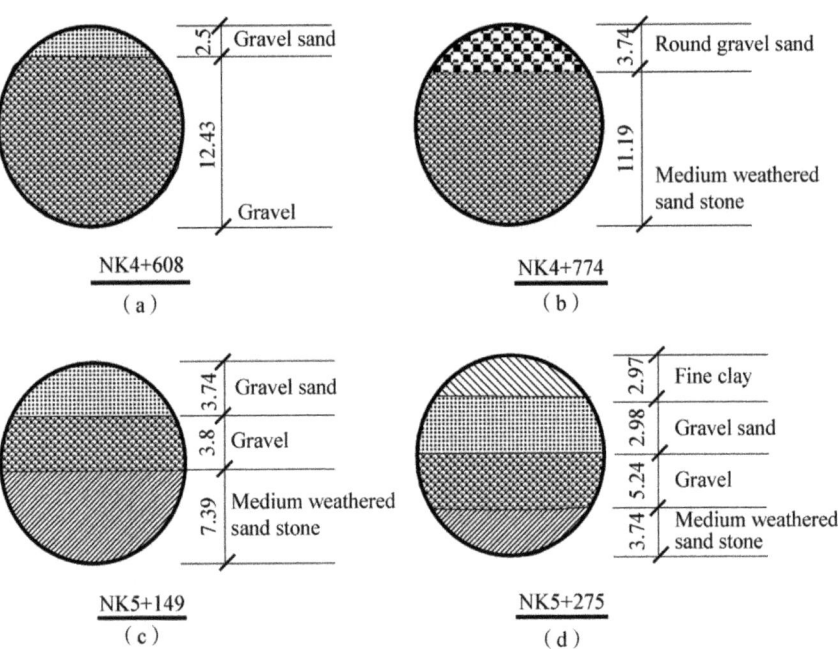

Fig. 5.45 Schematic diagram of tunnel crossing composite stratum

5.4 Example of Saturating Method of Pressure-Based …

technology in the Japan Tunnel Society survey In the development direction of the shield technology, the muddy water pressure type shield is used for construction when the water pressure is greater than 70 kPa, 74 μm, the soil particle content is less than 8%, and the unevenness coefficient is less than 6, the muddy water pressure type shield is used for construction. Excavation excavation is extremely prone to damage without assistance, and the construction carried out by the measures is more difficult without assistance. The penetration coefficient of the gravel formation through the three-way Yangtze River tunnel in Nanjing reaches 10^{-2} cm/s, the permeability is strong, the formation pores are large, and the mud in the pressure chamber can easily penetrate the square formation in front of the excavation under the action of high pressure. The layer makes the stable mud film difficult to form, the pressure of supporting the formation is difficult to guarantee, and the safety of construction has great variables.

The shield tunneling distance is long, and the lithology of the rock passing through the stratum is hard, which will lead to serious wear of the shield cutter. During the construction of the Nanjing Yangtze River tunnel, the cutter is seriously worn, and the shield fails to be properly drilled. The pressure-opening maintenance under the water pressure of 5 m high at the bottom of the river severely affected the normal construction period and increased the huge opening of the cabin.

The investment in cutting tools and cutter heads has brought huge construction risks and adverse social impacts. The shield section of the north line of the Yangtze River Tunnel in Nanjing is 3537 m; the shield section of the south line is 4134.8 m, It is more than 1,000 m longer than the Nanjing Yangtze River Tunnel, which has been completed and opened. Geological reports indicate that the quartz content of sand in the layers of the shield is generally higher (about 60%), Its quartz content in gravel and gravel formations is as high as 70% or more (commonly known as "topazite"). The rock is hard and easily causes serious shield cutters. Wear and shield construction was forced to stop.

The cross-section of the stratum has a soft and hard composite stratum section, which is not conducive to the formation of mud film. Under this engineering geological condition, the danger of shield tunneling is extremely high. The stratum through which the shield tunnel passes has a composite stratum of fine sand, sand pebble and sandstone, and a full-section pebble layer. The lithology of the stratum is unevenly soft and hard, and the difference is obvious. The tunnel is prone to uneven settlement, which causes local additional stress and stress concentration of the lining structure. At the same time, it is extremely difficult to control the excavation parameters, and it is easy to appear collapsed and muddy. These composite sections are high-permeability formations, and the formation of mud membranes is difficult. When the shield enters these composite formations, the mud formula used in the uniform formation can no longer meet the needs of construction in the formation, and the mud formula needs to be adjusted. A mud formulation suitable for various forms of composite sections is obtained. In addition, the shield may be used to open the knives during these composite sections. So it poses a huge threat to the safety of the entire tunnel and staff.

These composite formations contain large-diameter stones, which have less clay content, poor self-slurrying ability, poor mud stability, sedimentation is easy to occur at lower flow rates, and the slurry has a long transport distance and large particle size. The pebble and sandstone are easily blocked by mud tube sedimentation.

When the shield is shut down for opening, the stability and safety of the excavation face is also one of the difficulties of the project. Due to the large section diameter of the tunnel, the buried depth is large (the maximum buried depth is 7 m, the top of the tunnel is −62 m), the water pressure is large, and the geological conditions are complex. High permeability, these are unfavorable to the stability of the excavation face. When the shield is stopped in this condition, the excavation is in front of the excavation. Part of the mud will be replaced by gas, it is difficult to ensure long-term pressure stability. It is more difficult to maintain the pressure in the pressure chamber relative to the mud water pressure, and it is more prone to a sudden drop in pressure caused by gas leakage, which will cause the collapse of the excavation surface, groundwater breakdown, and also threaten the personal safety of the construction personnel. Retirement, therefore has great engineering risks.

5.4.2 Application of Mud Film Closed Gas in Pressurized Open Compartment

1) Overview of shield tunneling

When the S line of the Wei Three Road Crossing River Tunnel in Nanjing arrived at the Yangtze River Dike, a total of 468 rings were drilled, about 936 m. At this time, the density of the mud in the returning pool is 1.15 g/cm^3, the viscosity is 21 s, and the mud material is mainly bentonite. The mud can meet the normal conditions. Stability requirements for excavation surfaces during excavation. The shield has passed through the stratum mainly consisting of silty clay stratum, fine sand stratum, medium coarse sand and gravel sand stratum. The medium coarse sand and gravel sand stratum has higher quartz content, and occasionally more than 5 cm in the discharged muck. Stones, the wear of the tool is serious. After the machine stop, it is found that some atmospheric pressure replaceable tools can be replaced and checked by the replacement of normal pressure replaceable tools, which is damaged badly by Rub. In order to ensure the smooth tunneling of the shield at the bottom of the river, it is necessary to carry out the inspection and maintenance of the cutter of the mud shield between the rivers to maintain the good working performance of the shield.

The s-shi shield stop position of the Wei Three Road Crossing Tunnel in Nanjing is 468 ring, where the stratum is a composite stratum and the upper strata are. The gravel sand layer, which occupies half of the tunnel section, is composed of pebbles, gravel strata and some fine sand strata. The gravel sand and pebble gravel formation is a high permeability formation, and the permeability coefficient of the formation is as high as 3.47×10^{-2} cm/s and 4.05×10^{-2} cm/s. The particle gradation curve

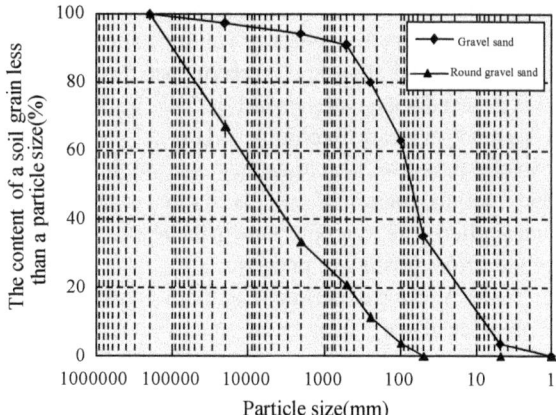

Fig. 5.46 Particle distribution curve of gravel and pebble gravel stratum

of the formation is shown in Fig. 5.46. It can be seen from the particle distribution curve that the gravel sand layer still has more fine particles. Exist, although this has certain advantages for mud film formation and mud film closure, the formation is still a high permeability formation, It is still difficult to form a mud film of higher quality. In the pebble gravel stratum, the fine particle content is very small, which is more challenging for mud film formation and mud film closure.

After the shield enters the bottom of the river, the stratum is mainly composed of coarse-grained strata and rock layers. The wear of the cutter is more serious, and when the shield is at the bottom of the river, the challenge of the gas-tightness of the mud is greater when the tank is opened. Therefore, it is necessary to inspect and maintain the shield cutter before the shield enters the river bottom.

The maximum pressure that the entry operator can withstand is 0.57 MPa. At this time, at the shield stop position, the depth of the shield center line is about 53.8 m. After calculation, the mud pressure to be set is about 0.7 MPa, so Cabin personnel need special protective measures to protect the safety of the incoming personnel.

2) Closed-air effect experiment of film formation after first penetration

In view of the problem that the formation pores are large and the fluid loss is large when the mud is formed, firstly, the mud with less density penetrates into the formation.

After the initial clogging of the pores of the formation, a high-quality mud-type mud film is formed on the excavation surface with a higher density mud, and then the stability under pressure is tested, and the non-permeable zone is only The airtightness of the mud film is compared, the feasibility of a mud film forming scheme for the mud film after the osmosis belt.

A total of three sets of comparative experiments were conducted this time: A, B, C. Among them, A_1, B_1, and C_1 experiments are to form the osmotic zone, first, then form the mud, and then test the stability of the mud film under pressure; A_2, B_2, C_2 is a mud-like type of skin, and then the stability of the film under air pressure is tested. The experimental mud properties and experimental results are shown in Table 5.4. Select ⑥$_1$ pebble gravel formation as the time experimental formation of mud membrane closed gas performance test.

The test device is shown in Fig. 5.47. In this experiment, the experimental column I is a mud permeation column, and the gas pressure of 0.6 MPa is set on the mud as the osmotic pressure; the experimental column II is used as the back pressure column, and a certain amount of water is charged in the lower part of the experimental column and applied. The water pressure in the formation allows for a more realistic simulation of the actual formation pressure.

In the A group experiment, in the A 1 group, the in-situ expanded bentonite slurry with a density of 1 0.06 g/cm^3 and a viscosity of 17.81 s was first infiltrated into the formation to form an osmotic zone (Fig. 5.48), followed by density. A slurry of 1.12 g/cm^3, viscosity 52 s, and strontium d_{85} of 45 μ m was infiltrated into a film. After 4 h of infiltration, a 2 mm thick muddy mud film was formed, as shown in Fig. 5.49. The air-tightness of the mud film was tested, and the gas-tight performance of the mud film was good, and the gas-tightness was maintained for 24 h under a pressure difference of 0.2 MPa. In the control group, a 3 mm mud film was formed in the experiment without forming the osmotic zone, and the mud film was under a pressure difference of 0.2 MPa. The time to keep the air tightness is only 2 h.

In the B group experiment, in the B1 group, the bentonite mud which was firstly expanded at a density of 1.06 g/cm^3 and a viscosity of 18.25 s was used. The slurry penetrates into the formation to form an osmotic zone, and then a slurry with a density of 1.15 g/cm^3, a viscosity of 36.65 s, and a TF5 of TF6 is used. After 4 h of penetration, a 3 mm thick muddy skin is formed. The mud film is shown in Fig. 5.50. The mud film can maintain a closed gas time of 12 h under a pressure difference of 0.2 MPa, and obvious cracks appear on the surface of the mud film after ventilation, as shown in Fig. 5.51. In the B2 group test, no osmotic zone was formed. After 4 h of infiltration, a 3 mm mud film was formed, and the mud film was able to maintain the gas shut-off time of 4 h under a pressure difference of 0.2 MPa.

The mud in the C group experiment was the mixed slurry mud of bentonite mud, which was first used in the C1 group of experiments with density 1.07 g/cm^3, viscosity is 29 s (Mars funnel viscosity), and thumping of d_{85} 43 μm on-site puffed soil mud to penetrate the formation, forming a permeable zone, followed by a density of 1.15 g/cm^3, viscosity is 45 s (in the first test) The mud of the Mars funnel viscosity forms a mud film, which, after 4 h penetration, forms a 5 mm thick mud film, as shown in Fig. 5.52. The mud film in the pressure difference of 0.2 MPa, can maintain the shut-air time of 3 h, mud film damage in the mud membrane formed a small hole, as shown in Fig. 5.53. In the C2 group of tests did not form a osmotic belt, after 4 h penetration, forming a 5 mm mud film, but the mud film at 0.2 MPa pressure difference can remain closed gas for 1.7 h.

Summary of test results: formulated bentonite mud (density 1.15 g/cm^3 or so), In the puffed soil pulp according to the ratio of 1:1 (puffed soil pulp than the slurry in the plasma tank) to add the slurry in the slurry pool, the density of 1.12 g/cm^3, viscosity 52 s, d_{85} for 45 μm of mud can form a mud skin. This mud can close at 24

5.4 Example of Saturating Method of Pressure-Based …

Table 5.4 Basic properties of mud and experimental results

Experimental content		Mud density (g/cm³)	Viscosity (s)	d_{85} (μm)	Penetration time (h)	Pressure difference (MPa)	Film thickness (mm)	Closed air time (h)
A1	Form a permeable belt	1.06	17.81	36	0.5	0.2	0	–
A2	Form a mud	1.12	52	45	4	0.2	2	24
	Form a mud	1.12	52	45	4	0.2	3	2
B1	Form a permeable belt	1.06	18.25	36	0.5	0.2	0	–
	Form a mud	1.15	35.65	40	4	0.2	3	12
B2	Form a mud	1.15	35.65	40	4	0.2	3	4
C1	Form a permeable belt	1.07	29	36	0.5	0.2	0	–
	Form a mud	1.15	45	43	4	0.2	5	3
C2	Form a mud	1.15	45	43	4	0.2	5	1.7

Fig. 5.47 Test device for airtightness of pressure limiting cement film

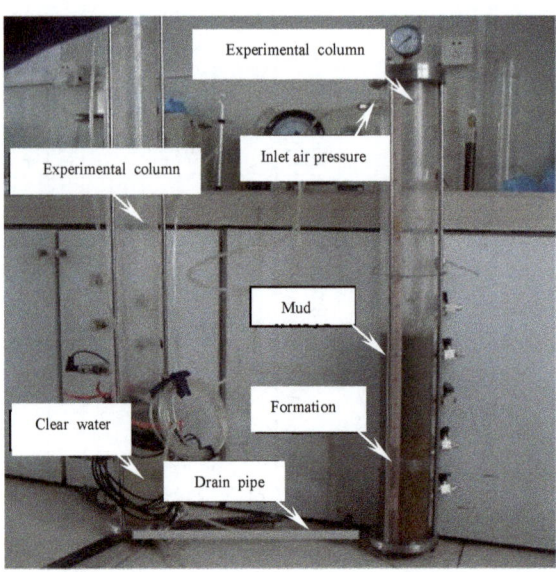

Fig. 5.48 A_1 group of experiments first formed an infiltration zone

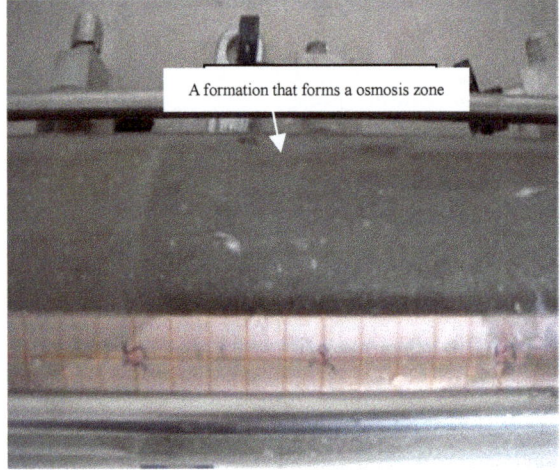

h at differential pressure (difference between air pressure and pore water stress) at 0.2 MPa.

3) Experiment on the effect of CMC addition on mud penetration and mud film closure

The gas-tightness of the mud membrane is the key to ensuring the success of pressure-opening in high-permeability formations. However, the gas-tight performance of the mud film is affected by various properties of the mud, especially the effect of the cmc

Fig. 5.49 Mud film formed in the A_1 experiment

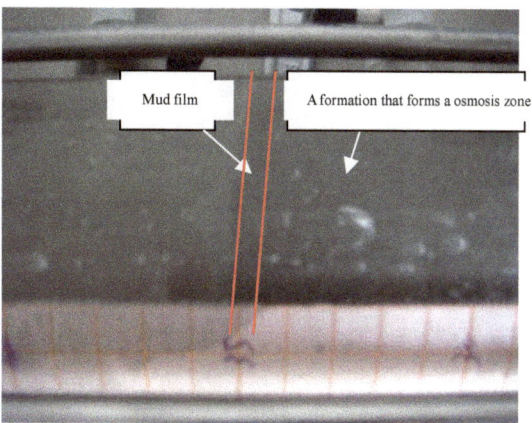

Fig. 5.50 Mud film formed in the B1 group experiment

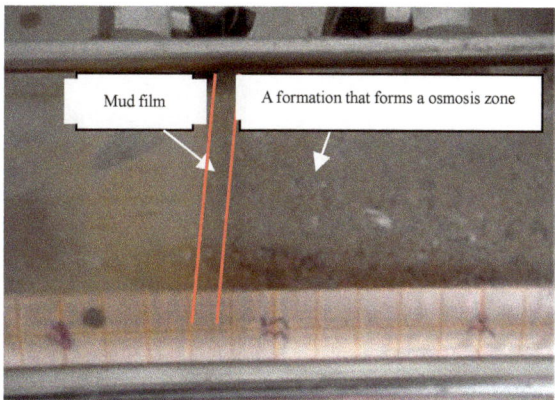

Fig. 5.51 Morphology of mud film destruction in group B1

Fig. 5.52 Mud film formed by mud in group C_1

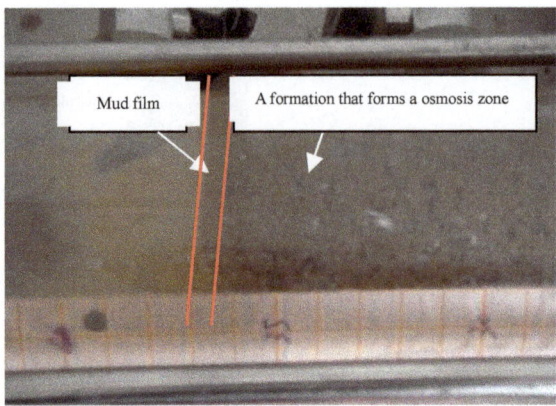

Fig. 5.53 C_1 group after the destruction of the mud film

material commonly used in mud preparation on the formation of mud film, which is a problem that must be clarified during construction. In the experiments of d, e, and f groups, the mud in the slurry pool was used as the base slurry. Add CMC dry powder with 0.5, 0.1 and 0.03% mud mass. The basic properties of the mud are shown in Table 5.5. The osmotic pressure difference when the slurry forms a mud film is 0.12 MPa, and the thickness of the mud film formed after 2 h of penetration is 3, 5 and 10 mm. Then, the mud was discharged, and the gas-tight effect of the mud film was tested by applying a pressure of 0.2 MPa. Under the experimental conditions, the gas-closing time of the three groups of mud films was 2 h, 6 h and more than 15 h, respectively. Figure 5.54 is formed by group E mud film 6.

Three sets of experiments show that the addition of CMC can increase the viscosity of the mud faster, reduce the water loss of the mud and the thickness of the mud film; as the amount of CMC increases, the thickness of the mud film rapidly decreases, and the gas shut-off time of the mud film also changes. Short, in order to form airtight

5.4 Example of Saturating Method of Pressure-Based … 259

Table 5.5 Basic properties and experimental results of the slurry in the slurry pool + CMC

Mud	The amount of CMC (%)	Relative density	Ma's viscosity (s)	Penetration time (h)	The difference of penetration pressure (MPa)	Mud thickness (mm)	The difference of closed air pressure (MPa)	Closed air time (h)
D	0.5	1.16	122	2	0.12	3	0.2	2
E	0.1	1.16	48	2	0.12	5	0.2	6
F	0.03	1.16	40	2	0.12	10	0.2	>15

Fig. 5.54 Mud formed by mud in group

better mud film, the amount of cmc should not be too large and should be controlled within 0.1%.

In order to check the wear of the cutter, the Nanjing Weisan Road Crossing Tunnel Project has been carried out twice. The cabin, through the observation of the mud film morphology during the opening process, also found the same phenomenon as the above test. The first pressurized open tank is in a coarse sand formation with a permeability coefficient of 1.4×10^{-2} cm/s. First, a bentonite clay with a higher density and viscosity (with a small amount of CMC added) is prepared in the mud pool. Replacing the mud in the mud chamber through the mud circulation to make the mud in the tank density is 1.2 g/cm^3, viscosity is 48 s, the set mud pressure is 50–100 kPa higher than the formation water pressure, about 8 h The mud penetrates and forms a dense mud film with a certain thickness on the excavation surface. Verify the mud by on-site pressure test. The film has a good gas shut-off effect. The shape of the mud film when entering the cabin is shown in Fig. 5.55.

The second pressurized open compartment is in a pebble formation with a permeability coefficient of 3.0×10^{-2} cm/s. First, the mud in the mud chamber is replaced with a pure bentonite slurry with a density of 1.2 g/cm^3 and a viscosity of 50 s. In order to further improve the viscosity of the mud, directly add about 2% of CMC dry

Fig. 5.55 The first time with pressure open cabin high quality mud film

powder to the mud chamber, adjust the viscosity of the mud to more than 10 s, the mud density is about 1.2 g/cm^3, and then use the same method as the first time. The mud penetrates into the film. However, the filming effect of this high-viscosity mud is not ideal. The part of the gravel in the formation is exposed outside the mud film, and the mud film in the pressure chamber is shown in Fig. 5.56. Comparison of mud film from two open cabins. It can be seen that the excessively high amount of CMC will make it difficult for the mud to penetrate into the formation, making it difficult for the mud particles to be on the surface of the formation. The accumulation of mud film has an adverse effect on the mud membrane.

4) Summary of mud filming experience during opening

According to the test and the actual opening situation, the experience of the pressure opening of the Weisan Road Tunnel in Nanjing can be summarized as follows:

Fig. 5.56 The second pressurized open mud film

(1) Implementation scheme: Firstly, a section of the permeable belt is formed, and then a muddy-type mud film with a certain thickness is formed, which can better achieve the mud film closed gas and improve the stability of the excavation surface.
(2) The addition of CMC can improve the viscosity of the mud more quickly, but with the increase of CMC addition, the mud permeation water loss will become more difficult, the formation of the mud film thickness is small, the mud film's closed gas effect is also worse.

5.4.3 Saturation Method with Pressure into the Cabin Change Process

Saturated method with pressure-opening tool changer is more effective than conventional compressed air-pressure open-cavity tool change technology rate. The operator lives in a living room with a certain pressure. When it is necessary to carry out the opening and changing operation, the operator is transported by the shuttle, and the shuttle is docked with the shield human cabin to balance the pressure difference, and then the operation can be started. Saturated method with pressure into the cabin. The tool change process is basically the same as the conventional compressed air belt opening and changing process, except for the docking and transportation of the living compartment and the shuttle compartment. The interior of the living compartment and the shuttle compartment are shown in Figs. 5.57 and 5.58, Figs. 5.59–5.65 shows the lifting of the shuttle cabin from the outside to the docking process with the human brake. Figure 5.66 is the flow of the saturation picture.

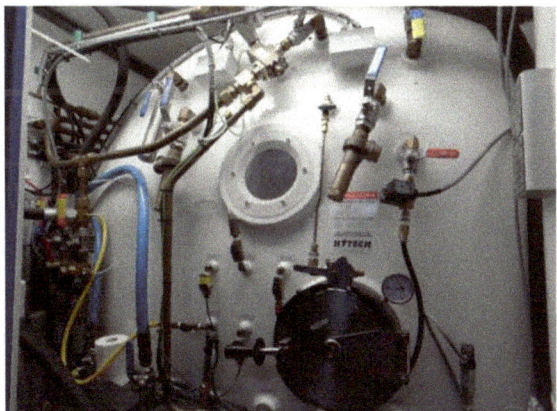

Fig. 5.57 Living compartment

Fig. 5.58 Shuttle cabin interior

Fig. 5.59 External lifting shuttle cabin

Fig. 5.60 Transport vehicle transports the shuttle cabin to the tunnel

5.4 Example of Saturating Method of Pressure-Based … 263

Fig. 5.61 Internal lifting shuttle cabin

Fig. 5.62 Running of the shuttle cabin on the track

Fig. 5.63 Shuttle compartment traverses and is connected to the assembly boom

Fig. 5.64 Rotating the shuttle cabin through the assembly machine

Fig. 5.65 Shuttle cabin docked with the human lock cabin

5.4.4 Summary of Saturation Method with Pressure Change into the Cabin

The use of saturated belt pressure opening technology has greatly improved the safety of the opening operation. This technology has become the current shield of China. One of the technical focus of the industry. The complete set of saturated pressure-opening equipment and technical methods equipped with saturated pressure-opening technology, including the living room, shuttle compartment and shuttle cabin transportation docking equipment, are the necessary premise and basis for the construction of saturated pressure-opening cabins. Pay attention to the following points when using the saturation method to open the cabin:

5.4 Example of Saturating Method of Pressure-Based …

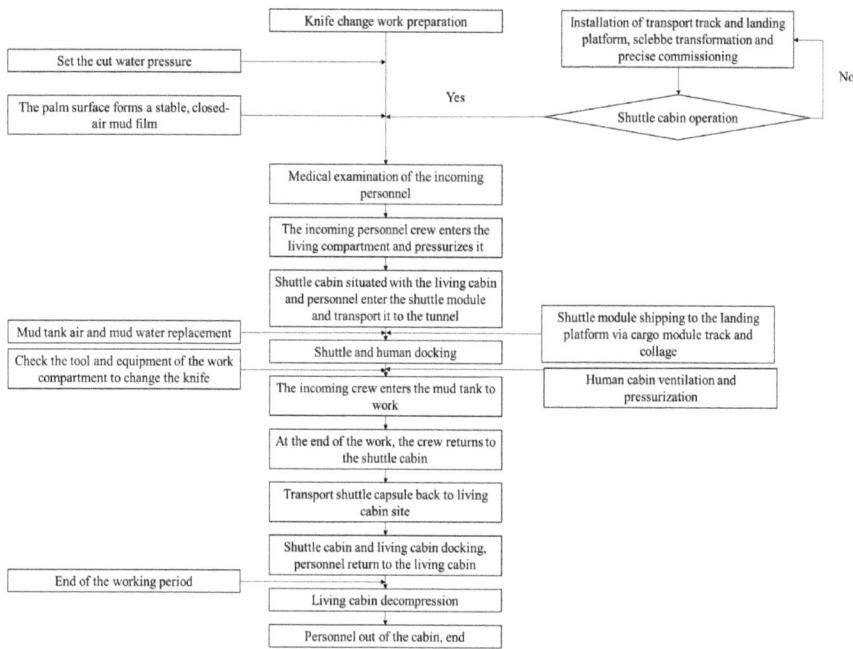

Fig. 5.66 Flow chart of the saturation method

(1) To ensure the gas supply and energy supply of the shuttle module, to ensure the safe stability of shuttle module transport and the fast docking of the shuttle module. The technical methods summed up through practical research have been well verified during the construction of the Three-Way River Tunnel project in Nanjing.

(2) When the saturation method opens the cabin, the mud liquid surface drops and the excavation surface is reinforced with compressed air support. The air pressure does not change along the vertical direction, that is, the pressure is the same at each point, so that the form of air pressure and the mud pressure form is quite different, the setting of the air pressure must be careful. The setting of compressed air pressure should refer to the limit support pressure under the condition of mud support in the position of the reduced mud surface, and consider a certain amount of surplus. The more the mud fluid drops, the greater the air pressure needed to keep the excavation surface stable. In addition, a certain amount of excess should be considered.

(3) When using saturation method with pressure-opening cabin change technology, the operators work in the cabin for a long time, during which the closed gas time and the gas-closed gas value of the mud film are required to meet the requirements of the open cabin operation. Therefore, it is particularly important to improve the gas-closed time and the gas-closed gas value of the mud film. The three methods mentioned above can be used to extend the gas-closed time

of the mud film, increase the closed gas value of the mud film, and ensure the stability of the excavation surface. The actual engineering effect shows that the formation is effectively silted with low density mud, and after the effective penetration zone is formed, and then the high density mud is permeable, which is more beneficial to the formation of a denser mud-type mud film.

(4) The process of saturation opening operation is more complex than compressed air operation, and its safety factors are more, and the analysis and evaluation of the safety factors of the saturation open cabin operation is helpful to control the construction risk and improve the safety and reliability of the saturated open-cabin construction.

Chapter 6
Other Special Tool Changing Techniques

From clay soil, sandy soil, gravel and cobble soil to soft rock, shield tunneling method is more and more applicable to all kinds of soil.

At the same time, as shield construction tends to develop in the direction of long distance and large depth, the demand for construction technology under the condition of composite strata is increasing day by day. Shield tool replacement has become a major difficulty in the construction, whether under normal pressure or with pressure will affect the construction progress and construction safety to some extent. When the shield cutter needs to be replaced in the tunneling construction of shield tunnel, it is usually used to open the cabin at normal pressure or with pressure cabin. The operator shall change the cutter to the front of the shield cutter head. This approach not only has a high risk of accidents, but also increases the time and cost of construction. In order to ensure the safety of tool change, tool change technology has been researched and developed in many countries.

6.1 Tool Replacement Technology While Encountering Riprap and Bedrock Shield

6.1.1 Project Summary

Ma Liu Zhou Traffic Tunnel (Hengqin third Channel) is located in the north end of Hengqin Island in Zhuhai City, connecting northward to Nanwan City, Zhuhai City. The project starts from Hengqin Middle Road to the south, and then goes north through Huandao North Road, Qinhai North Road, and north along the planned Baozhong Road to Nanwan Avenue after crossing the Ma Lizhou waterway. The total length of the project is about 2834.6 m, which is divided into three sections: Hengqin tunnel south connection (south bank and north bank construction general contract section), Hengqin tunnel (shield section design and construction general contracting section), Hengqin tunnel north connection (south bank and north bank construction general contract section 2). The route direction is shown in Fig. 6.1.

The shield section of the tunnel is 1082 m in length, the tunnel is a double tube circular section, and a single tube is set up in one direction and three lanes. The combination of the two tubes forms a double direction six lanes, the length of the west line is about 1.09 km, the length of the east line is about 1.08 km, the outer diameter of the tunnel is 14.5 m, and a mud water pressure balance shield with an outer diameter of 14.93 m is used. The segment is a general cuneiform ring segment with a wedged weight of 20 mm, a width of 2 m and a thickness of 0.6 m. Each ring segment is composed of 10 pieces. The buried depth of the tunnel is 7.7–23.0 m, and the excavated section is 175.07 m 2. The tunnel section is shown in Fig. 6.2.

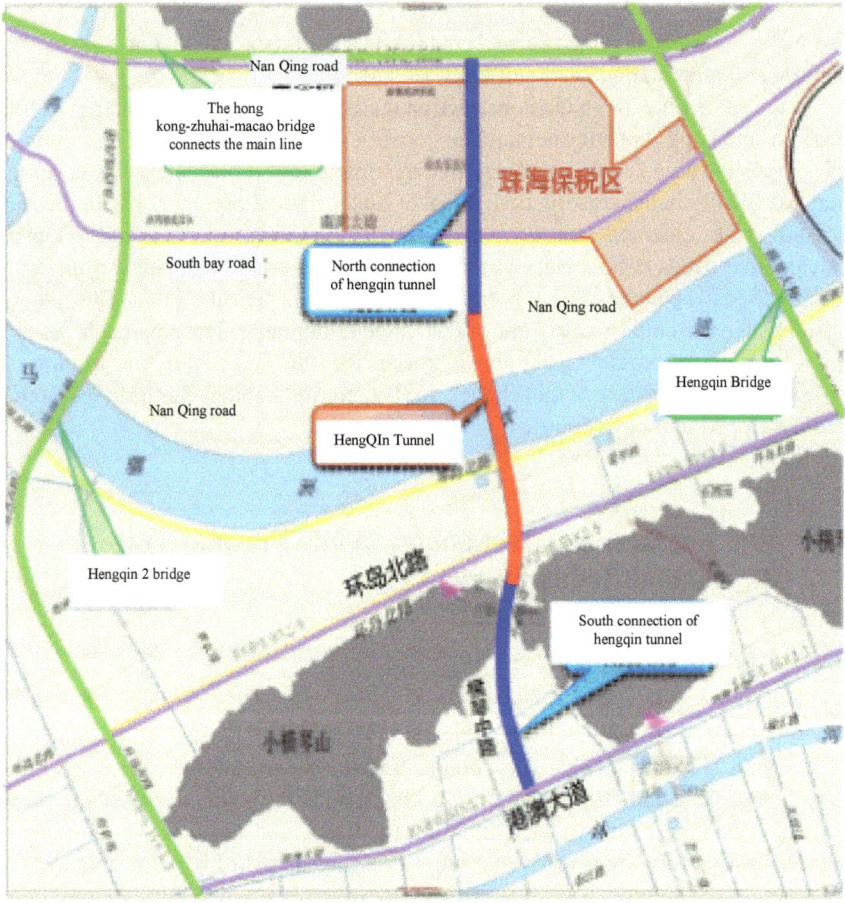

Fig. 6.1 Schematic diagram of project location

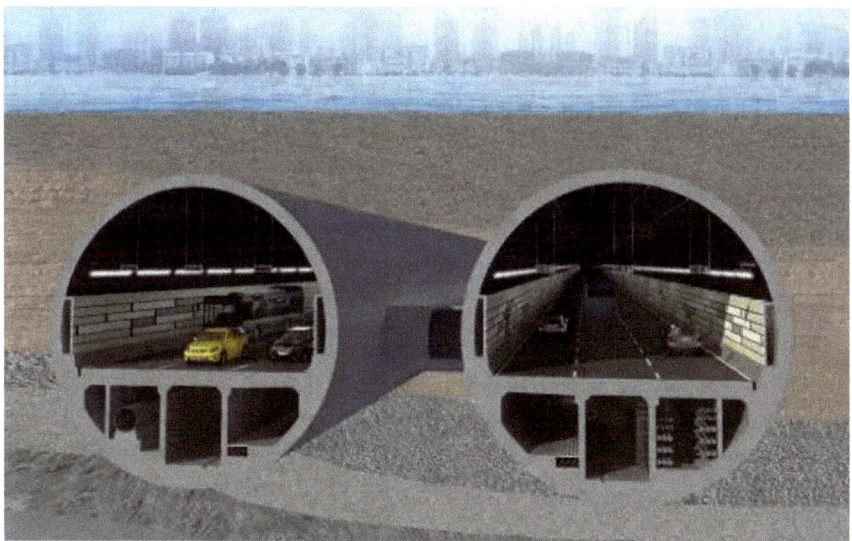

Fig. 6.2 Tunnel section rendering

The foundation soils are Quaternary loose sediments (Q4) and Yanshanian weathered granites in the middle of the survey depth within the depth of reconnaissance, mainly are the artificial fill layer (Qml), intersections between sea and land, marine sedimentary mud, muddy soil, clay, medium coarse sand layer (Q^{mc}_4), granite residual soil (Q^{cl}_4) and underlying total weathered, strong weathered and apoplectic granite (γ 2–3 5). The ground standard is about +3.7 m, and the tunnel is mainly excavated in silt, clay, coarse sand with clay and silt clay; the local shield is buried deep or the bedrock is shallow, involving medium coarse sand, gravel clay soil, fully weathered granite and strongly weathered granite; the shield is exposed to weathered granite in some areas. In the process of propulsion, 70% of the range is carried out in the silt layer, and the most complex area in the river straddles 7 soil layers from the silt layer to the middle weathered granite layer. It has distinct characteristics of upper, soft and lower hard.

The surface water is mainly Ma Lizhou waterway, the width of the river is 500–600 m, the river is relatively flat, and the current is also slower. Under the influence of seawater tide, the water level difference is generally about 1.5 m, which can reach 3 m in rainstorm season. In typhoon season, the tide can overflow over the embankment during the maximum tide, resulting in a large area of flooding, and the highest flood level is about 3.7 m. According to the regional hydrological data, the most wonderful part of the water level in 50 years and 100 years is 2.26 m and 2.42 m respectively (based on the elevation of the Pearl River). The groundwater types along the route are mainly Quaternary loose rock pore diving, confined water and bedrock crack water.

6.1.2 Cutter Head Design

In order to deal with riprap and bedrock that may be encountered in the process of shield tunneling, this project adopts SSP advanced detection system, and makes corresponding design arrangement for shield tunneling cutter head arrangement.

In order to ensure the excavation radius, the excavation trajectory of 80# is 2 17.5in hob, 81# ~ 83# excavation track is 7 18in hob, and the other excavation trajectory is 17in hob. When you get out of the hole, the shell knife can be used instead of the hob, and a total of 102 hob can be replaced with the shell knife in the front of the cutter head of 1# ~ 71#. The hob is 175 mm higher than the surface of the cutter disk. A total of 182 scrapers, 140 mm above the disk surface, 43 leading knives, 150 mm above the disk surface, 16 peripheral scrapers, 8 diameter-keeping knives and 1 copying knife. The wear detection system is distributed on the cutter head steel structure, which consists of five tool wear detection devices and one steel structure wear detection device. Each detection point consists of a sensor bolt and a welded sensor bolt cage. The sensor bolt is screwed into the cage at a specific angle. When the sensor bolt is worn through by the normal wear and tear in front, the pressure sensor detects the pressure drop in the system and outputs a warning signal. The cutter head arrangement is shown in Fig. 6.3.

At the same time, according to the characteristics of the shield tunneling through the bedrock layer of this project, it is specially equipped with a high-performance stone crusher to meet the requirement that the diameter is less than 0.8 m of stone can be discharged normally after crushing, ensuring the tunneling speed of shield. The stone crusher device is shown in Fig. 6.4.

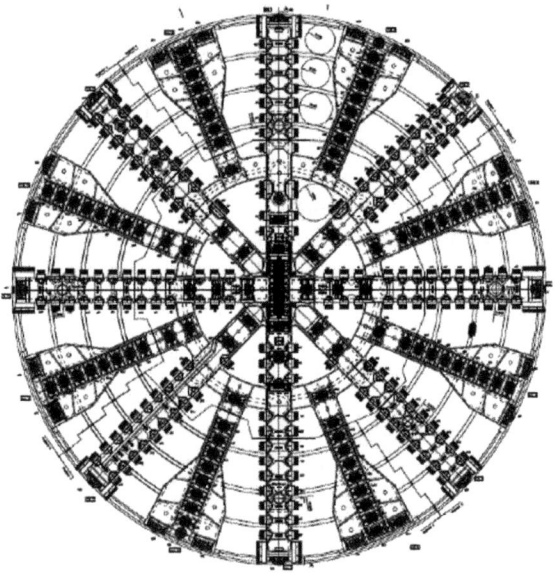

Fig. 6.3 Schematic diagram of cutter head arrangement

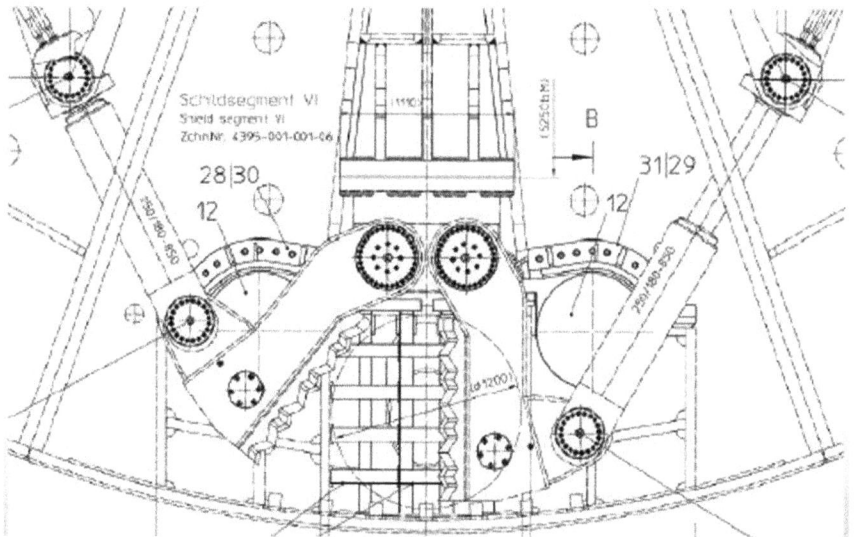

Fig. 6.4 Schematic diagram of stone crusher installation

6.1.3 The Influence of Riprap and Bedrock on Shield Tunneling and Its Location Survey

According to previous construction experience, in the process of tunneling, when there is riprap on the advance line, the riprap will rotate with the cutter head, which leads to very difficult and frequent jamming of the cutter head, and difficult control of the attitude and direction of the shield. Uneven stress on the cutter head leads to main bearing damage or seal damage; Increase the wear of the cutter head and the cutter, so that the strength and stiffness of the cutter head and the cutter unable to dig; The falling of riprap will lead to uneven settlement of ground and structures, and damage existing roads and pipe lines. The uplift of bedrock will cause the shield structure to be in the state of fast turning and slow pushing when it meets the rock layer, which will easily lead to the loss of upper soil mass and ground collapse. Although the hob can deal with a certain strength of rock, but in high strength, a large range of weathered rock excavation will lead to serious tool wear damage, need to open the cabin for several times to change the knife, and open the cabin to change the knife in the river with pressure state, construction difficulty, high risk, with pressure change efficiency is low, will seriously affect the time limit; It will be more difficult to adjust the attitude of shield, and the precision of tunnel axis is difficult to control. In order to effectively avoid the serious adverse effects of riprap and bedrock on shield tunneling and engineering risks, it is necessary to detect their positions and make pretreatment before shield tunneling.

1) Riprap location

According to the geological data, there is a small amount of riprap in the range of 45–65 m from the east–west line to the working well on the south bank, and the buried depth is about 8 m (the surface standard height is about +3.5 m), which is close to the tunnel structure. The shear wave reflection method is used to detect the progress of the south bank embankment and the north bank embankment. The results are shown in Figs. 6.5 and 6.6.

According to the geophysical results, there are a large number of riprap in the upper section of the south bank (420 m north of the working well), the buried depth is-2.90~7.97 m (the tunnel depth is-4.63~11.04 m), and some of the riprap has entered the shield propulsion section. The shear wave reflection method can only detect the top of the riprap and cannot determine the specific buried depth of the riprap. In order to ensure the accuracy of geophysical results, and the relative position relationship between riprap and tunnel is found out more accurately, and the way of drilling and coring is selected for supplementary investigation. The hole spacing is 5 m, the hole depth is 25 m, and the drilling hole is blocked with cement slurry. The hole arrangement is shown in Figs. 6.7 and 6.8.

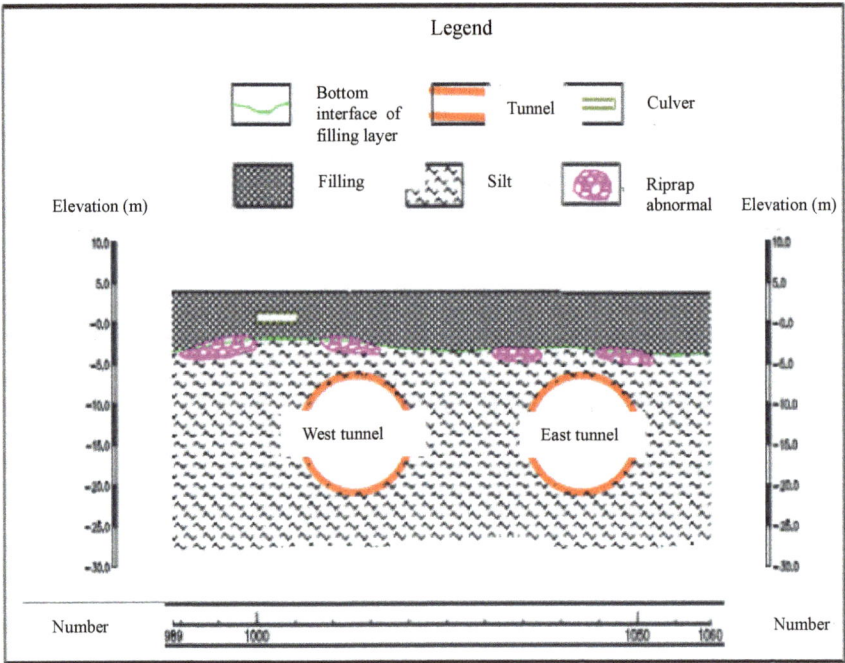

Fig. 6.5 Schematic diagram of representative cross section of south bank embankment

6.1 Tool Replacement Technology While Encountering Riprap … 273

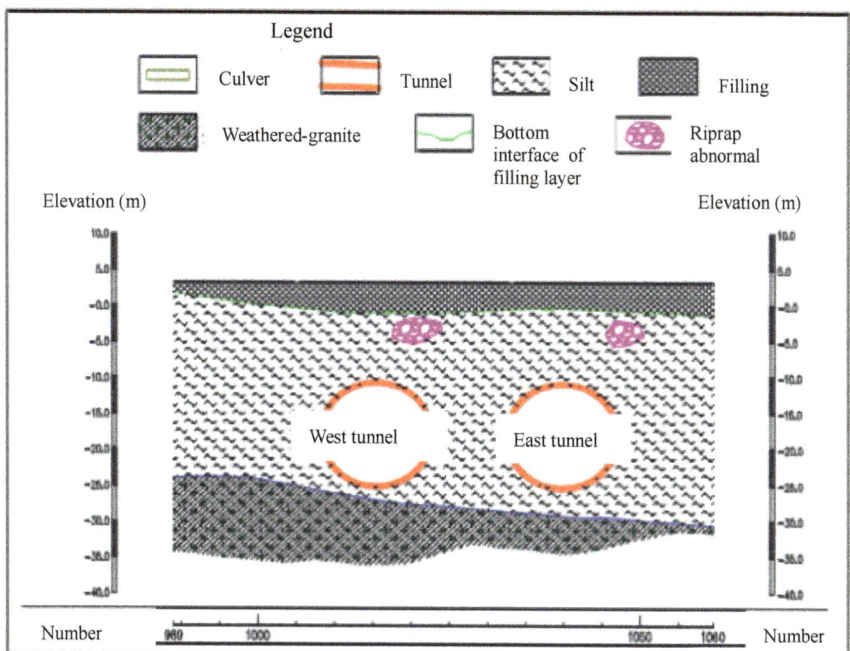

Fig. 6.6 Schematic diagram of representative cross section of north bank embankment

Fig. 6.7 Schematic diagram of the hole layout of the north bank dike

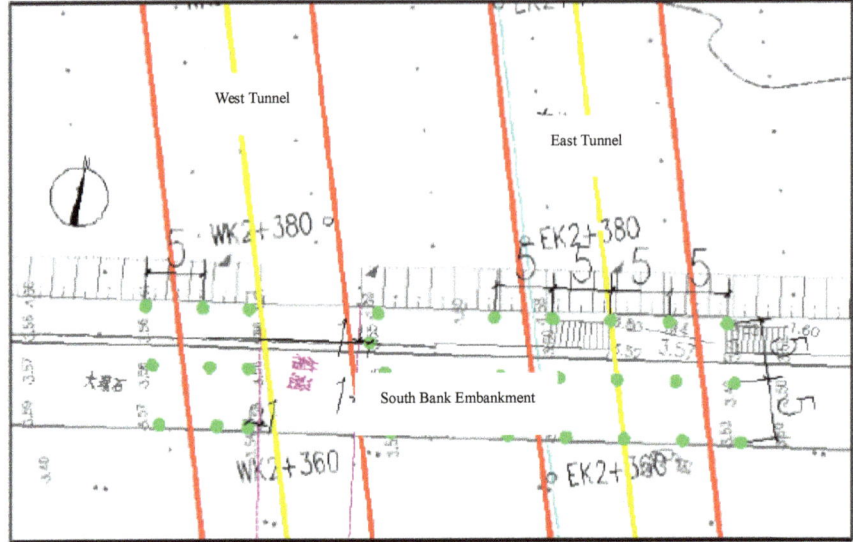

Fig. 6.8 Schematic diagram of hole layout of the south bank dike

2) Uplift position of bedrock

The results of design exploration data, onshore shear wave reflection method and SSP seismic scattering show that the bedrock protrusions encountered in the propulsion section of circular tunnel shield are divided into two parts, one is located about 155 m south of the south bank embankment, and the uplift is about 3.6 m in the shield propulsion range, in which the weathered strata enter the shield propulsion section about 1.30 m. The other is located in the south bank embankment at a position of 140 m (located in the area of Maruzhou waterway). The mileage pile number WK2+506, enters the shield propulsion section range of about 3.0 m, in which weathered granite enters the shield propulsion section range of about 1.25 m.

Bedrock uplift plane and longitudinal sections are shown in Fig. 6.9 and Fig. 6.10, respectively. See Table 6.1 for bedrock bulge distribution.

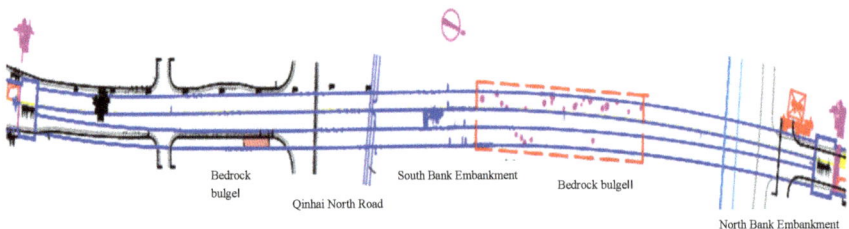

Fig. 6.9 Schematic diagram of uplift plane of bedrock

6.1 Tool Replacement Technology While Encountering Riprap ...

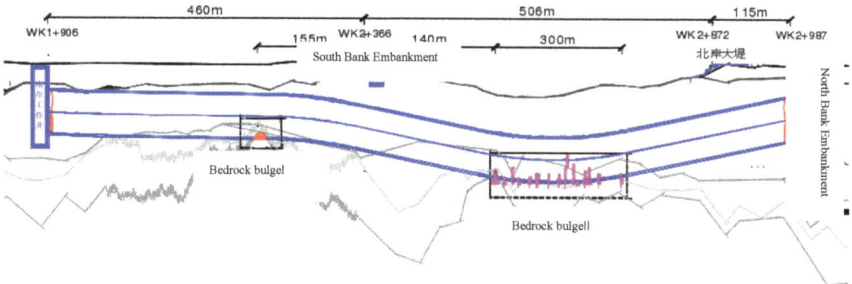

Fig. 6.10 Schematic diagram of longitudinal section of bedrock uplift

Table 6.1 Uplift distribution of bedrock

Order number	Mileage range	length	Rock top elevation	Protruding height	Preparation note
1	Wk2+201-WK2+234.5	33	−19.25	2.7	On the east line, the tunnel section enters the middle weathered rock range of about 2.70 m
2	Wk2=506-Wk2+727	221	−32.6	1.25	East–west line, in which the projection height of moderately weathered rock is 1.25 m

6.1.4 The Construction Plan

6.1.4.1 Stone Processing

1) Shore section

First of all, the pick machine is used to break the road surface in the range of shield propulsion line, and the social pipeline is removed, and the excavator and loading vehicle are organized to remove the old pavement debris centrally. After the pavement breaking is completed, the underground rock throwing is explored by high pressure rotary jet grouting pile machine, and then the rock throwing area is cleaned by full rotary drilling rig supplemented by steel casing and two special machine tools such as crawler crane, and then the foundation is strengthened by high pressure rotary jet machine for 50 m later. The whole obstacle removal process adopts the form of flowing water operation to carry on the construction. The clearance range is from

420 m north of the working well to the embankment on the south bank, and widen 0.5 m outward from both sides of the shield propulsion line. Firstly from south to north construction to clean up the west line, after the west line clean up from north to south clean up the east line. The obstacle removal operation of the two lines is shown in Figs. 6.11 and 6.12. The construction process is shown in Fig. 6.13.

2) Embankment section

From the exploration results, it is found that there are a large number of stone throwing under the south bank embankment (located directly above the propulsion section), and it is easy to fall to the propulsion excavation surface during the propulsion process. At the same time, there is a drainage plate in the propulsion section, the existence of the drainage plate will increase a large number of seepage paths, seriously affect the stability of the excavation surface, easy to lead to great land subsidence and embankment structure damage. Considering the situation of drainage plate and

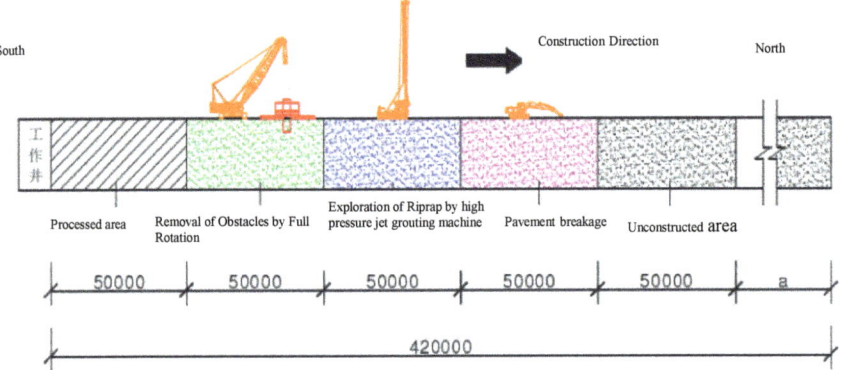

Fig. 6.11 Schematic diagram of barrier clearance flow on the western route

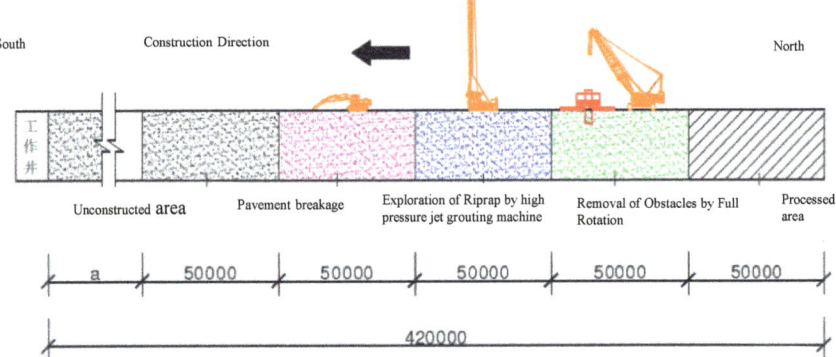

Fig. 6.12 Schematic diagram of barrier clearance flow on the eastern route

6.1 Tool Replacement Technology While Encountering Riprap ...

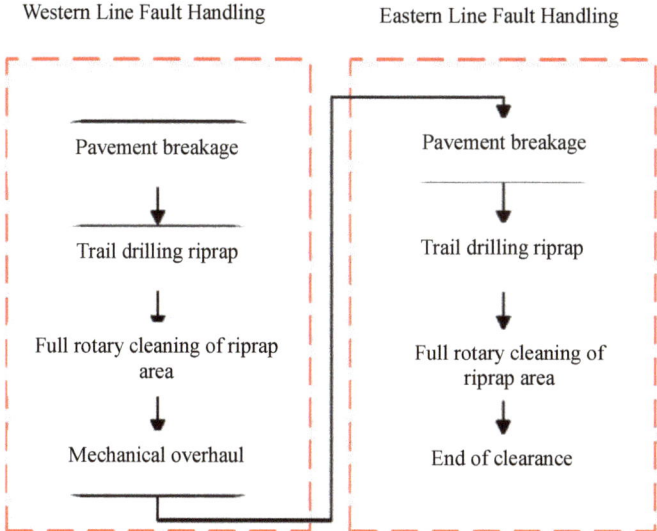

Fig. 6.13 Flow chart of obstacle removal operation

stone throwing, on the one hand, the shield axis is lowered and the throwing stone is avoided; on the other hand, the cement and water glass rapid consolidation double liquid slurry is used to reinforce the rock throwing above the section of the circular tunnel at the embankment, and the seepage path of the drainage plate is blocked. The stone throwing distance from the propulsion section at the north bank embankment is far from the propulsion section, so it is not treated according to the actual situation.

(1) Axis adjustment. In order to minimize the impact of riprap on the advance section at the embankment, after communication with the design, it was decided to lower the circular tunnel section at the embankment by 3 m (considering the bedrock uplift comprehensively), so as to ensure the advance section away from the riprap area as far as possible, and then carry out reinforcement treatment. The following plans are considered after the axis reduction.

(2) Reinforcement area. The length and width of the reinforcement area are 39 m and 21 m (3 m respectively on both sides of the advancing section), and the depth is 10–16 m below the ground, as shown in Figs. 6.14 and 6.15. After the reinforcement, the strength of the reinforcement zone shall not be less than 1 MPa.

(3) The construction process is as follows:

① Drill to design depth with engineering rig.
② Connect the power supply and data lines, start the oil pump, and place the pile machine.
③ he bit is connected to the in-ground pressure monitoring display to confirm that the bit is cleared without load.

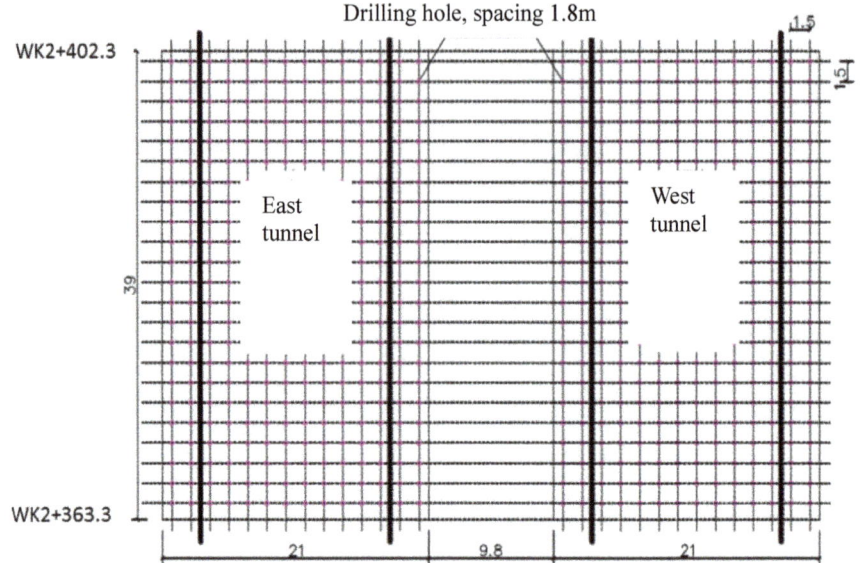

Fig. 6.14 Double slurry reinforcement plan

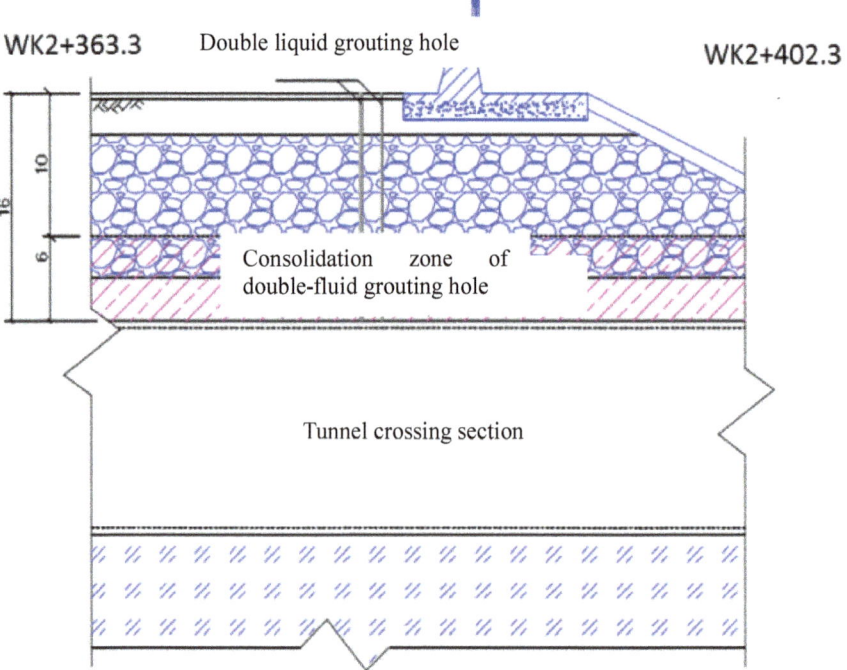

Fig. 6.15 Longitudinal section diagram of double liquid slurry reinforcement

6.1 Tool Replacement Technology While Encountering Riprap ...

④ ocking drill pipe and drill bit, docking carefully check the sealing ring, see whether missing or damaged, and to determine whether the pressure in the ground is normal.
⑤ he equipment drills down the pipe.
⑥ Repeat steps ④ and steps ⑤ until the drill reaches a predetermined depth.
⑦ When the bit reaches the predetermined depth, the pressure should not be too high when the high pressure cement pump is turned on, and the pressure should be pressurized step by step until the specified pressure is reached. After the specified pressure has been reached and the pressure in the ground has been confirmed to be normal, the lifting can begin.
⑧ We should closely monitor the pressure in situ during construction. When the pressure is abnormal, it must be adjusted in time.
⑨ when lifting a drill pipe, the drill pipe is removed, pay attention to the process of removing the drill pipe, carefully check the sealing ring and the condition of the data line, to see whether it is damaged, and whether the pressure display in the ground is normal. Eliminate any problems in time. Remove the drill pipe
After that, the drill pipe shall be rinsed and maintained in time.
⑩ Repeat step ⑨ until the construction is finished.
⑪ After the construction, wash and maintain the equipment.
⑫ Strengthen construction monitoring during construction.

6.1.4.2 Rock Blasting

For the rock layer and the combination of upper soft and lower hard geology, shield can use hob and stone crusher to break the rock, but because the rock is difficult to fix in the stratum, the hob can not do effective treatment, once encountered in the advance, the effect of the hob breaking rock is very poor. The uplift of bedrock will bring the following direct risks: the shield tunneling is slow and seriously affects the construction period; Although the hob can handle a certain strength of rock, but in high strength, a large range of weathered rock excavation will lead to serious tool wear, need to open the compartment for several times to change the knife; It is difficult and risky to change the tool with pressure, and the efficiency of changing the tool with pressure is low, which will seriously affect the working period. When encountering rock strata, the shield is in a state of fast rotation and slow pushing, which is easy to cause the loss of upper soil and ground collapse. It will be more difficult to adjust the attitude of shield, and the precision of tunnel axis is difficult to control.

    The rocks in the tunneling line can be broken through blasting, full rotation barrier clearance and artificial pressure chamber. The blasting method is suitable for shield tunneling line with isolated stone and bedrock protruding locally and other geological conditions. On the basis of detailed survey, the blasting quantity of isolated stone is determined, and the detailed construction plan is made to minimize the impact of blasting method on the surrounding environment. Compared with other methods, use blasting to break isolated stones and bedrock is more reliable and safer, saves time and has high cost performance.

Blasting is used to treat the bedrock on the bank and in the middle of the river, and the rock layer entering the tunnel section is broken to fragments with a straight diameter less than 30 cm, so as to reach the treatable state of shield. After the blasting is completed, the blasting area is sampled by full rotation. After the blasting on the shore is evaluated, the blasting parameters and technology in the middle of the river are adjusted to ensure the effect of the rock treatment in the middle of the river. After the completion of drilling and blasting in the middle of the river, the blasting area should be reinforced by grouting.

1) Drilling detection

   (1) The bedrock on the shore. According to the previous geophysical exploration results, the moderately weathered strata enter the tunnel Sect. 300 m north of the working well of the eastern tunnel. The actual drilling depth is 11.2 m × 33.56 m, the depth is about 25 m, and the drilling spacing is 1.6 m × 1.6 m, as shown in Fig. 6.16.

According to the preliminary geophysical survey results, the 300 m of moderately weathered rock strata north of the working shaft of the east line tunnel were put into the tunnel section, and the actual drilling depth of rock strata was detected. The drilling range is 11.2 m × 33. 56 m, depth approximately 25 m, drill hole spacing press 1.6 m 6 m, as shown in Fig. 6.16. If the edge probing hole detected weathered rock layer is 0. Within the 2 m limit, press 1 outward. 6 m interval increases drilling hole, and so on, to ensure that all into the tunnel section of the range of weathered rock. Drilling pipe with an aperture of 90 mm is adopted for shore drilling, and 75 mm PVC casing is lowered, as shown in Fig. 6.17. If the moderately weathered rock strata in the edge probing are located within the limit of 0.2 m below the advancing

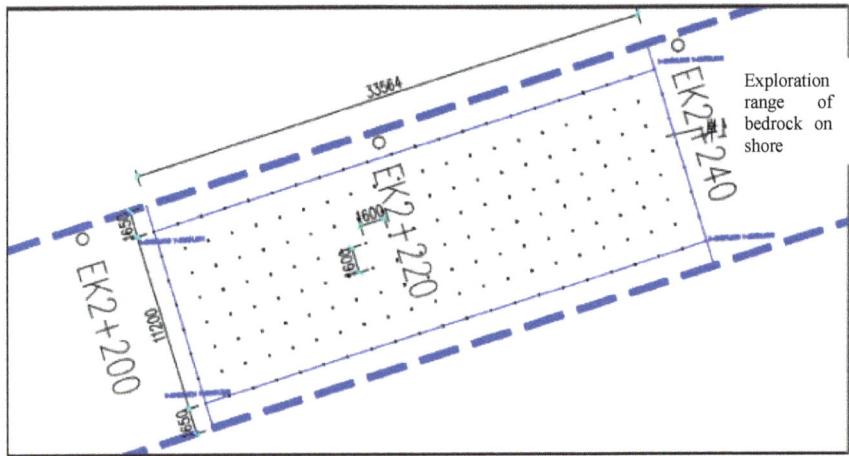

Fig. 6.16 Plan of borehole layout on shore section

6.1 Tool Replacement Technology While Encountering Riprap ...

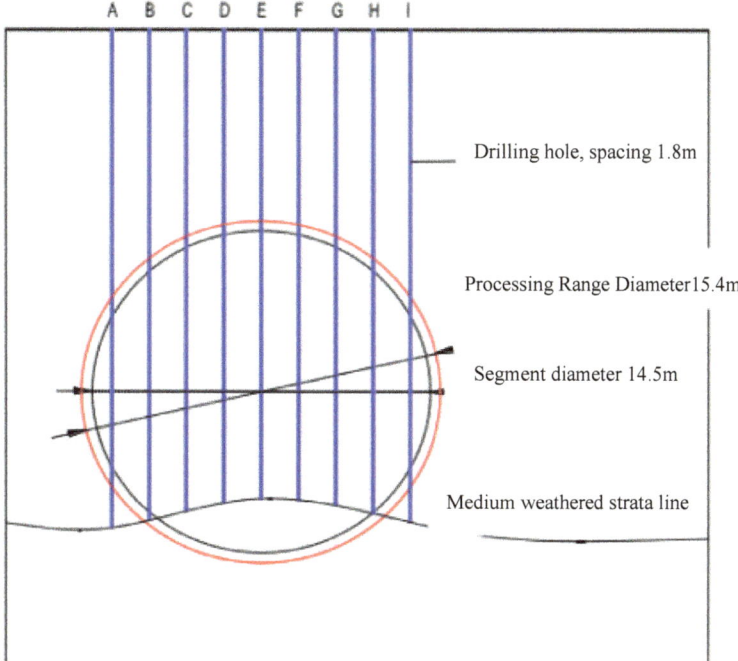

Fig. 6.17 Cross section diagram of drilling on shore

tunnel section, the probing holes will continue to be increased at a distance of 1.6 m to the outside, and so on to ensure that all the moderately weathered rock strata in the tunnel section are clear. The hole diameter 90 mm drill pipe is used to drill on the shore, and the 75 mm PVC casing is lowered, as shown in Fig. 6.17.

(2) The bedrock in the middle of the river. In the course of this construction, the preliminary positioning and drilling of each bedrock are based on the geophysical exploration report provided by geophysical exploration units, and the treatment range is considered according to the external expansion of shield propulsion section. For bedrock with diameter greater than 2 m, the hole spacing is 2 m, and the hole spacing is 1 m when the diameter is less than 2 m bedrock. Holes shall be drilled in the scope of geophysical prospecting bedrock. If the bedrock entering the tunnel section is not found or the bedrock is incomplete, 2 m hole shall be expanded outward in the tunnel advancing range. After each bedrock drilling detection is completed, it is necessary to report the drilling results to the geophysical exploration unit to confirm whether the bedrock detection is completed here (finding the bedrock in the geophysical report or determining its absence) or hole detection is required. Geophysical exploration units need to confirm the construction results in time, and follow-up construction can be carried out after the confirmation is completed. Use geological drilling rig drilling, aperture 130 mm, lower 110 mm diameter PVC pipe. The drill arrangement is shown in Figs. 6.18 and 6.19.

Fig. 6.18 Layout of drilling plane in the middle part of the river

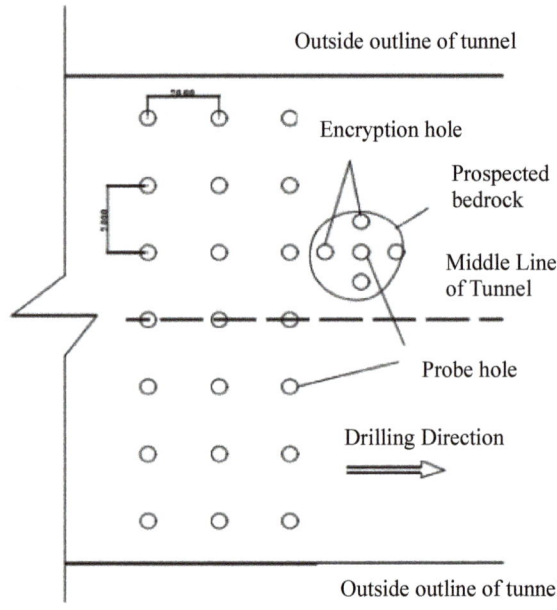

Fig. 6.19 Cross section diagram of drilling hole in the middle part of the river

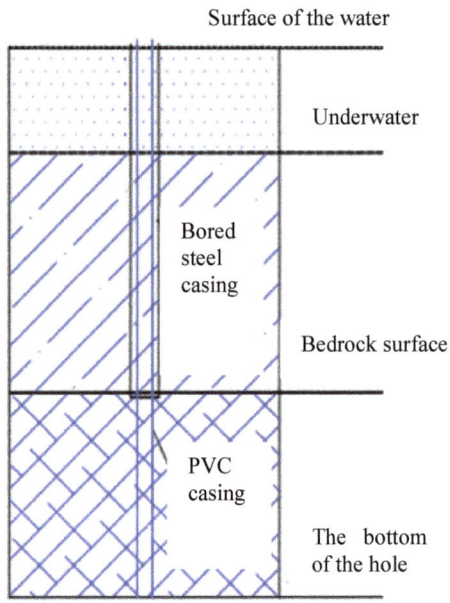

6.1 Tool Replacement Technology While Encountering Riprap ...

In order to facilitate construction and accurately control the direction of drilling, vertical drilling is adopted. In this project, the drilling surface is drilled, because it is necessary to keep the hole free of sundries for a certain period of time, so there are strict requirements in the drilling method. In order to ensure the verticality of drilling, single layer casing or multi-layer casing should be drilled while drilling until the micro-weathered rock layer is in contact with each other. When the drilling is completed, it is necessary to PVC casing up and down, and the steel casing that has been laid cannot be moved at the same time. After the blasting construction of the hole is completed, the steel casing in the hole is pulled out, which not only ensures the quality of the hole, but also releases the concrete mortar blocked in the pipe, thus completing the sealing of the hole.

(3) Main parameters of blasting

According to the result of drilling and the position of tunnel after reduction, the position and quantity of blasting hole are analyzed and determined, and the blasting construction is carried out in zoning section.

(1) Selection of firework equipment. The millisecond detonator is used in the hole detonator, the detonator is used in the initiation detonator, and the emulsion explosive is used in the explosive selection. The standard diameter is Φ 60 mm, which is processed according to the needs of the field.

(2) Charge structure and initiation network. Because of the deep hole depth and the deep buried depth of the rock that needs blasting treatment, the explosive package is suspended at the blasting point by soft rope, and one end is fixed at the hole position, and the elevation error should not be greater than 10 cm. The medicine is packaged in a special PVC tube, and the detonator must have good waterproof performance. Because there is a water column about 15 m high above the detonator and the pressure is more than 0.1 MPA, an anti-floating solid hoist needs to be suspended below the detonator. The gun hole is priming with forward charge, priming with electric and non-electric combination, priming with two instantaneous generation detonators, which belong to two electric detonation networks, and priming with parallel connection of two sets of networks. The initiation network is shown in Fig. 6.20.

(3) Charge parameters and hole distribution mode. Considering that it is difficult to pull out the casing after blasting in the river, and even because of the influence of rising and falling tide and water flow, the casing will break or shift. If the casing can not be pulled out, there is a direct connection between the water surface and the tunnel, which may lead to grouting or leakage of the shield when the shield passes through. In order to solve this problem, a small charge package is arranged at the same time during blasting, the charge quantity is controlled at about 1 kg, which not only does not destroy the stability of the original formation, but also breaks the PVC casing so that there is no direct connection, and it can also compaction the sand and the original formation in a certain area under the charge. The position of the medicine bag is 4 m below the mud surface, which can not affect the thickness of the overlying soil.

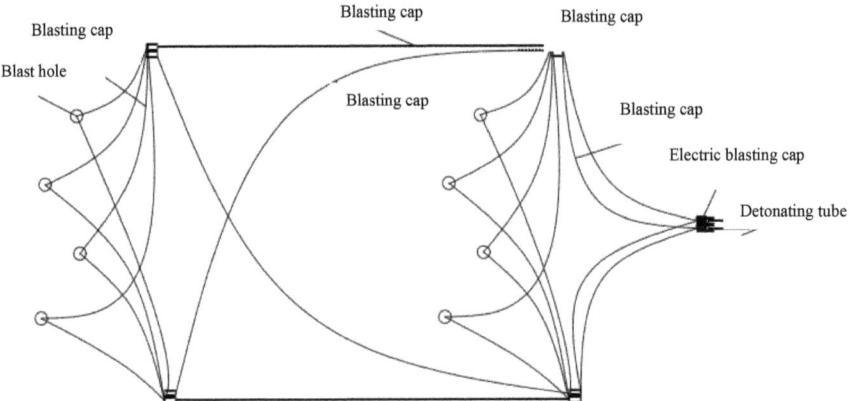

Fig. 6.20 Schematic diagram of initiation network

3) Blasting process

(1) Technical disclosure. Before construction, the technical personnel first carry on the technical exchange to the drilling worker, convey the technical requirements such as the principle of hole laying and the allowable deviation of drilling to all the construction personnel, so that the staff can understand and master the standard and standard of the project ideologically and consciously. Draw up the plan and implementation rules of quality control before construction, and operate in strict accordance with the plan and implementation rules in the construction.

(2) Construction positioning of drilling and blasting ship. Six anchors are used to control the forward and rear movement of the ship, and the lateral movement of the ship is controlled by using the left and right four-door octave anchors and two main anchors. The front and rear main anchor wire cables control the longitudinal movement of the ship. RTK-DGPS positioning system (accuracy: horizontal ± 2 cm $+ 2 \times 10^{-6}$, elevation ± 4 cm $+ 4 \times 10^{-6}$) and total station positioning are adopted for the positioning of borehole. The deviation between the measured hole position and the design hole position shall not exceed 20 cm.

(3) Control of drilling elevation. lead the reference point provided by the field engineer representative to the construction area, and set up a water gauge, set up a temporary working point next to the water gauge, and check it with a level. During drilling and blasting construction, RTK DGPS and total station instrument are used to observe the change of tide level, and water gauge is used to calibrate regularly.

(4) Drilling and acceptance. The hole spacing is 1 m, because of the fixed position of the drilling rig on board, the hole can be drilled by misplaced drilling. After drilling, the hole should be accepted by technical personnel,

the deviation should be too large, the hole with large deviation of resistance line should be abandoned, and the charge construction can only be carried out after acceptance.

(5) Charge construction alert. For the safety of the site machinery and equipment and construction personnel, the charge explosion area must be initially alerted, Party A must assist in the site cleaning work.

(6) Hole charging. After drilling, gunners should follow the following procedures:

① Use the sounding line to check the depth of the hole, hole bottom elevation if not up to the construction design bottom elevation requirements, should be required to re-drill.
② According to the design requirements of processing the detonator and loading explosives.
③ Use the sounding line to check whether the explosive has reached the bottom of the hole, if not, the gun should be pressed to the bottom of the hole.
④ After the explosive is packed, fill the upper part of the gun hole with a sandbox.
⑤ Pull up the casing and take out the detonator tube.
⑥ A blasting of the gun hole all loaded with explosives, connected to the initiation network.

(7) Security alert and initiation. After the above work is completed, the experienced operators will be connected to the network, and after repeated inspection, they will begin to alert correctly. Blasting warning signals include visual signals (flags, etc.) and auditory signals (alarms and whistles). The warning ring with explosion point as the center, underwater blasting radius 150 m and land blasting radius 300 m are set up, the warning ring is set up at the boundary of blasting warning circle, and the land and water guard posts are set up respectively. A blasting warning command system centered on drilling and blasting ship is established, and radio interphone and mobile phone are used as contact methods to ensure the smooth communication of information in the process of warning and to ensure the safety of blasting. The blasting alert is shown in Fig. 6.21.

4) Grouting reinforcement

In order to prevent the pulping phenomenon in the process of shield propulsion, the pierced area of Ma Lizhou waterway was strengthened by high pressure swirl spray reinforcement by reforming the rotary jet drilling rig carried on board the ship. The reinforcement range is 2 m outside the drilling area, the reinforcement depth is 9–12 m below the water surface, the cement content is 20%, and the water-cement ratio is 0.8%. High pressure rotary spray reinforcement is shown in Fig. 6.22.

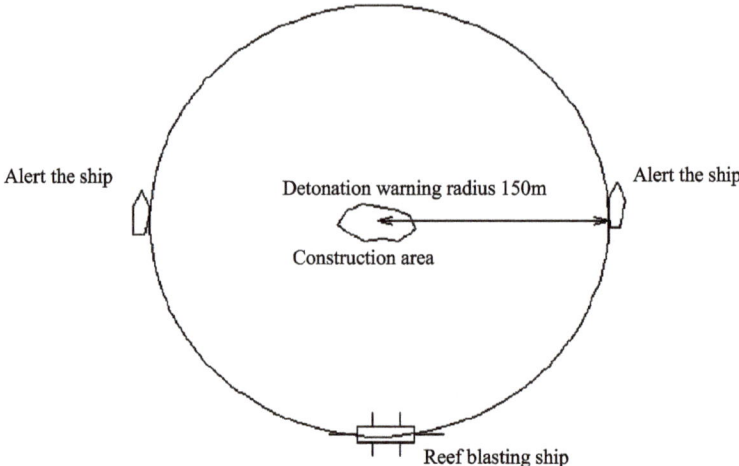

Fig. 6.21 Blasting alert

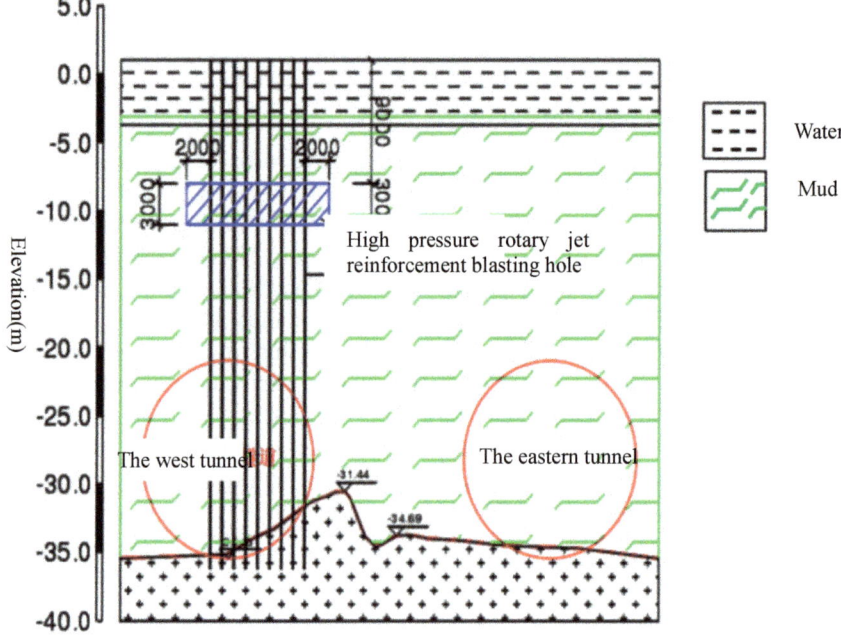

Fig. 6.22 Schematic diagram of high pressure rotary spray reinforcement

The other technological parameters are as follows:

(1) The spacing of holes is 1.5 m × 1.5 m square arrangement, and the grouting depth is 3.0 m.
(2) Grouting pressure: 0.5–4 MPA.
(3) Pile verticality error: ≤1/100.
(4) Lifting speed: 40 min/ m.

(1) Slurry flow rate: 85–100L/ min.
(2) Materials: P.O.42.5 cement, tap water. Considering that the reinforcement effect of cement slurry in silt layer is not good, double liquid slurry grouting is used when necessary, and 38–41 water glass is added. Slurry ratio: A liquid cement slurry: B liquid water glass = 1: (0.8–1). The curing time of double liquid slurry is 30–45 s.

6.1.4.3 Tips for Changing Cutters

The tool must be replaced between the drainage plate at the south bank reinforcement area and the drainage plate at the south bank embankment and the bedrock area in the middle of the river, including the following two points:

In this project, tool replacement should be carried out between the south bank tool change reinforcement area and the drainage plate at the south bank embankment and the bedrock area in the middle of the river, mainly including the following two points:

(1) Tool type replacement. Before the working well on the south bank is out of the hole, 102 hob cutters have been replaced with cutting cutters. After the shield is pushed to the south bank embankment to replace the cutter reinforcement area, 102 cutters must be replaced with hob to ensure the propulsion ability of the middle part of the river.
(2) Tool wear and replacement. In the whole process of shield propulsion, according to the feedback data of tool wear detection system, if the tool wear is serious and the shield cannot advance normally, the open cabin tool change (pressure tool change) is carried out. Because there are a large number of bedrock protrusions in the river, some bedrock may still remain after blasting treatment. If the opening cabin is carried out in the river to change knives, the blasting drilling area should be avoided as far as possible. The pressure holding test must be carried out before opening and changing the knife. The inlet pressure should be controlled at about 2.6 MPa, and no obvious pressure drop can be observed for 4 h before entering the cabin to change the knife. If the pressure drop is obvious, the shield should be pushed forward for a distance, and then the pressure holding check should be carried out. If the pressure still drops (indicating that there is a seepage path in the bottom layer), the upper sandbag should be blocked, and then the pressure holding inspection and the cabin inspection tool should be carried out.

6.2 Encounter Large Boulder Shield Cutter Protection and Replacement Technology

6.2.1 Project Overview

The scope of the No. 1 project of Xiamen Railway Line 2 is from the starting point to Dongdu Road Station (inclusive). The project mainly includes four stations, including Lukeng Station, Haicang CBD Station, Haicang Avenue Station and Dongdu Road Station, which are four shield construct intervals. Among them, Haicang Avenue Station-Dongdu Road Station section starts from Haitang Avenue Station, and then the coastal road is laid to the north, then enters Haitang Bay Park with a radius of 500 m and enters the sea. It will land on the No. 1 berth of the International Terminal, and then pass through the second phase of the cruise city at a radius of 350 m to reach Dongdu Road Station. The line plane spacing is 14–17 m. The interval tunnel is constructed by the mining method and shield method, and the line direction is shown in Fig. 6.23.

At present, the geological encryption and supplementary survey of the coastal section of haicang avenue station—haicang side seawall has been completed. along the center line of the left and right lines at 2.5 m, geological coring was carried out for supplementary prospecting hole, and 2 bedrock bulges and 4 isolated rocks were found. Of which, solitary stone 1 is located on the left line DK18+687.5 – DK18+690, solitary stone 2 is located on the left line DK18+747.5, solitary stone 3 is at DK18+745 on the right line, and solitary stone 4 is at DK18+752.5 on the right line. Solitary stone 1 and 4 are mostly in the tunnel, solitary stone 3 is only partly in the tunnel, and solitary stone 2 is all in the tunnel. From top to bottom, the left line is composed of plain fill, silt, silty clay, loose or fragmentary weathered granite and medium weathered granite. The right line is composed of plain earth, silt, silty clay or eluvial sandy clay, fully weathered granite and moderately weathered granite from top to bottom. The offshore section has completed the geological filling and prospecting of haicang side seawall—daduyu section. The geological coring has

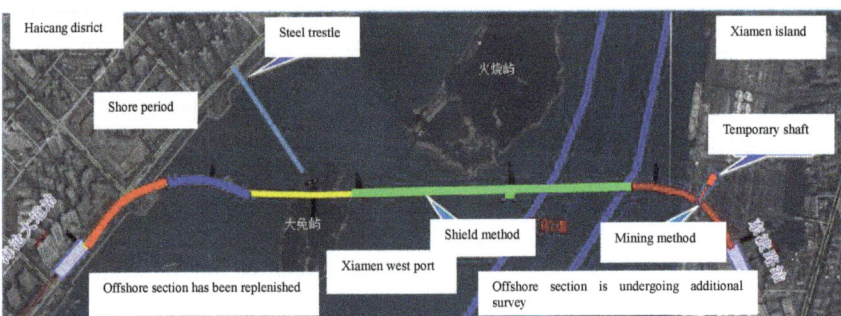

Fig. 6.23 Shield tunnel line trend of station 1 dongdu road, haicang

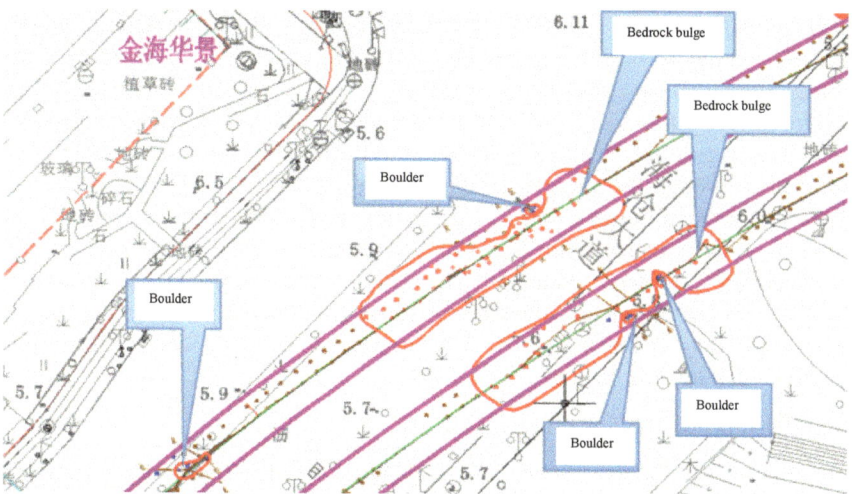

Fig. 6.24 Plane diagram of isolated rock and bedrock uplift

been carried out with a filling and prospecting hole of 5 m along the center line of the left and right lines. A total of 5 bedrock bulges and 9 isolated rocks have been found. Solitary stone 1 is at T4-49 hole on the right line 1, solitary stone 2 is at T4-63 hole on the right line, solitary stone 3 is at T4-65 hole on the right line, solitary 4 is located at M2Z3—THD—B48 hole on the right line, solitary stone 5 is at M2Z3—THD—B23 hole on the right line, solitary stone 6 is at THD14G hole in the left line, solitary stone 7 is at T4-30 hole in the left line, solitary stone 8 is at T4-36 hole in the left line, solitary stone 9 is at T4-37 hole on the left line. With the exception of isolated rocks 2, 3, 5, 6, other isolated rocks are basically located in the tunnel. The upper to lower strata of the left line are silt, fully weathered andesite or granite, and strongly weathered andesite or granite, while the upper to lower strata of the right line are silt, fully weathered granite, and dispersed strongly weathered granite. The isolated stone and bedrock bulges found in the shield tunnel interval in the early stage need blasting treatment before the shield tunnel passes, as shown in Fig. 6.24.

6.2.2 Analysis of Influence of Boulder and Bedrock Bulging on Shield Tunneling Construction and Blasting Construction Technology

According to the experience of shield construction at home and abroad, the soil layer in the soft soil layer cannot provide the reaction force of the hob to break the rock. The lone stone will move with the destruction of the soil or be bounced off by the tool, blocking the front of the cutter head and damaging it. The knives cause large disturbances in the stratum, which may cause catastrophic consequences such as tunnel collapse or roof collapse. The bedrock bulge is mainly medium-weathered granite, and the average strength is up to 92 MPa, which will cause frequent tool change during shield construction, and there is a great risk of tool change in soft soil layer. According to the geological conditions of the cryptographic exploration, before the start of the shield construction, the existing boulder and bedrock bulges are drilled through the geological rig, and the explosives are placed in the rock body to be blasted from the surface. The energy generated by the explosion breaks down the rock by using explosives. The rock with stronger hardness is divided and disintegrated into pieces, and the particle size of the piece meets the requirements of the slag outlet of the shield, so that the shield can pass smoothly during construction, thereby avoiding engineering risks. The blasting project is divided into two parts: the upper part and the sea part.

1) Blasting construction process on the shore

   (1) Drilling construction. The geological rig is used to take holes in the boulder and bedrock ridges in the shield line. The construction adopts the 108 mm diameter drill pipe to take the hole vertically. After drilling to the design depth, all the gravel, silt and mud in the hole must be removed, and the blasthole should be kept clear. 75 mm PVC pipe protection hole is provided below. Record the rock face height, that is, the thickness of the cover layer and the length of the blasthole in the rock. Charge construction can only be carried out after acceptance.

   (2) Cannon hole acceptance. After the blasthole is drilled, it will be inspected by the technician. The blasthole can be installed into the casing and the medicine package to pass the upper and lower venting. The wall protection and the hole protection are well protected.

   (3) The blasthole charge. The emulsion explosive was placed in a PVC pipe with an inner diameter of 75 mm, and 4 detonators were placed, and the length of the charge was greater than the thickness of the rock layer to be blasted. The total length of the rope of the medicine package is larger than the hole depth of 2 m, and the obvious mark such as a small red cloth strip is attached to the rope for marking. The length of the mark to the bottom of the PVC pipe is equal to the full blast hole (cover thickness + rock blast hole depth) depth.

   (4) The blasthole is blocked and covered. After the medicine bag is in place, slowly pour the sand into the blasthole, and the length of the blockage is more than 5 m. Since the medicine package is below the ground level of 16 m, no flying stones will be produced, but the high pressure gas generated after the explosion may push the mud water in the blasthole out of the hole. In order to prevent the splash of muddy water and the rise of the PVC pipe, it is necessary to protect the hole of the hole and the heavy-duty gland, and the overhead height is 0.6 m. The steel frame cage of the whole welding is used for overhead, and the steel mesh is welded or riveted at the bottom of the bracket cage, two layers of bamboo are bundled on the steel mesh. The concrete block is placed on the bamboo

6.2 Encounter Large Boulder Shield Cutter Protection ...

Fig. 6.25 Gun hole coverage

piece, and the weight of each bracket after the weight is more than 2000 kg. The hole cover is shown in Fig. 6.25.

(5) Blasting network. The multi-stage high-precision millisecond digital electronic detonator is used, and the interval delay interval is 25–50 ms. The millisecond deferred blasting between the holes or the rows is performed, the interval interval is 25–50 ms, and the detonation is performed by holes or rows by row. 4 detonators are installed in each blasthole, and a dedicated detonator is detonated. The number of blasting holes per blast is determined by the maximum amount of blast that is allowed to detonate. The hole-by-hole or row-by-row detonation mode is shown in Fig. 6.26.

(6) Detonation. In order to ensure the safety of the blasting construction, the blasting construction announcement is posted at the main location before

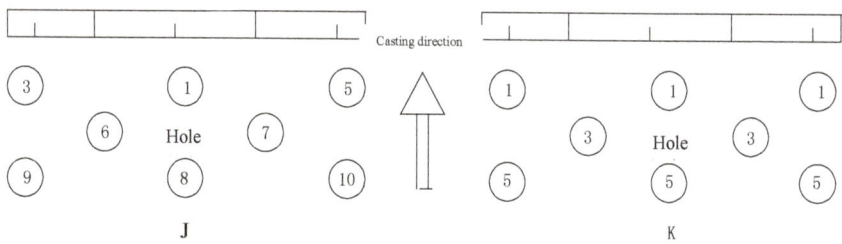

Fig. 6.26 Schematic diagram of hole by hole and row by row initiation. J hole by hole initiation mode; K one row by row initiation mode; number a detonator segment number, that is, the sequence of detonation

the blasting operation, and other warning measures such as evacuation personnel are taken. Once all the alert work is completed, contact each alert point again to confirm the correct order and issue the detonation order. After the blasting is completed, the technician will check the scene and make a warning order.

(7) Detection of blasting effect. After blasting, the blasting area should be subjected to coring detection and blasting effect. The test section takes three cores, and the cores of other areas are cored at two places, one of which needs to be located at the center of the boulder. The number of detection holes of the bedrock protrusion is 3% of the total number of blast holes, and the core 21 is taken at each place, and the core depth reaches the design depth. After the core is taken out, the core crack is analyzed and judged, and the unidirectional length of the complete core extracted is ≤ 25 cm. If the core is unqualified, the hole is filled for secondary blasting.

(8) Ground sealing. The blasting holes are arranged directly above the tunnel, and the quality of the sealing must be strictly guaranteed. The cement mortar is used for sealing, and the original blasting hole is sealed to the ground level. The quality of the sealing must be checked and accepted by the site management personnel.

(9) Grouting. After the blasting is completed, the casing is grouted with a sleeve valve to reinforce the soil above the blasting area. The grouting of the sleeve valve tube is cement slurry. The grouting range is 3 m for the front, rear, left and right of the blasting area. The bottom mark height is the tunnel center line, and the top mark height is 3 m above the top of the tunnel. Segmented grouting is adopted, and after each grouting is completed, the length of the heart tube of one step is moved up or down. It is advisable to use lifting equipment to move, or manually clamp the heart tube symmetrically with 2 pipe clamps, and apply force evenly on both sides to move the heart tube. Each time a 3–4 m grouting length is completed, a section of the grouting heart tube is to be removed. After the grouting is finished, the grouting pipe is covered with a stuffy cover to facilitate the re-injection construction. After the completion, the grouting effect is tested, and those who fail are re-constructed.

(10) Grouting effect detection. At 28d after grouting, the reinforcement area should be corroded and tested. The core is taken at 2 places. The number of detection holes of bedrock is 3% of the total number of grouting holes. Those who fail will be re-constructed. Drill the core in the constructed consolidated body and make it into a standard test piece for indoor physical and mechanical properties test, check the internal pile body.

(11) The original ground is restored. After the completion of the construction, the original road surface will be restored according to the design requirements.

6.2 Encounter Large Boulder Shield Cutter Protection ...

2) Blasting construction process at sea

   (1) Drilling construction. The auxiliary ship will tow the drilling and blasting ship to the blasting area, and assist the drilling and blasting ship to throw six anchors. One of the bow and the stern will throw a medium anchor, the anchor cable will be 150–300 m long; With side anchors, the anchor cable is 150–200 m long. When drilling underwater positioning, the RTKDGPS global satellite positioning system is used for drilling positioning. After that, a down-hole impact drill is used, which requires a drill to the bottom of the design hole (including the depth of the drill) and a plum-shaped hole. During the process of drilling the downhole hole into the hole, the steel casing is followed up and buried in the rock face. The diameter of the casing is 219 mm and the wall thickness is 10 mm. Due to the inaccuracy of the boundary between the boulder and bedrock, the blasthole layout during construction must exceed the range of the boulder and bedrock to ensure the blasting effect.

   (2) Cannon hole acceptance. After the blasthole is drilled, it will be checked by the technician. The blasthole can be filled up and down to be qualified, and the hole protection can be done. After the acceptance is passed, the charge can be applied.

   (3) The blasthole charge. The special round plastic cylinder is used to charge the column. The diameter of the column at the lone stone is 60 mm, and the diameter of the column at the base rock is 100 mm. The explosive is a high-density, high-performance anti-water emulsion explosive. Each hole is filled immediately after drilling. The weight of 50–100 cm fine sand is bundled on the upper part of the blasthole column to meet the weight of the blasthole hole, so that the medicine bag can not float up in the water hole.

   (4) The blasthole is blocked. After the medicine bag is in place, slowly pour the sand into the blasthole, and the length of the blockage is more than 5 m.

   (5) Sealing and extubation. The protective cylinder is pulled up and removed section by section. At the same time, the cement slurry (water-cement ratio is 0.8:1) is sealed into the sleeve, and the sealing quality must be checked and accepted by the site management personnel.

   (6) Blasting grid. The 1–7-stage detonating tube millisecond detonator is used, and the interval delay time is 25–50 ms. The inter-hole or inter-row millisecond delay blasting is carried out. The interval time is 25–50 ms, and the detonation is performed by holes or rows by row. 4 detonators are installed in each blasthole, and a dedicated detonator is detonated. The number of blasting holes per blast is determined by the maximum amount of blast that is allowed to detonate.

   (7) Be alert, confirm that all personnel and vessels are outside the danger zone, and after the passage to the adjacent and explosive zone has been temporarily blocked, and confirm that there is no white dolphin in the

danger zone, the detonation signal is issued before it can detonate. The blasting warning range is 150 m around the explosion area. The maritime alert uses a warning ship to patrol around the blast, and screams and dissuaces nearby ships.

(8) Detonation. Once all the warning work is completed, contact each security point again and confirm the correctness. After the blasting is completed, the technician will check the scene and make a warning order.

(9) Detection of blasting effect. After blasting, the blasting area should be subjected to coring detection and blasting effect. The test section takes 3 cores, and the cores of other areas are cored at 2 places, one of which needs to be located at the center of the boulder. The number of detection holes of bedrock protrusions is 3% of the total number of blast holes, and each core is taken at 7 places, and the core depth reaches the design depth. After the core is taken out, the core crack is analyzed and judged, and the unidirectional length of the complete core extracted is ≤ 25 cm. If the core is unqualified, the hole is filled for secondary blasting.

(10) Grouting. After the blasting is completed, the blasting area is grouted and reinforced. The reinforcement range is larger than the blasting range of 2 m, the spacing of the cloth holes is 0.8–1 m, the slurry is cement slurry, the water-cement ratio is 1:1, the appropriate amount of quick-setting admixture is added to the slurry, and the amount of quick-setting agent is blended on site. The slurry uses UJW150 mixer on board. The grouting pressure is controlled at 1.5–3 MPa. According to the actual situation on site, the grouting parameters can be adjusted appropriately. Drilling with a rig, the casing is placed down to the inside of the blasting stone. A sleeve valve tube is installed inside the sleeve, and a gap is filled between the two tubes with a double slurry, and then grouting is performed. When the grouting pressure of each hole reaches the design final pressure and should be stabilized for 10 min, and the feed rate is less than 1/4 of the initial feed rate, the grouting is considered to be completed.

(11) Grouting effect detection. At 28d after the end of grouting, it is necessary to carry out the coring detection and reinforcement effect on the reinforcing area, and the core is taken at two places. The number of detection holes of bedrock protrusions is 3% of the total number of grouting holes, and each core is taken at 7 places. If it is unqualified, it should be reconstructed. The core is drilled in the constructed consolidated body and made into a standard test piece for indoor physical and mechanical performance test to check the uniformity of the internal pile.

6.2.3 Construction Parameter Design

There are boulder and bedrock bulges in the shield construction section. Blasting must be carried out before the shield construction. After blasting, the longest side of the rock is required to be less than 25 cm.

1) Design of construction parameters of the upper shore

    (1) Drilling diameter and hole detection. Drilling with geological drilling rig, the drilling hole diameter is 108 mm, and 75 mm PVC casing is set underneath. The drilling of the boulder and bedrock can be used to check the position of the blasthole until the boulder and bedrock bulge boundaries are detected.
    (2) Selection of fire equipment. The hole detonator of the rock blasting on the shore uses digital electronic detonator, and the explosive is selected as emulsion explosive, which is processed according to the actual situation on site.
    (3) Arrangement of the blasthole. In order to reduce large rocks and fully break the rock, a plum-shaped cloth hole is made. Take the base rock bulge on the land $a = b = 70$ cm, the lone stone $a = b = 60$ cm, and the blasthole density coefficient is $m = a/b = 1\sim 1.2$, to ensure that the "crack areas" between the blastholes overlap as much as possible.
    (4) Explosive unit consumption. Since the blasted rock is below the surface of the overburden (underground) 16 m, that is, the overburden rock has a cover layer with a thickness of 16 m or more, and the blasted rock is directly thickly pressed, so that the blasted rock has no vacant surface, and it is difficult to carry out the slag throwing. For the phenomenon of "internal action drug pack", the fracture range of the rock is mainly "explosion cavity, crushing zone and crack zone". In order to maximize the "burst, crushing zone, crack zone", the explosive consumption of the underground rock is greater than the explosive consumption of the same type of rock on the ground (with the airfront), with reference to the rock blasting test and application results of Xiamen Rail Transit Line 1 Generally, it can be taken from 5.0 to 9.0 kg/m^3. In order to ensure that the covering layer and the underground pipeline are not damaged or disturbed, the explosive unit consumption relative to the covering layer is not more than 0.05 kg/m^3 on land.
    (5) The size requirements of the blasthole super deep and super boundary. In order to completely break the rock, there is no under exploitation rock ridge, the shield excavation section is over-expanded to 1.0 m, and the blasthole needs to be 1 m deep.
    (6) The parameters of the blasthole and the charge of the boulder and bedrock. The blasthole parameters of the bedrock bulge in the full section of the shield and its charge: It is assumed that the bedrock bulge occupies the full section and the distance from the tunnel vault to the ground is 16 m. The

plum-shaped cloth hole is made, and the distance between the blastholes is $a = b = 70$ cm, and then the depth of the adjacent two rows of blastholes and other blasting parameters are obtained. The blasthole parameters of the bedrock bulge of the shield 1/2 section and its charge: It is assumed that the bedrock bulge occupies the 1/2 section of the shield, and the distance from the tunnel vault to the ground is 16 m, that is, the thickness of the cover layer of the blasted rock is 19.52 m and the plum-shaped cloth hole is made. The hole spacing of the middle-weathered rock is taken as $a = b = 70$ cm, and then the depth of the adjacent two rows of blastholes and other blasting parameters are obtained. Since the boulder may exist in any part of the shield excavation range, and the shapes are different and the sizes are different, the blasting parameter design is required for this case. Make a plum-shaped cloth hole, which is a vertical blast hole. And corresponding to the different thickness of the rock on the shore, calculate the corresponding blasting parameters.

(7) Set the underground surface. Because the blasting of the underground shield boulder and the bedrock bulge is carried out under the condition that the covering layer is thick and not excavated, that is, without the empty surface, it is difficult to use a large explosive unit consumption. The rock is fully broken to achieve the blasting effect of the gravel with a longest side of less than 25 cm. To this end, the underground surface is created to create favorable conditions for the full fragmentation of the rock. The specific practices are as follows:

① Bedrock raised on the air surface. When the first blast of each bedrock bulge, it is necessary to take a hole in the bedrock bulge (no charge), the position is near the middle row of the blasthole, 15–20 cm away from the charge blasthole, evenly arranged 2–4 holes, the depth is more than 1.0 m, and the hole diameter is more than 100 mm.

② Lone stone surface setting. In the boundary layer of the lone stone, 2–3 blastholes are arranged. These soil blastholes are parallel with the first row of boulder holes, and the depth exceeds 1.0 m of the boulder hole. The diameter of the blasthole in the soil layer is as small as possible. The single-hole hole charge of the soil layer is 20–40% of the single-row single-hole charge of the lone stone. The soil blasthole uses the first-stage detonator to detonate at the same time, thereby forming a cavity at the lone stone. This cavity is the empty surface of the first row of blastholes.

(8) Calculation of blasting safety distance. In order to ensure that the blasting construction of this project does not affect surrounding buildings and pipelines, the maximum amount of drug must be strictly limited. To this end, according to the blasting vibration speed calculation formula of the Blasting Safety Regulations (GB6722-2014), the maximum amount of medicine allowed for different building (structures) at different distances is calculated. That is to say, only the blasting construction according to

the maximum charge amount of each blasting of the rock at different distances can make the blasting vibration speed less than the blasting vibration speed value allowed by the blasting safety regulations, thereby ensuring the safety of buildings and facilities.

> The maximum amount of drug used in blasting construction shall be adjusted according to the actual observed blasting vibration velocity data. For the setting of safety distance, consider the following two aspects:
> ① The blasthole plugging method and the underground "internal action pack" method are used for blasting. Therefore, under the condition that the plugging length and the plugging quality are sufficient, the blasting air shock wave will not be generated, and the hazard can be ignored.
> ② The blasting is below the ground level of 16 m, and the explosive consumption of the relative covering layer is less than 0.05 kg/m³, which belongs to the ultra-weak loose blasting, which is in the form of "internal action drug pack". There is no change in the ground, so there will be no blasting flying stone phenomenon. A warning distance of 50 m is safe enough.

(9) Monitoring program. On the one hand, the seismic velocity is detected by arranging one or several surface lines along the radial or circumferential direction of the blasting center. In the radial arrangement, the distances of the measuring points are arranged in a logarithmic curve, and each measuring point is preferably capable of simultaneously three vectors. Observe the influence of blasting vibration on the building (structure), and the measuring points are arranged on the surface and buildings near the building to be tested. On the other hand, monitoring can be carried out by deploying surface settlement monitoring points around the blasting area.

2) Construction parameters of the offshore section, the main blasting parameters can be dynamically adjusted according to the results of the blasting test.

(1) Drilling diameter and hole detection. Drilling with a down-the-hole drilling rig with a borehole diameter of 138 mm, the boulder hole and bedrock bulge can be re-checked during drilling until the boulder and bedrock bulge boundaries are detected.

(2) Selection of fire equipment. The detonator in the blasting hole is selected from the detonator tube millisecond detonator, and the explosive is selected from high-density and high-performance water-resistant emulsion explosive. The reliability of underwater operation is verified before use.

(3) Arrangement of the blasthole. In order to reduce the large pieces and fully break the rock, a plum-shaped cloth hole is made. The hole distance and row spacing of the blasthole are small, and the base rock of the seabed is $a = b = 120$ cm, the seabed boulder $a = b = 60$ cm, and the blasthole

density coefficient $m = a/b = 1{\sim}1.2$ to guarantee the "crack areas" of the blasthole overlapping as much as possible.

(4) Explosive unit consumption. Since the blasted rock is below the surface of the covering layer by 16 m, that is, the covering layer with a thickness of 16 m or more on the blasted rock. The blasted rock is directly thickly pressed, so that the blasted rock has no vacant surface, there is almost no concession space, and it is difficult to carry out the slag throwing. It is characterized by the phenomenon of "internal action drug pack". In order to maximize the "explosion cavity, crushing zone and crack zone", the single consumption of explosives should meet the requirements of "internal action drug pack", that is, it can guarantee no damage or disturbing the cover layer; the explosive unit consumption relative to the cover layer should generally be less than 0.10 kg/m^3 in the sea.

(5) The size requirements of the blasthole super deep and super boundary. In order to completely break the rock, there is no under exploitation rock ridge, the shield excavation section is over-expanded to 1.0 m, and the blasthole needs to be 1 m deep.

(6) The parameters of the blasthole and the charge of the boulder and bedrock. The blasthole parameters of the bedrock bulge of the shield 1/2 section and its charge: It is assumed that the bedrock bulge occupies the 1/2 section of the shield, and the distance from the tunnel vault to the bottom of the sea is 16 m, that is, the cover of the blasted rock The thickness is 19.52 m, and the plum-shaped cloth hole is taken. The distance between the holes of the middle-weathered rock is $a = b = 120$ cm, and then the depth of the adjacent two rows of blastholes and other blasting parameters are obtained. The blasthole parameters of the bedrock bulge of the shield 1/3 section and its charge: It is assumed that the bedrock bulge occupies the 1/3 section of the shield, and the distance from the tunnel vault to the bottom of the sea is 16 m, that is, the cover of the blasted rock The thickness is 20.69 m, and the plum-shaped cloth hole is made. The abutment distance of the middle-weathered rock is $a = b = 120$ cm, and then the depth of the adjacent two rows of blastholes and other blasting parameters are obtained.

Since the boulder may exist in any part of the shield excavation range, and the shapes are different and the sizes are different, the blasting parameter design is required for this case. If the lone stone is within the scope of the shield excavation, the whole hole is blasted, and the blasthole penetrates the boulder in the vertical direction, and the hole is drilled in the horizontal direction beyond the boulder; if the boulder is only partially in the shield Within the excavation range, the rock in the excavation area is fully blasted and the blasthole penetrates the boulder. The rock within 1.0 m beyond the excavation boundary also needs to be drilled and blasted, but the blasthole does not need to penetrate.

(7) Set the underground surface. Because it is under the condition that the cover layer is thick and not excavated, that is, the blasting of the shield rock

and the bedrock bulge is carried out without the air surface, it is necessary to set the underground air surface to fully break the rock. Create favorable conditions as follows:

① Bedrock raised on the air surface. When the first blast of each bedrock bulge, it is necessary to take a hole in the bedrock bulge (no charge), the position is near the middle row of the blasthole, 15–20 cm away from the charge blasthole, evenly arranged 2–4 holes, the depth is more than 1.0 m, and the hole diameter is more than 100 mm.

② Lone stone surface setting. Two to three blastholes are arranged in the boundary layer of the boulder. These blastholes are parallel to the first row of boulder holes, and the depth exceeds 1.0 m of the boulder hole. The diameter of the blasthole in the soil layer is as small as possible. The single-hole hole charge of the soil layer is 20–40% of the single-row single-hole charge of the lone stone. The soil blasthole uses the first-stage detonator to detonate at the same time, thereby forming a cavity at the lone stone. This cavity is the empty surface of the first row of blastholes of the lone stone; the detonation time of the lone stone hole should be more than 100 ms.

(8) Calculation of blasting safety distance. In order to ensure that the blasting construction of this project does not affect surrounding buildings and pipelines, the maximum amount of drug must be strictly limited. To this end, according to the blasting vibration speed calculation formula of the Blasting Safety Regulations (GB6722-2014), the maximum amount of medicine allowed for different building (structures) at different distances is calculated. That is to say, only the blasting construction can be carried out according to the maximum charge of each blasting of the rock at different distances, so that the blasting vibration speed can be less than the blasting vibration speed value allowed by the blasting safety regulations, so as to ensure the safety of buildings and facilities.

The maximum amount of the dose used in the blasting construction shall be adjusted according to the actual observed blasting vibration velocity data. Shield rock blasting is an "internal action pack form" that does not generate shock waves in water, but for safety reasons, it is regarded as underwater reef blasting, and thus determined according to the "Blasting Safety Regulations" (GB6722-2014) to ensure the safe distance of the shock wave in the shield rock blasting water.

(9) Monitoring program. During the blasting process, a monitoring point is buried beside the Haitang side seawall and the No. 9 restaurant building in Haijiao. The impact of blasting vibration on the building (structure) is observed. The measuring points are arranged on the surface and buildings near the surface building. Applying the formula and the one-way regression method to carry out regression analysis on the measured data of the building, and obtain the coefficients K and α related to the medium and the terrain, so as to obtain the attenuation law of the vibration velocity of the

particle. And then allow the maximum vibration velocity and the explosion distance according to the formula to calculate the maximum amount of drug. The obtained vibration speed is compared with the safety criterion (predetermined permissible vibration speed) to determine whether the building or the structure is safe. If the measured vibration speed is greater than the allowable vibration speed, the damping measures are strengthened.

6.2.4 Construction Guarantee Measures

(1) Noise control measures. The boulder treated by this project is mainly concentrated in the underground of 16–22 m deep, and there is water in the blasthole. Similar to underwater blasting and deep hole blasting, only the muffled sound will occur during blasting, and there is protection above the blasthole. It will affect the surrounding residents and the environment.
(2) Shock absorption technical measures.

> ① Control the maximum dose of a burst, strictly control the amount of explosives according to different distances.
> ② Using millisecond delay blasting technology, design a reasonable starting sequence. The time interval between the segments is controlled at 50-100 ms, and the large source generated by the simultaneous explosion of all charges is divided into small earthquake sources with several millisecond delays.

(3) Treatment of construction mud. In the construction of the upper section of the shore, the geological drilling rig needs to use the mud to protect the wall into holes. When drilling, it is necessary to make the enclosure, prepare the woven bag and the mud tank, and use the method of slag bagging and mud loading to transport the road in time and flush the road surface.
(4) Treatment of construction dust.

   ① The main road of the transportation vehicle is regularly sprinkled and cleaned to keep the road surface of the vehicle clean to reduce the dust pollution caused by the vehicle.
   ② Set a protective net to reduce the impact of dust and dross that may occur during construction.
   ③ Close the compartments of vehicles with easy-to-make materials such as construction materials and dregs to avoid the spreading, flowing and polluting environment along the transportation process.
   ④ The construction site should be sprinkled in time to prevent dust. In the case of windy weather, the muck should be covered.

(5) Safety management measures for blasting equipment. The quantity of blasting equipment to be used shall be consistent with the practical and remaining

quantity. Undocumented personnel shall not be exposed to blasting equipment; blasting equipment must be handed over to the operation site, and detonators and explosives shall be stored separately; at noon, blasting equipment shall be temporarily provided as required. Put it in a burglar-proof and explosion-proof container (safety box) that is supervised; fill it in time after loading the medicine; the blasting equipment should not be stacked in a dangerous area and exposed to the sun.

(6) Navigation related measures. In view of the frequent characteristics of the passing ships around the project, the following measures have been formulated: all units involved in the operation, ship certificates, personnel certificates, etc., all personnel must be certified to work; safety disclosure, education and training must be carried for all participating ships to implement safety responsibilities to everyone, reduce potential safety hazards, and prevent the occurrence of safety accidents.

(7) Construction safety measures during the foggy season. All construction work vessels shall strictly implement the "Rules for Navigation in the Sea" and the regulations for the harbor fog. The self-propelled work ship should stop sailing when the fog is heavy, and choose the underground anchor. The blasting vessel shall be arranged by the relevant departments such as the port management office and the traffic control station, and may be anchored or towed to the anchorage.

(8) Fire prevention and anti-skid measures.

① The heat treatment of electricity, gas welding, etc. on board is subject to the approval of the chief engineer, report to the master of the ship, fill out the "fire application report", and report to the port supervisor for approval. Fire-fighting equipment must be prepared before welding, check the nearby area, confirm that there are no inflammable and explosive materials, and send someone to take care of them. After the operation, the site should be carefully inspected to confirm that there is no possibility of resurgence before leaving the site.

② Strictly observe the operating procedures of the electric welder. When the welding is not applied, the welding clamp should not be grounded. It is absolutely forbidden to carry out the welding and painting work at the same time. It is forbidden to heat or weld pressurized containers such as steel cylinders and oil pipes.

③ Oxygen, acetylene, propane, etc. should be isolated from each other to prevent collision and keep away from fire and oil. When it is hot, take measures to cool down. After use, close and wear a safety cover.

④ Inflammable and explosive materials should be placed at designated locations. It is strictly prohibited to be stored in boilers, chimneys and other high temperature places.

⑤ Rubbing oil rags, cotton yarn heads, etc. should be handled at any time, not allowed to be placed indiscriminately, to avoid fire.

⑥ Smoking is not allowed in machines, pump warehouses, material warehouses and battery rooms.
⑦ When the deck or sidewalk has oil, mud and water, it should be removed or laid in time.

(9) Measures against tropical cyclones.

① The project manager department formulates the site defense plan, including the anti-typhoon organization, the anti-typhoon organization job responsibilities, the anti-typhoon work procedure, the anti-typhoon communication network, and the anti-typhoon duty system.
② When it is necessary to use the anchorage to avoid wind and prevent the platform, the project department plans to use the anti-satellite anchorage in advance, and applies for the anchorage to the local maritime department 24 h in advance. After obtaining the instructions agreed by the local maritime department, the vessel enters the sheltered anchorage.
③ To ensure the safety of the ship, personnel should be inspected for ship equipment, navigation equipment, navigation instruments, mooring equipment, communications, life-saving, fire prevention, watertight devices, plugging and drainage equipment before the typhoon season to ensure that the equipment is in good condition. All crew members learn to prevent typhoon.
④ During the construction period, the construction vessel shall be well protected from wind. Receive weather forecasts every day, master the meteorological dynamics at sea, be alert to tropical cyclone signs and sudden attacks of tropical depression, and if necessary, arrive in the harbor or shelter from the wind to avoid the wind.

6.3 Japan's Kumagai Group SunriseBite Method of Tool Change

Japan's Kumagai Group and tunnel excavator company JIMT jointly released a rotary tool change technology for the shield cutter replacement technology, SunriseBit, at the end of October 2017 (Fig. 6.27). The method is to install a rotating device (rotating body) equipped with a plurality of cutting blades in a shield spoke, and rotate the device by a hydraulic cylinder to replace the worn tool (Fig. 6.28). This method is suitable for ultra-long-distance excavation. It can work under various conditions such as large depth and high water pressure corresponding to various geological conditions. It is a safe and efficient tool change technology through remote operation.

The Sunrise Bite method is to arrange a rotating device with a plurality of spare tools in the spokes of the shield, and rotate the hydraulic jack to perform the tool change. Up to 8 reinforced shell knives can be equipped according to the necessary number of tool changes. Since the hydraulic jack and the ratchet are used to rotate, the operator does not need to enter the tool change position, and can perform remote

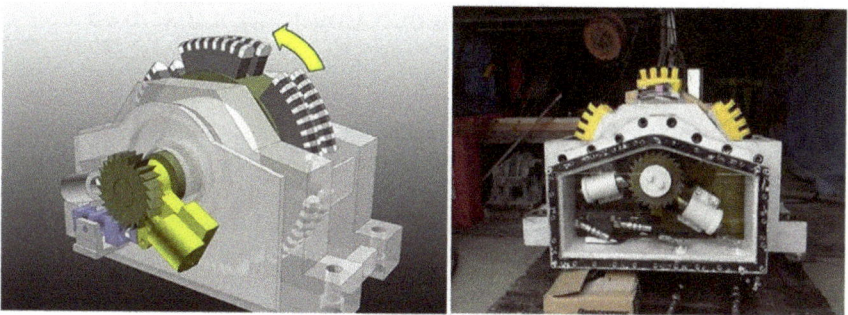

Fig. 6.27 Schematic diagram and full-size model of Sunrise Bit

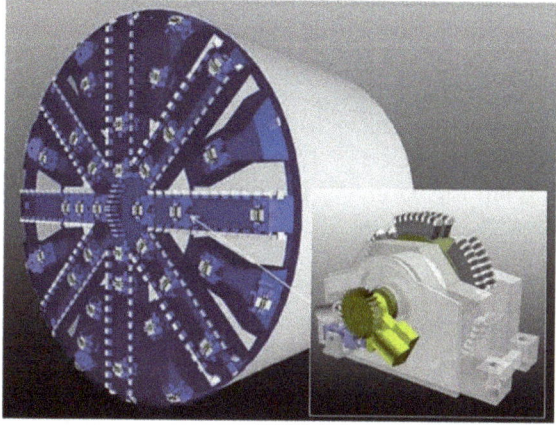

Fig. 6.28 Schematic diagram of replacement device

operation in a safe place. The application method is for ultra-long-distance excavation with a total length of about 10 km, a composite stratum with soft soil layer to hard rock stratum change, a large-diameter pebble stratum and other layers with severe tool wear, and environment with large ground depth and high water pressure that are difficult to improve and enter. The advantages of the new tool change method are: the tool can be changed under the premise of safety and no influence on the construction progress; it is suitable for the conditions that the foundation improvement, high water pressure and the like cannot be performed, and the tool can be changed in the environment that the personnel cannot enter; Special tools such as temporary wall cutting or cutting obstacles are directly cut.

6.4 Japan's Kashima Construction Company's Disc Hob Replacement Technology

The disc hob replacement technology uses the tool holder sliding method to change the tool so that the tool change operation can be carried out in the shield. The worn disc hob is pulled along with the seat by a hydraulic jack into the spokes, and then the sprinkler gate mounted on the spokes is closed to prevent groundwater from flowing into the shield from the face. Figure 6.29 shows the replacement procedure for the disc hob.

This technology is applicable in principle to rock tunnels with large depth and high water pressure, as well as mountain tunnels of gravel and pebble formations. It can be applied to the replacement of disc hobs in shield cutters (Fig. 6.30). In December 2016, in a simulation experiment with a depth of about 100 m and a maximum water pressure of about 1 MPa, the safety of the tool change operation and the water repellency under pressure were confirmed. Compared with the rotary tool change method used in the past, the amount of soil sand inflow during the tool change is controlled. The technology can replace four disc hobs in one day, eliminating the need for foundation improvement or the use of auxiliary methods such as grouting, which can significantly shorten the construction period and reduce construction costs.

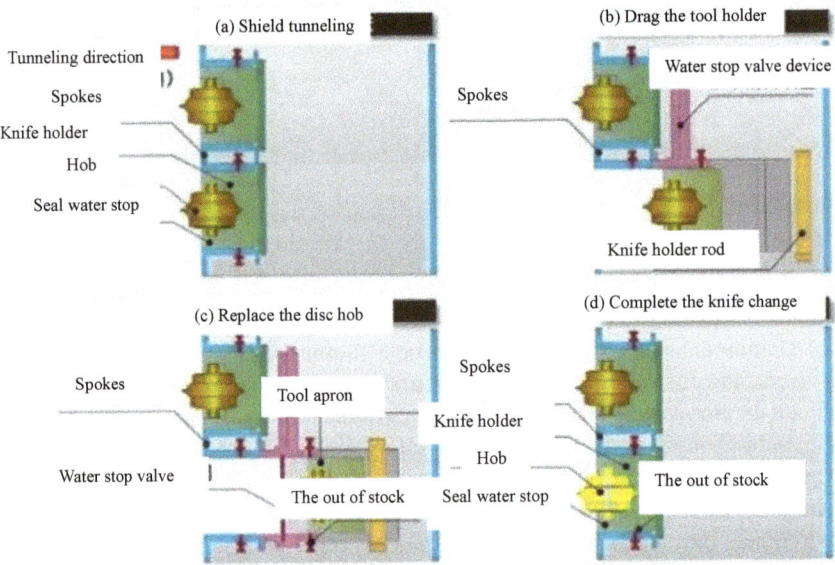

Fig. 6.29 Replacement steps of disc hob (sliding mode of knife seat)

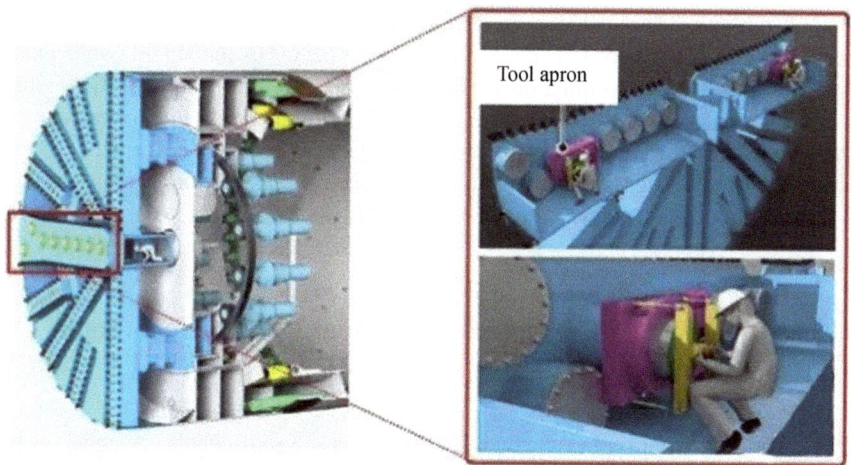

Fig. 6.30 Schematic diagram of changing position of disc hob

6.5 Japan Dacheng Construction Company Central Hob Replacement Technology

The central hob replacement technology is a new technology developed based on the spherical rotary tool change technology. It mainly uses a movable annular seal to stop the water while rotating the ball hob bracket with the built-in hob from the inside of the shield to the soil side of 180°, then remove the back cover and replace the new hob. The central hob replacement procedure is shown in Fig. 6.31.

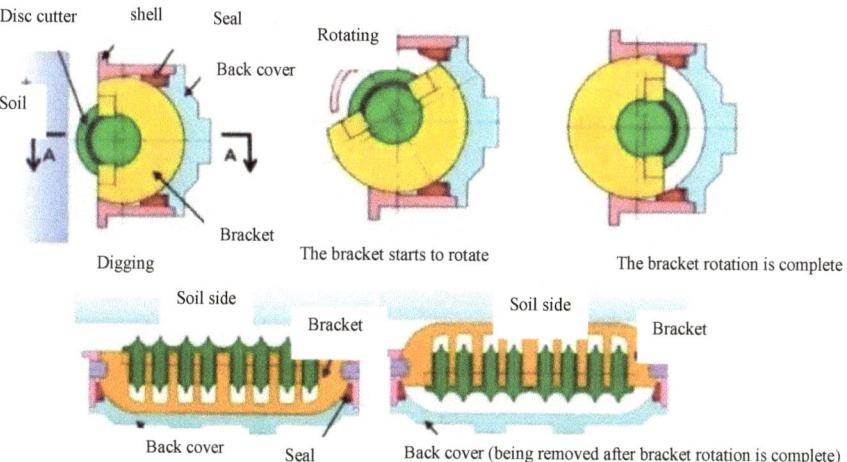

Fig. 6.31 Central hob replacement step

This method is applicable to hard rock tunnels under large depth and high water pressure, and is used for hob replacement in the center of the shield cutter panel. In April 2016, in the simulation experiment with a depth of about 100 m and a maximum water pressure of 1.1 MPa, the safety of the tool change operation and the water stop under pressure were confirmed. The method can also replace the tool when the groundwater level cannot be lowered during construction, and the series of hobs can be replaced at one time, which shortens the working time.

6.6 Japan Feidao Construction Company Chameleon Tool Replacement Technology

The chameleon tool replacement technology sets the tool change chamber in the center of the shield. The cutter module is set in the cutter head without affecting the operation of the shield. These cutter modules are automatically installed and replaced by the jack. The normal tooling in soil sand and hard rock formations is accommodated by replacing the center cutter and spokes of the cutter. The chameleon tool replacement technology is shown in Fig. 6.32.

The fishtail knife is arranged in the center of the shield cutterhead, and the hob that can move back and forth is arranged at the center without affecting the position of the fishtail knife. The position of the hob is controlled by the jack (if it is higher than the fishtail knife, the cutter is used for cutting, if it is lower than Then use a fishtail knife to cut), as shown in Fig. 6.33. In the soft formation, the fishtail knife is used for excavation. In the hard layer, the hob behind the fishtail knife is pushed out by the jack, and the cutting height is higher than the cutting height of the fishtail knife. The amount of push is determined by the cutter's cutter rotation speed and the tunneling speed. When the soil becomes a soft formation again, the hob is retracted and the fishtail knife is used for cutting.

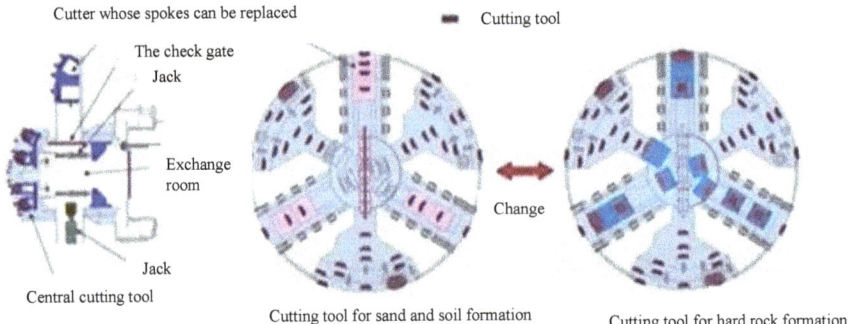

Fig. 6.32 Chameleon tool replacement technology

6.6 Japan Feidao Construction Company Chameleon Tool ...

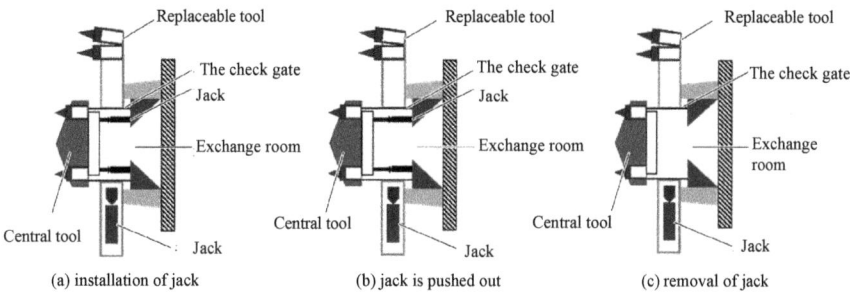

Fig. 6.33 Central tool replacement steps

The module with a plurality of large cutters is arranged on each spoke of the cutter head, and the lost cutter module is pulled to the center replacement chamber by the jack, and then the new cutter module is installed in the fixed position through the jack, as shown in Fig. 6.34.

The method is adapted to the construction of a composite stratum shield with a water pressure lower than 1.0 MPa and a geological condition of clay, silt, sand, gravel, and the like. This method can replace the manual automatic tool change by

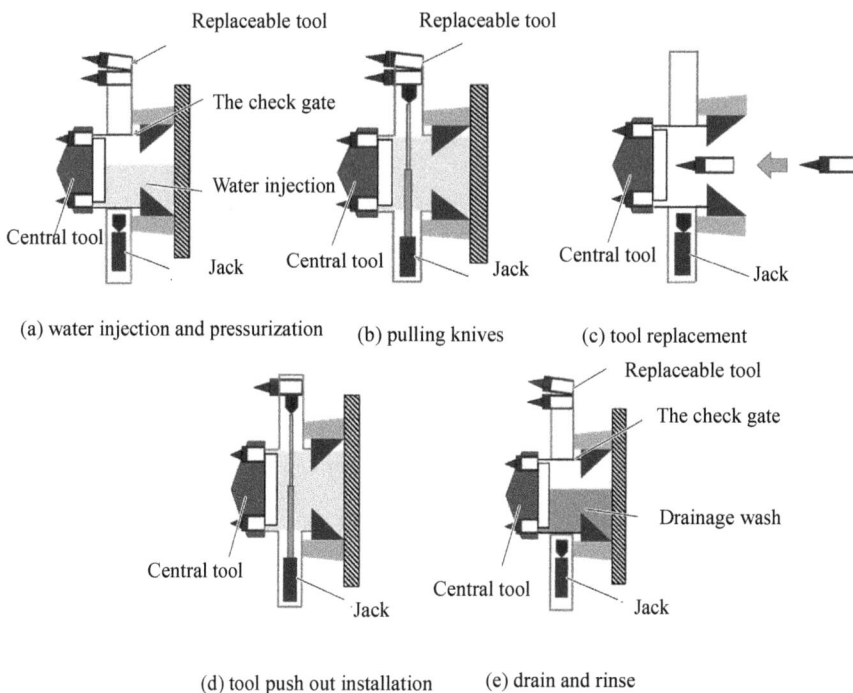

Fig. 6.34 Spoke part tool replacement steps

mechanical substitution in the actual operation process, which increases the safety of the project. The method was developed in August 2015, but there is no actual construction verification yet.

It can be seen from the above several new technologies that the Japanese shield enterprise has innovated the tunnel tool replacement technology under the rock formation under high depth and high water pressure and the composite formation, which ensures the safety of the construction tool replacement process, improves the tool, speeds up replacement, making the tool replacement process more specialized and refined.

Lightning Source UK Ltd.
Milton Keynes UK
UKHW022024270822
407877UK00001B/3